Fundamentals of Programming in SAS®

A Case Studies Approach

James Blum
Jonathan Duggins

S.sas.

sas.com/books

The correct bibliographic citation for this manual is as follows: Blum, James and Jonathan Duggins. 2019. *Fundamentals of Programming in SAS®: A Case Studies Approach*. Cary, NC: SAS Institute Inc.

Fundamentals of Programming in SAS®: A Case Studies Approach

Copyright © 2019, SAS Institute Inc., Cary, NC, USA

978-1-64295-228-5 (Hardcover)
978-1-63526-672-6 (Paperback)
978-1-63526-671-9 (Web PDF)
978-1-63526-669-6 (epub)
978-1-63526-670-2 (kindle)

Contents

Foreword

To Readers

This book is designed to help you develop an understanding of the SAS programming language and to help you develop good programming practices. It is intended as a learning guide and a skill builder, not as a reference book. To that end, it introduces sets of topics within each chapter that are connected through a single case study. Concepts are introduced on an as-needed basis to complete required tasks, so you are immediately exposed to writing complete programs. As further concepts are introduced they might be new topics, or they might revisit previously introduced topics at a more complex level. This reflects how many of the best SAS programmers have built their talents—by continually adding layers of knowledge onto a base set of skills. The book mimics this type of experience by increasing the complexity of the case study, requiring the addition of newer skills, or more complex versions of earlier skills, as they are needed.

Because of this circling back to content from previous chapters, a pedagogical concept known as a *spiral curriculum*, you will not learn everything this book covers on a topic in any single chapter. Of course, no one book could serve the purpose of giving a complete treatment of all concepts included; therefore, you will often be referred to outside resources, such as SAS Documentation, for a full description of syntax or for more detailed commentary. SAS Documentation is the standardized name used in this text for the collection of help files and examples provided by SAS. This documentation is available via the Help menu in SAS or online. Reading such references is a strategy commonly used by the best SAS programmers to refine their abilities and is an important habit for you to develop to build your skills as a SAS programmer and to expand on those skills in the future.

Due to the introduction of concepts in spiral fashion, it is important to begin with the setup material in Chapter 1 and then to proceed through the book sequentially. For easy reference, the numbering on all output in Chapters 1 through 7 directly corresponds to the number of the program that generates it. However, not all programs generate output. In all chapters, the programs, output, tables, and figures are numbered sequentially within each section. The same case study is used to provide continuity through the narrative when building on earlier concepts. Other case studies are also available for use to build continuity for additional programming activities and exercises. This includes a case study located in Chapter 8 for which the sections are aligned with the learning objectives of each chapter. Additional case studies are available online by visiting the author page for either author.

To Classroom Instructors

As stated in the previous section, this book is designed to tap into some of the best practices in educational theory as it spirals back onto topics throughout your course. It is designed by instructors with over 25 years of combined experience in teaching SAS either in the classroom or in industry. The more technical details are isolated in their own sections so that you can easily include or exclude them to fit the needs of your course. Multiple case studies are also provided so that the case study assignments can be customized for your students' interests and to the content presented in your course. The case study provided in Chapter 8 ensures students have immediate access to a case study for reference while reading the text. Additional case studies are made available through the author pages for either author to ensure you get the benefit of updated materials on a regular basis. Any instructional materials will also be available either via the author pages (for public resources) or by contacting SAS to verify your status as an instructor (for instructor-only materials). These resources will be regularly updated.

About the IPUMS CPS Data

The IPUMS CPS data includes the Integrated Public Use Microdata Series (IPUMS) and Current Population Survey (CPS) beginning in 1962. These data sets provide person- and household-level information about a variety of demographic variables. A cross-section of recent data (2001, 2005, 2010, and 2015) was released for this publication and is included here as the main case study in the narrative. Visit https://cps.ipums.org to learn more about the IPUMS CPS or to extract newer data to continue honing your SAS programming skills.

About This Book

What Does This Book Cover?

This text covers a wide set of topics available in the Base SAS software including:

- DATA step programming including:
  - Reading data sets from non-SAS sources
  - Combining and restructuring SAS data sets
  - Functions and conditional logic
  - DO loops and arrays
- Basic analysis procedures: MEANS, FREQ, CORR, and UNIVARIATE
- Reporting procedures: CONTENTS, PRINT, and REPORT
- Restructuring data with PROC TRANSPOSE
- Visualization with the SGPLOT and SGPANEL procedures
- SAS formats and the FORMAT procedure
- Output Delivery System

While this book covers the foundations of the topics listed above, additional details are often beyond the scope of this text. References are provided for those interested in further study.

Is This Book for You?

Are you trying to learn SAS for the first time? Are you hoping to eventually earn your Base SAS certification and become a SAS Certified Professional? Are you already comfortable with some SAS programming but are looking to hone your skills? If the answer to any of those questions is "yes," then this is the book for you! This book takes a novel approach to learning SAS programming by helping you develop an understanding of the language and establish good programming practices. By following a single case study throughout the text and circling back to previous concepts, this book aids in the learning of new topics through explicit connections to previous material. Just as the best SAS programmers expand their capabilities by continually adding to their already impressive skill sets, as you read this text you will gain the skills and confidence to take on larger challenges with the power of SAS.

This book does not assume any prior knowledge of the SAS programming language. However, an understanding of how file paths function in your operating system is necessary to facilitate the storage and retrieval of data sets, raw data files, and other files such as documents and graphics.

What Should You Know About the Examples?

This book includes tutorials for you to follow to gain hands-on experience with SAS. The majority of the examples are based on a case study using real data. Some examples use subsets of the case study data or introduce smaller data sets to help illustrate a topic. Chapters 2 through 7 contain a wrap-up activity that uses the case study to tie together concepts from the current chapter, and every chapter references another case study contained in Chapter 8 for further practice. You need access to the software listed in the next section to complete the exercises.

For easy reference, the numbering on all output in Chapters 1 through 7 directly corresponds to the number of the program that generated it. However, not all programs generate output. In all chapters, the programs, output, tables, and figures are numbered sequentially within each section.

Software Used to Develop the Book's Content

SAS 9.4TS1M3 and higher were used to develop the examples and exercises. To follow along with the examples simultaneously or to complete the exercises, you only need the Base SAS software except for the portions of Chapters 7 and 8 that use SAS/ACCESS to connect to Microsoft Excel workbooks and Microsoft Access databases.

Example Code and Data

You can access the example code and data for this book by linking to its author page at https://support.sas.com/authors.

SAS University Edition

This book is compatible with SAS University Edition. If you are using SAS University Edition, then begin here: https://support.sas.com/ue-data.

Where Are the Exercise Solutions?

Readers: Exercise solutions to selected exercises are posted on the author page at https://support.sas.com/authors.

Classroom Instructors: To obtain the full solutions, contact saspress@sas.com.

We Want to Hear from You

Do you have questions about a SAS Press book that you are reading? Contact us at saspress@sas.com.

SAS Press books are written *by* SAS Users *for* SAS Users. Please visit sas.com/books to sign up to request information on how to become a SAS Press author.

We welcome your participation in the development of new books and your feedback on SAS Press books that you are using. Please visit sas.com/books to sign up to review a book

Learn about new books and exclusive discounts. Sign up for our new books mailing list today at https://support.sas.com/en/books/subscribe-books.html.

Learn more about these authors by visiting their author pages, where you can download free book excerpts, access example code and data, read the latest reviews, get updates, and more:
http://support.sas.com/blum
http://support.sas.com/duggins

About These Authors

James Blum is a Professor of Statistics at the University of North Carolina Wilmington where he has developed and taught original courses in SAS programming for the university for nearly 20 years. These courses cover topics in Base SAS, SAS/SQL, SAS/STAT, and SAS macros. He also regularly teaches courses in regression, experimental design, categorical data analysis, and mathematical statistics; and he is a primary instructor in the Master of Data Science program at UNC Wilmington, which debuted in the fall of 2017. He has experience as a consultant on data analysis projects in clinical trials, finance, public policy and government, and marine science and ecology. He earned his MS in Applied Mathematics and PhD in Statistics from Oklahoma State University.

Jonathan Duggins is an award-winning Teaching Professor at North Carolina State University, where his teaching includes multiple undergraduate and graduate programming courses. His experience as a practicing biostatistician influences his classroom instruction, where he incorporates case studies, utilizes large data sets, and holds students accountable for the best practices used in industry. Jonathan is a member of the American Statistical Association and is active with the North Carolina chapter. He has been a SAS user since 1999 and has presented at both regional and national statistical and SAS user group conferences. Jonathan holds a BS and MS in mathematics from the University of North Carolina Wilmington and an MS and PhD in statistics from Virginia Tech.

Learn more about these authors by visiting their author pages, where you can download free book excerpts, access example code and data, read the latest reviews, get updates, and more:
https://support.sas.com/blum
http://support.sas.com/duggins

Acknowledgments

I would like to thank my family—my mother Jo Ann, my father Robert, and my sister Amy—for their unwavering support in all of my academic endeavors over the years. A big thank you to all of the students and colleagues I have interacted with who, through their desire to learn more, also pushed me to learn more and bring more ideas back into the classroom. Thanks to all of the people at SAS who helped make this book a reality, particularly those in SAS publishing, and to all of the instructors I had for SAS Training who helped me transition from a mediocre SAS user to an actual SAS programmer. And a very special thank you to my love, Lilit, whose patience and encouragement gave me the extra strength to finish this project.

– Jim

I want to thank the hardworking staff at SAS who helped make this book a reality—it has been a learning experience and I am already looking forward to working with you on the next project! While many friends and colleagues supported me during this process, I especially want to thank Dr. Ellen Breazel for providing invaluable feedback from the perspective of a fellow SAS instructor and to Jordan Lewis for helping ensure the needs of the novice SAS user remained a primary concern during my writing process. Thank you to my wonderful and supportive wife, Katherine, for keeping me from getting too engrossed in writing this book. (Also for helping me write this acknowledgment!) Finally, I am eternally grateful for my parents, Bill and Teresa, who have provided enthusiastic encouragement during this project, just like they have done for all of my endeavors for longer than I can remember. This book would not have been possible without all of you.

– Jonathan

Chapter 1: Introduction to SAS

1.1 Introduction

This chapter introduces basic concepts about SAS that are necessary to use it effectively. This chapter begins with an introduction to some of the available SAS environments and describes the basic functionality of each. Essentials of coding in SAS are also introduced through some pre-constructed sample programs. These programs rely on several data sets, some provided with SAS, others are provided separately with the textbook, including those that form the basis for the case study used throughout Chapters 2 through 7. Therefore, this chapter also introduces SAS data sets and libraries. In addition, an introduction to debugging code is included, which includes a discussion of the SAS log where notes, warnings, and error messages are provided for any code submitted.

1.2 Learning Objectives

This chapter provides a basis for working in SAS, which is a necessary first step for successful mastery of the material contained in the remainder of this book. In detail, it is expected upon completion of this chapter that the following concepts are understood within the chosen SAS environment:

- Demonstrate the ability to open, edit, save, and submit a SAS program

- Apply the LIBNAME statement to create a user-defined library—including the BookData library that contains all files for this text, downloadable from the Author Page

- Demonstrate the ability to navigate through libraries and view data sets

- Think critically about all messages SAS places in the log to determine their cause and severity

- Apply ODS statements to manage output and output destinations

- Explain the basic rules and structure of the SAS language

- Demonstrate the ability to apply a template to customize output

Use the concepts of this chapter to solve the problems in the wrap-up activity. Additional exercises and case-studies are also available to test these concepts.

1.3 SAS Environments

Interacting with SAS is possible in a variety of environments, including SAS from the command line, the SAS windowing environment, SAS Enterprise Guide, SAS Studio, and SAS University Edition; with most of these being available on multiple operating systems. This chapter introduces the SAS windowing environment, SAS Studio, and SAS University Edition on the Microsoft Windows operating system and points out key differences between those SAS environments. For further specifics on differences across SAS environments and operating systems, consult the appropriate SAS Documentation. In nearly all examples in this book, code is given outside of any specific environment and output is shown in generic RTF-style tables or standard image formats. Output may vary somewhat from the default styles across SAS environments on various operating systems, and examples later in this chapter demonstrate some of these differences. Later chapters give information about how to duplicate the table styles.

1.3.1 The SAS Windowing Environment

The SAS windowing environment is shown in Figure 1.3.1 with three windows visible: Log, Explorer, and Editor (commonly referred to as the Enhanced Program Editor). The Results and Output windows are two other windows commonly available by default, but are typically obfuscated by other windows at launch. When code that generates output is executed, these windows (and possibly others) become relevant.

Figure 1.3.1: SAS Windowing Environment on Microsoft Windows

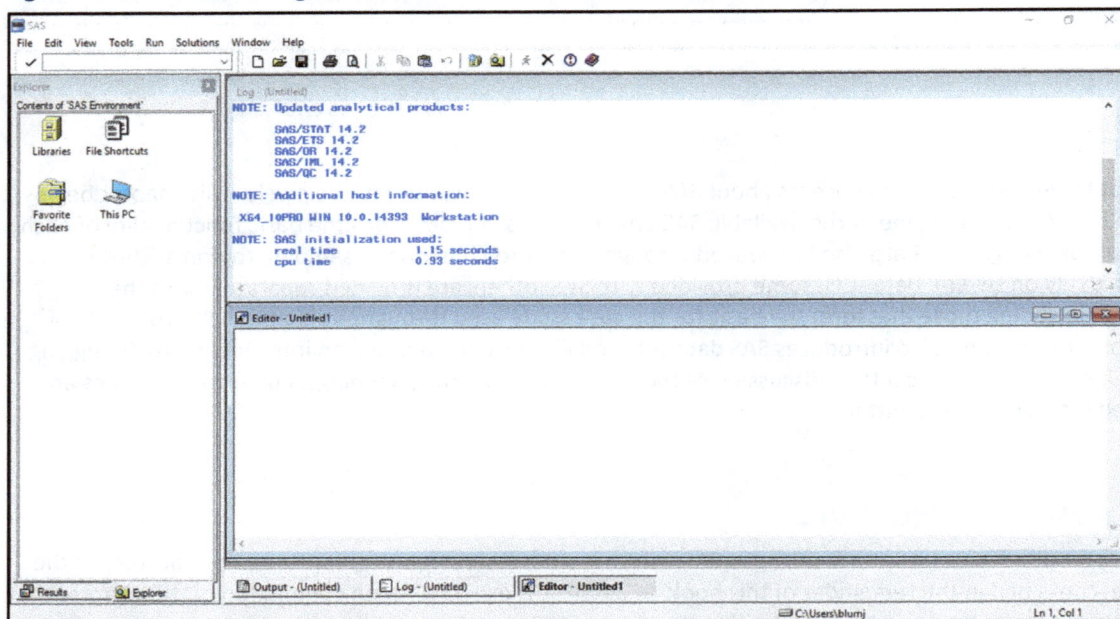

In the Microsoft Windows operating system, the menu and toolbars in the SAS windowing environment have a similar look and feel compared to other programs running on Windows. Exploring the menus reveals standard options under the **File, Edit,** and **Help** menus (such as **Open, Save, Clear, Find**). The **View, Tools, Solutions,** and **Window** menus have specialized options related to windows and utilities that are specific to SAS. The **Run** menu is dedicated to submissions of SAS code, including submissions to a remote session. As is typical in most applications, toolbar buttons in SAS provide quick access to common menu functions and vary depending on which window is active in the session. Some menu and toolbar options are reviewed below during the execution of the supplied sample code given in Program 1.3.1. This sample code is available from the author web pages for this book.

Program 1.3.1: Demonstration Code

```
options ps=100 ls=90 number pageno=1 nodate;

data work.cars;
  set sashelp.cars;

  mpg_combo=0.6*mpg_city+0.4*mpg_highway;
```

```
select(type);
  when('Sedan','Wagon') typeB='Sedan/Wagon';
  when('SUV','Truck') typeB='SUV/Truck';
  otherwise typeB=type;
end;

label mpg_combo='Combined MPG' typeB='Simplified Type';
run;

title 'Combined MPG Means';
proc sgplot data=work.cars;
  hbar typeB / response=mpg_combo stat=mean limits=upper;
  where typeB ne 'Hybrid';
run;

title 'MPG Five-Number Summary';
title2 'Across Types';
proc means data=cars min q1 median q3 max maxdec=1;
  class typeB;
  var mpg:;
run;
```

After downloading the code to a known directory, there are multiple ways to navigate to and open this code. Figure 1.3.2 shows two methods to open the file, each requiring the Editor window to be active.

Figure 1.3.2: Methods for Opening SAS Code Files in the SAS Windowing Environment

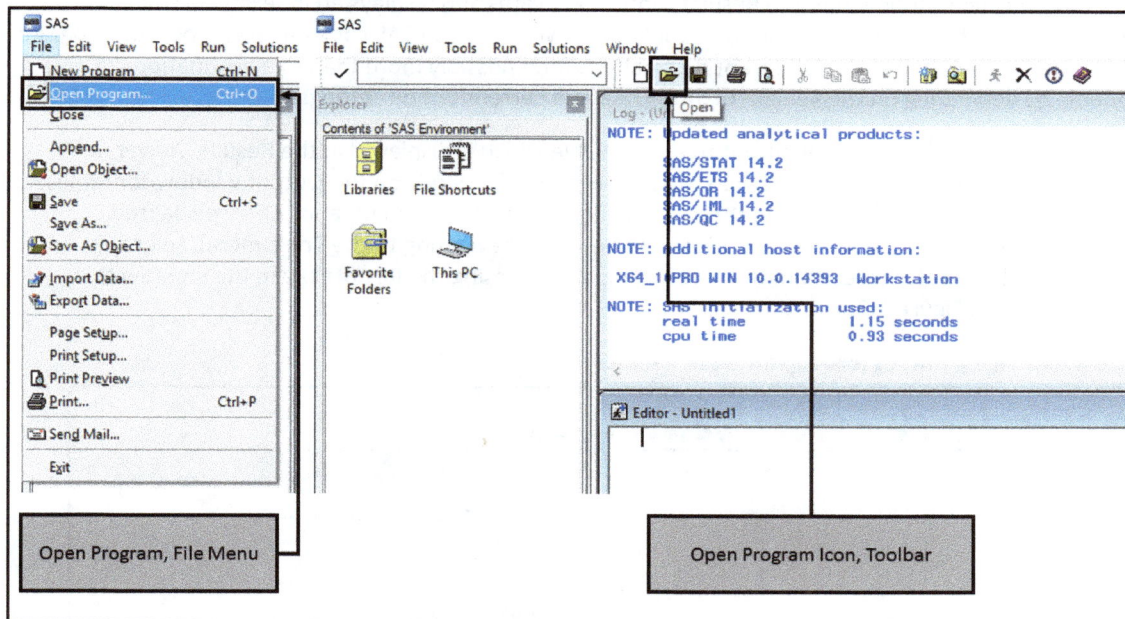

Either of these choices launches a standard Microsoft Windows file selection window, which is used to navigate to and select the file of interest. Upon successful selection of the code, it appears in the Editor window, and is displayed with some color coding as shown in Figure 1.3.3 (assuming the Enhanced Program Editor is in use, the Program Editor window provides different color coding). It is not important to understand the specific syntax or how the code works at this point, for now it is used simply to provide an executable program to introduce some SAS fundamentals.

Code submission can also occur in multiple ways, two of which are shown in Figure 1.3.3, again each method requires the Editor window to be the active window in the session. If multiple Editor windows are open, only code from the active window is submitted.

Figure 1.3.3: Submitting SAS Code in the SAS Windowing Environment

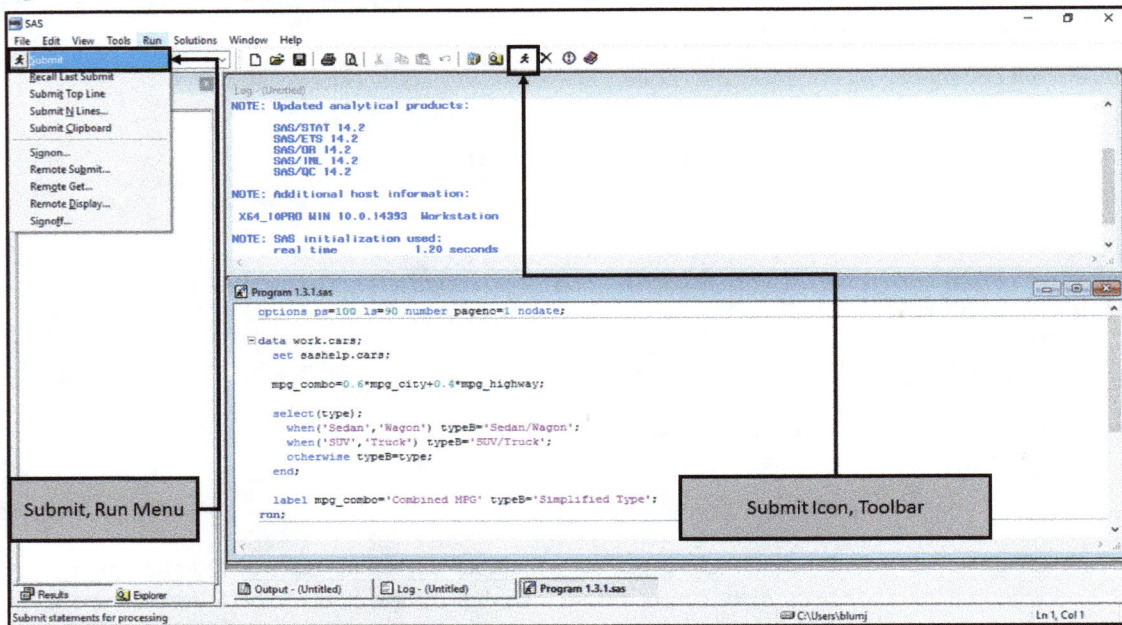

Typically, after any code submission the Results window activates and displays an index of links to various entities produced by the program, including output tables. While not all SAS code generates output, Program 1.3.1 does, and it may be routed to different destinations (and possibly more than one destination simultaneously) depending on the version of SAS in use and current option settings.

In SAS 9.4, the default settings route output to an HTML file which is displayed in the Results Viewer, a viewing window internal to the SAS session. Previous versions of SAS rely on the Output window for tables, an option which remains available for use in the SAS 9.4 windowing environment, and other specialized destinations for graphics. Default output options can be set by navigating to the **Tools** menu, selecting **Options**, followed by **Preferences** from that sub-menu, and choosing the **Results** tab in the window that appears, as shown in Figure 1.3.4.

Figure 1.3.4: Managing Output for Program Submissions

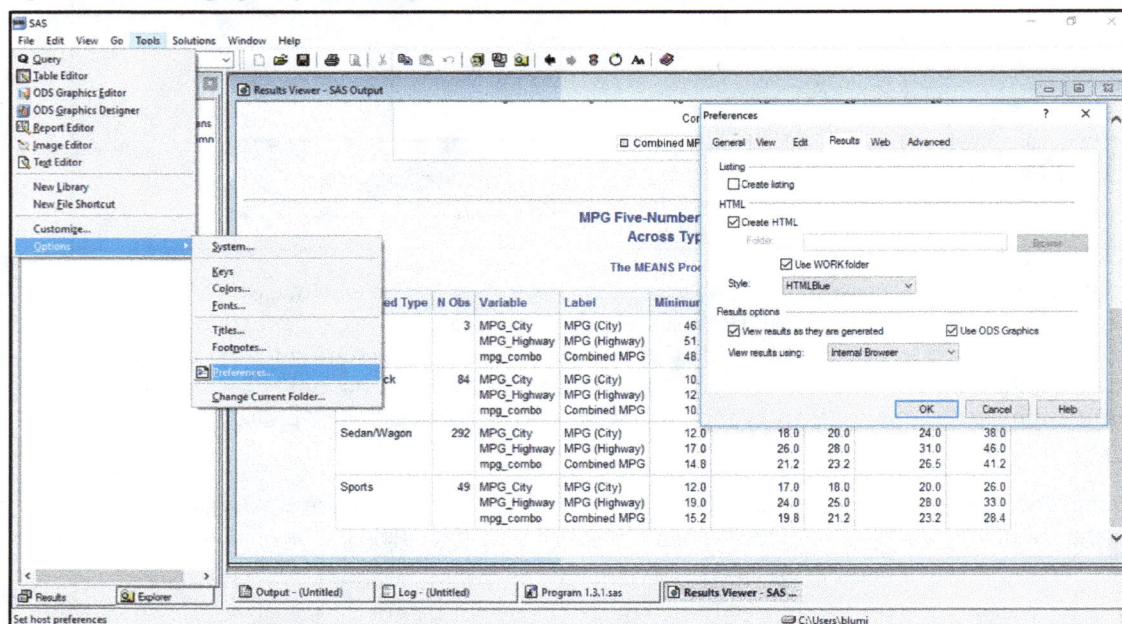

Among other options, Figure 1.3.4 shows the option for **Create HTML** checked and **Create Listing** unchecked. For tables, the listing destination is the Output window, so when **Create Listing** is checked, tables also appear in the Output window in what appears as a plain text form. It is possible to check both boxes, and it is also possible to check neither, whichever is preferred.

In the remainder of this book, output tables are shown in an RTF form embedded inside the book text, outside of any SAS Results window. Appearance of output tables and graphs in the book is similar to what is produced by a SAS session, but is not necessarily identical when default session options are in place. Later in this chapter, the ability to use SAS code to control delivery of output to each of these destinations is demonstrated, along with use of the listing destination as an output destination for graphics files.

1.3.2 SAS Studio and SAS University Edition

SAS Studio and SAS University Edition (which are, for the remainder of this text, singularly referred to as SAS University Edition) interface with SAS through a web browser. Typically, the browser used is the default browser for the machine hosting the SAS University Edition session, but this is not a requirement. Figure 1.3.5 shows a typical result of launching SAS University Edition (in this case using the Firefox browser on Microsoft Windows), launching in visual programmer mode by default. A closer match to the structure of the SAS windowing environment is provided by selecting SAS Programmer from the toolbar as shown.

Figure 1.3.5: SAS University Edition

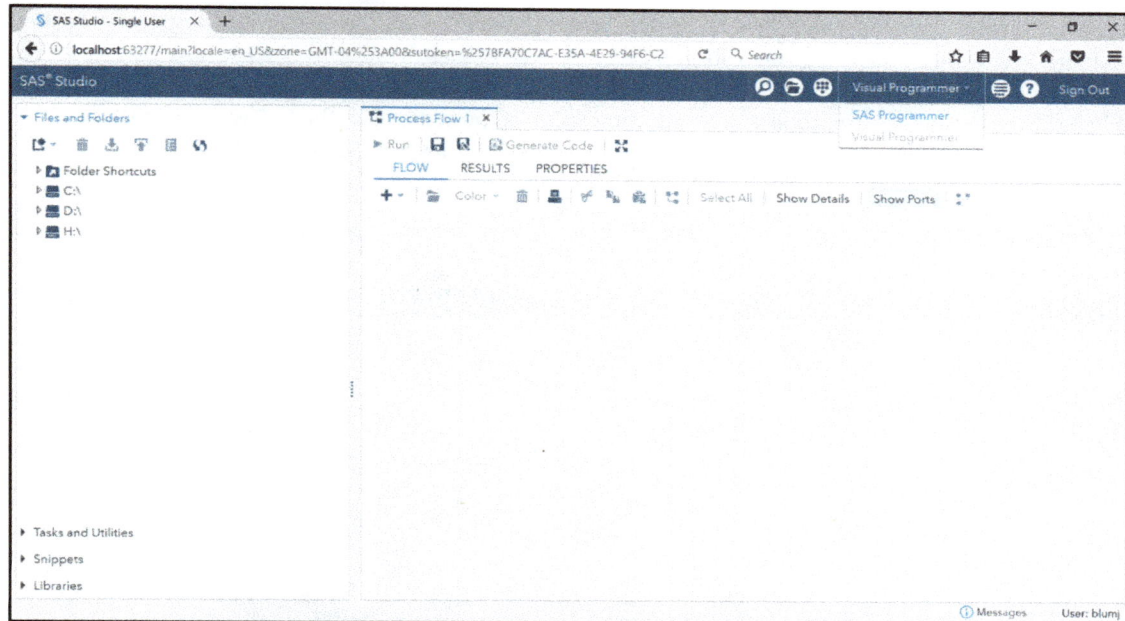

Opening a program is accomplished via the **Open** icon on the toolbar, as illustrated in Figure 1.3.6, and the opened code is displayed in a manner very similar to the that of the Enhanced Program Editor display shown in Section 1.3.1.

Figure 1.3.6: Opening a Program in SAS University Edition

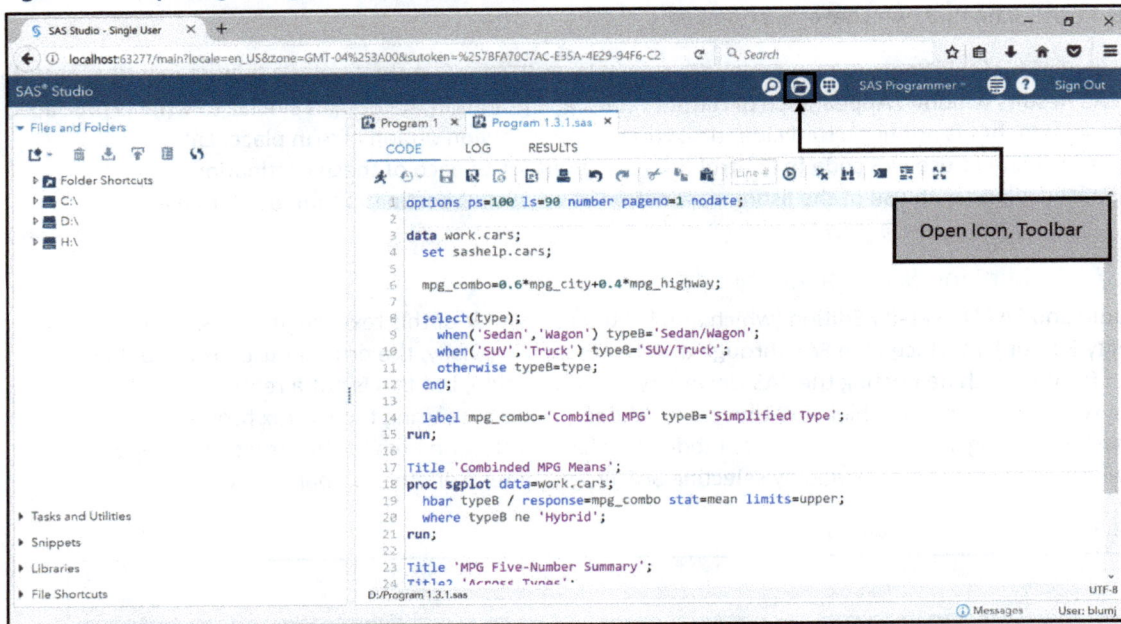

Though in a different position, the toolbar icon for submission is the same as in the SAS windowing environment, and selecting it produces output in the Results tab as shown in Figure 1.3.7.

Figure 1.3.7: Execution of a Program in SAS University Edition

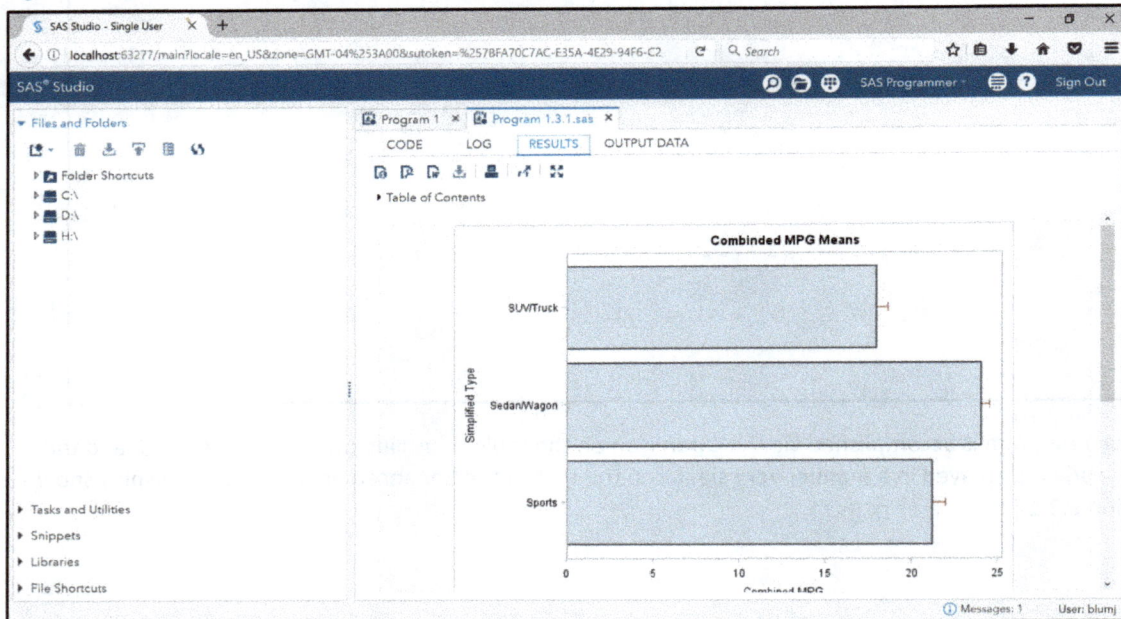

A few important differences to note in SAS University Edition: first, the output is displayed starting at the top, rather than at the bottom of the page as in the SAS windowing environment. Second, there is an additional tab for Output Data in this session. In Section 1.4.3, libraries, data sets, and navigation to each are discussed; however, SAS University Edition also includes a special tab whenever a program generates new data sets, which aids in directly viewing those results. Finally, note that the Code, Log, Results, and Output Data tabs are contained within the Program 1.3.1 tab, and each program opened is given its own set of tabs. In contrast, the SAS windowing environment supports multiple Editor windows in a single session, but they all share a common Log window, Output window, and (under default conditions) output HTML file. As discussed in other examples and in Chapter Notes 1 and 2 in Section 1.7, submissions from any and all Editor windows in the SAS windowing environment are cumulative in the Log and Output windows; therefore, managing results in each environment is quite different.

1.4 SAS Fundamentals

To build an initial understanding of how to work with programs in SAS, Program 1.3.1 is used repeatedly in this section to introduce various SAS language elements and concepts. For both SAS windowing environment and SAS University Edition, the features of each environment and navigation within them are discussed in conjunction with the language elements that relate to them.

1.4.1 SAS Language Basics

Program 1.4.1 is a duplicate of Program 1.3.1 with certain elements noted numerically throughout the code, followed by notes on the specific code in the indicated position. Throughout this book, this style is used to detail important features found in sample code.

Program 1.4.1: Program 1.3.1, Revisited

```
options ps=100 ls=90 number pageno=1 nodate; ❶

data work.cars; ❷
  set sashelp.cars;

  MPG_Combo=0.6*mpg_city+0.4*mpg_highway;

  select(type);
    when('Sedan','Wagon') TypeB='Sedan/Wagon';
    when('SUV','Truck') TypeB='SUV/Truck';
    otherwise TypeB=type;
  end;

  label mpg_combo='Combined MPG' typeB='Simplified Type';
 run;

title 'Combined MPG Means'; ❶
proc sgplot data=work.cars; ❸
  hbar typeB / response=mpg_combo stat=mean limits=upper;
  where typeB ne 'Hybrid';
run;

title 'MPG Five-Number Summary'; ❶
title2 'Across Types'; ❶
proc means data=work.cars min q1 median q3 max maxdec=1; ❸
  class typeB;
  var mpg:; ❹
run; ❺
```

❶   SAS code is written in statements, each of which ends in a semicolon. The statements indicated here (OPTIONS and TITLE) are examples of global statements. Global statements are statements that take effect as soon as SAS compiles those statements. Typically, the effects remain in place during the SAS session until another statement is submitted that alters those effects.

❷   The SAS DATA step has a variety of uses; however, it is primarily a tool for creation or manipulation of data sets. A DATA step is generally comprised of several statements forming a block of code, ending with the RUN statement, the role of which is described in ❹.

❸   Procedures in SAS are used for a variety of tasks and, like the DATA step, are generally comprised of several statements. These are generically referred to as PROC steps.

❹   The PROC MEANS result includes the variables MPG_City, MPG_Highway, and MPG_Combo even though none of these are explicitly written in the procedure code. The colon (:) at the end of a variable name acts as a wildcard indicating that any variable name starting with the prefix given is part of the designated set, this shortcut is known in SAS as a name prefix list. For other types of variable lists, see Chapter Note 3 in Section 1.7.

⑤ With DATA and PROC steps defined as blocks of code, each of these blocks is terminated with a step-boundary. The RUN statement is a commonly used as a step boundary, though it is not required for each DATA or PROC step. See Section 1.4.2 for details.

1.4.2 SAS DATA and PROC Steps

SAS processing of code submissions includes two major components: compilation and execution. In some cases, individual statements are compiled and take effect immediately, while at other times, a series of statements is compiled as a set and then executed after the complete set is processed by the compiler. In general, statements that compile and take effect individually and immediately are global statements. Statements that compile and execute as a set are generally referred to as steps, with the SAS language including both DATA steps and procedure (or PROC) steps.

The DATA step starts with a DATA statement, and a PROC step starts with a PROC statement that includes the name of the procedure, and all steps end with some form of a step boundary. As noted in Program 1.4.1, a commonly used step boundary in the SAS language is the RUN statement, but it is technically not required for each step. Any invocation of any DATA or PROC step is also defined as a step boundary due to the fact that DATA and PROC steps cannot be directly nested together in the SAS language. In general, it is considered a good programming practice to explicitly provide a statement for the step boundary, rather than implicitly through invocation of a DATA or PROC step. The code submissions in Figure 1.4.1 and Program 1.4.2 provide illustrations of the advantages of explicitly defining the end of a step.

In either the SAS windowing environment or SAS University Edition, portions of code can be compiled and executed by highlighting that section and then submitting. Having clear definitions from beginning to end for any DATA or PROC step aids in the ability to submit portions of code, which can be accomplished by using the RUN statement as an explicit step boundary. Figures 1.4.1A and 1.4.1B show submissions of the two PROC steps from Program 1.4.1 along with their associated TITLE statements.

Figure 1.4.1A: Submitting Portions of Code in SAS University Edition

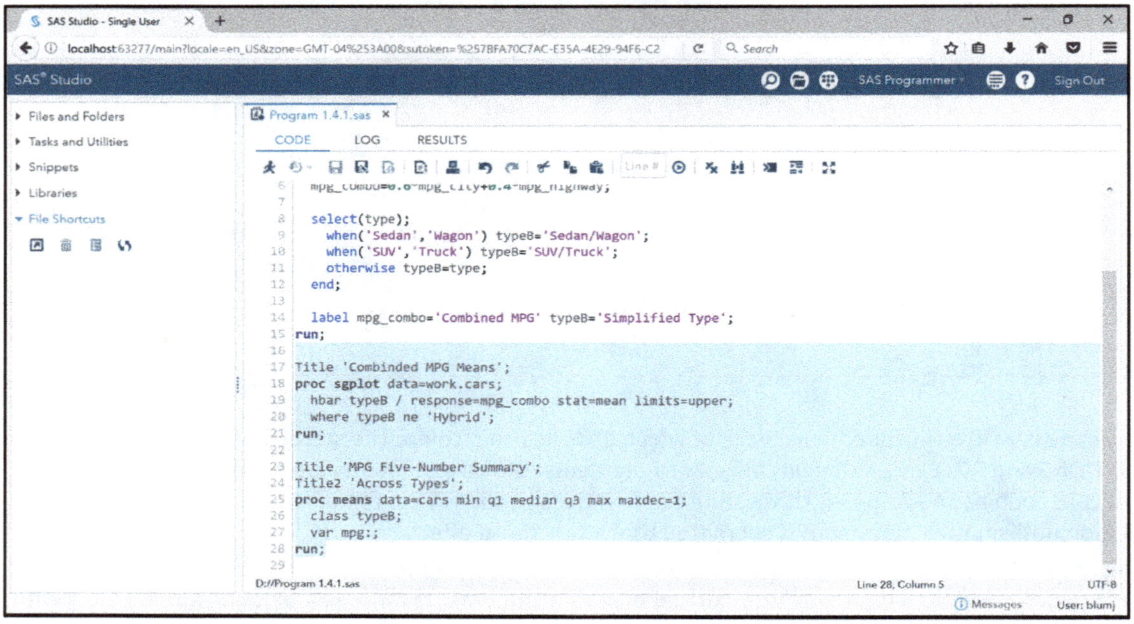

Figure 1.4.1B: Submitting Portions of Code in the SAS Windowing Environment

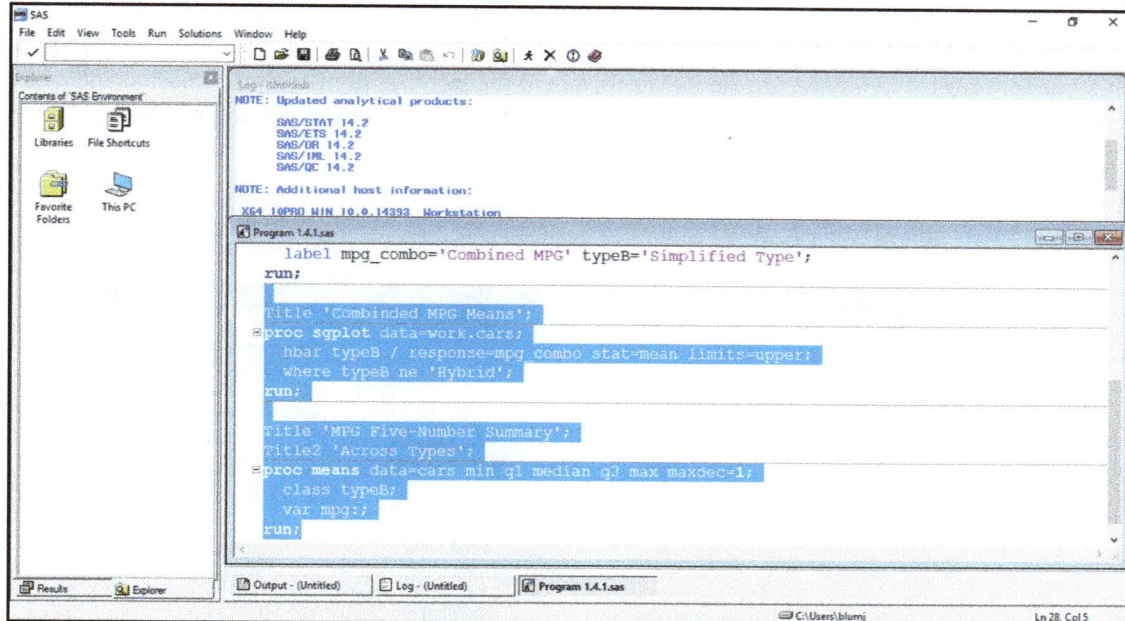

This submission reproduces the bar chart and the table of statistics produced previously in Figure 1.3.7. However, notice that the result is somewhat different in the SAS windowing environments and SAS University Edition. In the SAS windowing environment, the output is added to the output from the previous submission (and the log from this submission is also added to the previous log information). In SAS University Edition, the output is replaced, and the sub-tab for Output Data is not present because the DATA step did not run. With default settings in place, submissions are cumulative for both log and output in SAS windowing environment; conversely, replacement is the default in SAS University Edition. For more information about managing results in either environment, see Chapter Note 1 in Section 1.7.

Program 1.4.2 shows the code portion submitted in Figure 1.4.1 with the first RUN statement removed. Delete the RUN statement and re-submit the selection, review the output (Figure 1.4.2) and details below for another example of why explicitly ending steps in SAS is a good programming practice.

Program 1.4.2: Multiple Steps Without Explicit Step Boundaries

```
title 'Combined MPG Means'; ❶
proc sgplot data=work.cars; ❷
  hbar typeB / response=mpg_combo stat=mean limits=upper;
  where typeB ne 'Hybrid';
❸

title 'MPG Five-Number Summary';
title2 'Across Types'; ❹
proc means data=work.cars min q1 median q3 max maxdec=1; ❺
  class typeB;
  var mpg:;
run;
```

❶ The first statement compiled and executed is this TITLE statement, which assigns the quoted/literal value as the primary title line.

❷ The SGPLOT procedure is invoked for compilation and execution by this statement. Subsequent statements are compiled as part of the SGPLOT step until a step boundary is reached.

❸ This is the position of the RUN statement in Program 1.4.1 and, when it is compiled in that program, it signals the end of the SGPLOT step. Assuming no errors, PROC SGPLOT executes at that point; however, with no RUN statement present in this code, compilation of the SGPLOT step is not complete and execution does not begin.

❹   These two TITLE statements, which are global, now compile and take effect. Since the SGPLOT procedure still has not completed compilation, nor started execution, this TITLE statement replaces the first title line assigned in ❶.

❺   This statement starts the MEANS procedure which, due to the fact that steps cannot be nested, indicates that the SGPLOT statements are complete. Compilation of the SGPLOT step ends and it is executed, with the titles in ❹ now placed erroneously on the graph.

Figure 1.4.2: Failing to Define the End of a Step

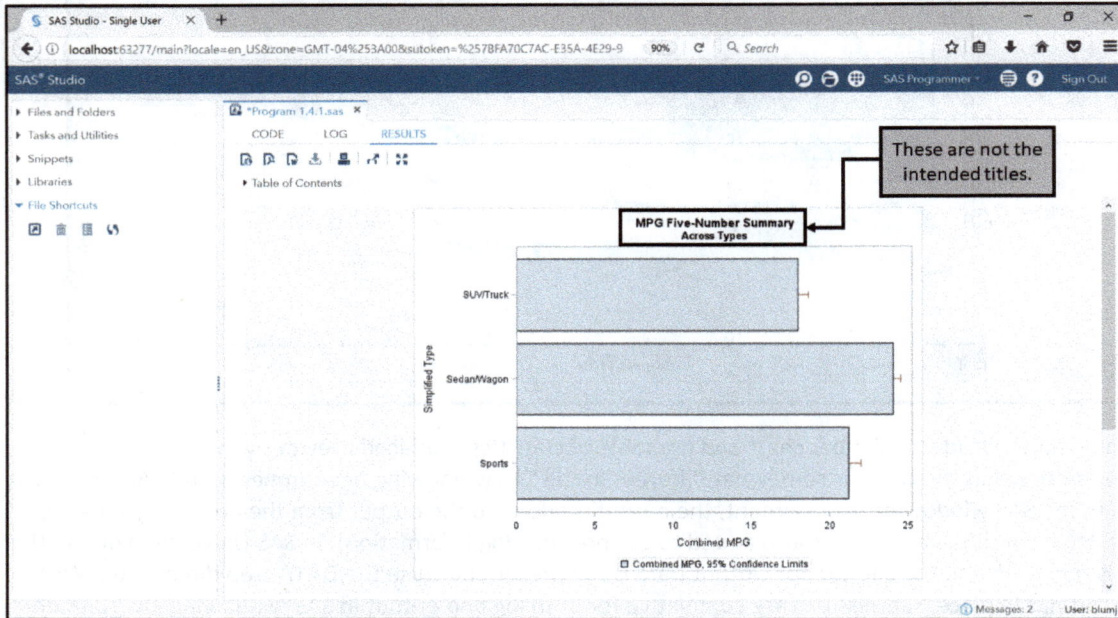

In any interactive session, the final step boundary must be explicitly stated. For a discussion of the differences between interactive and non-interactive sessions in the SAS windowing environment and SAS University Edition, see Chapter Note 2 in Section 1.7.

The remainder of Program 1.4.1 is a DATA step, which is shown as Program 1.4.3 with a few details about its operation highlighted. The DATA step is a powerful tool for data manipulation, offering a variety of functions, statements, and other programming elements. The DATA step is of such importance that it is featured in every chapter of this book.

Program 1.4.3: DATA Step from Program 1.4.1

```
data work.cars❶;
  set sashelp.cars❶;

  MPG_Combo=0.6*mpg_city+0.4*mpg_highway;❷

  select(type);
    when('Sedan','Wagon') TypeB='Sedan/Wagon';
    when('SUV','Truck') TypeB='SUV/Truck';
    otherwise TypeB=type;
  end;❸

  label mpg_combo='Combined MPG' typeB='Simplified Type';❹
run;
```

❶   The DATA statement that opens this DATA step names a SAS data set Cars in the Work library. Work.Cars is populated using the SAS data set referenced in the SET statement, also named Cars and located in the Sashelp library. Data set references are generally two-level references of the form `library.dataset`. The exception to this is the Work library, which is taken as the default library if only a data set name is provided. Details on navigating through libraries and data sets are given in Section 1.4.3.

❷   MPG_Combo is a variable defined via an arithmetic expression on two of the existing variables from the Cars data set in the Sashelp library. Assignments of the form `variable = expression;` do not require any explicit declaration of variable type to precede them, the compilation process determines the appropriate variable type from the expression itself. SAS data set variables are limited to two types: character or numeric.

❸   The variable TypeB is defined via assignment statements chosen conditionally based on the value of the Type variable. The casing of the literal values is an exact match for the casing in the data set as shown subsequently in Figures 1.4.4 and 1.4.5—matching of character values includes all casing and spacing. Various forms of conditional logic are available in the DATA step.

❹   Naming conventions in SAS generally follow three rules, with some exceptions noted later. Names are permitted to include only letters, numbers, and underscores; must begin with a letter or underscore; and are limited to 32 characters. Given these naming limitations, labels are available to provide more flexible descriptions for the variable. (Labels are also available for data sets and other entities.) Also note that references to the variables MPG_Combo and TypeB use different casing here than in their assignment expressions; in general, the SAS language is not case-sensitive.

1.4.3 SAS Libraries and Data Sets

Program 1.4.1 involves data sets in each of its programming steps. The DATA step uses one data set as the foundation for creating another, and the data set it creates is used in each of the PROC steps that follow. Again, data set references are generally in a two-level form of `library.dataset`, other than the exception for the Work library noted in the discussion of Program 1.4.3. The PROC steps in Program 1.4.1 each use one of the possible forms to reference the Cars data located in the Work library.

Navigation to data sets in various libraries is possible in either the SAS windowing environment or SAS University Edition. In the SAS windowing environment, the Explorer window permits navigation to any assigned library, while in SAS University Edition, the left panel contains a section for libraries. In either setting, opening a library potentially reveals a series of table icons representing various SAS data sets, which can be opened to view the contents of the data. As an example, navigation to and opening of the Cars data set in the Sashelp library is shown below for each of the SAS windowing environment and SAS University Edition. Figures 1.4.3 and 1.4.4 demonstrate one way to open the Cars data in the SAS windowing environment.

Figure 1.4.3: Starting Points for Library Navigation, SAS Windowing Environment

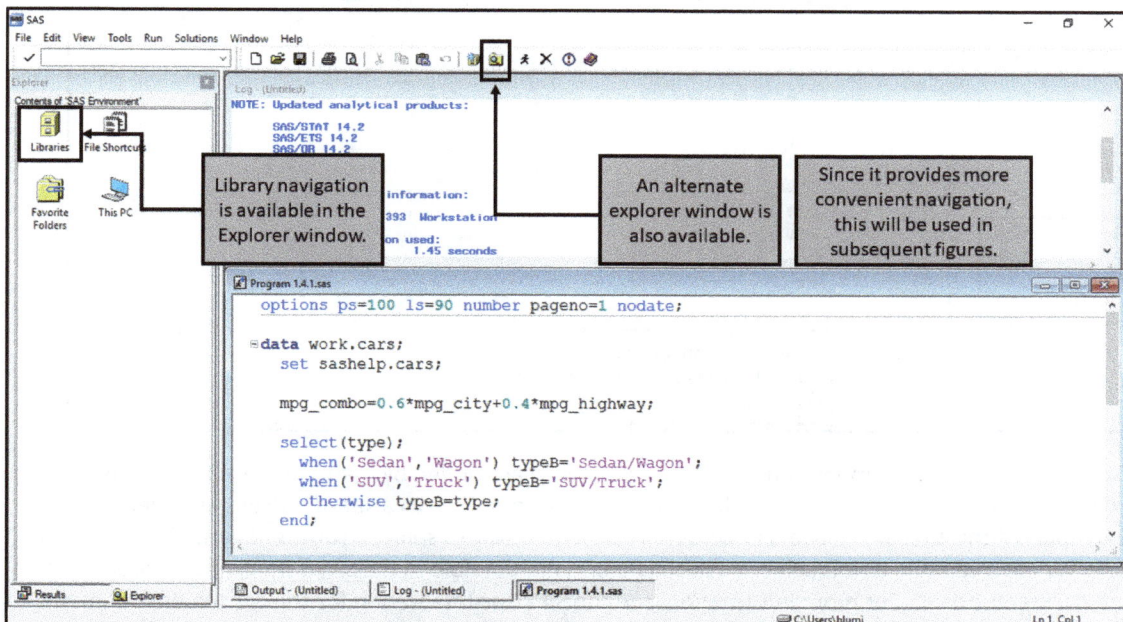

Figure 1.4.4: Accessing the Cars Data Set in the Sashelp Library in the Windowing Environment

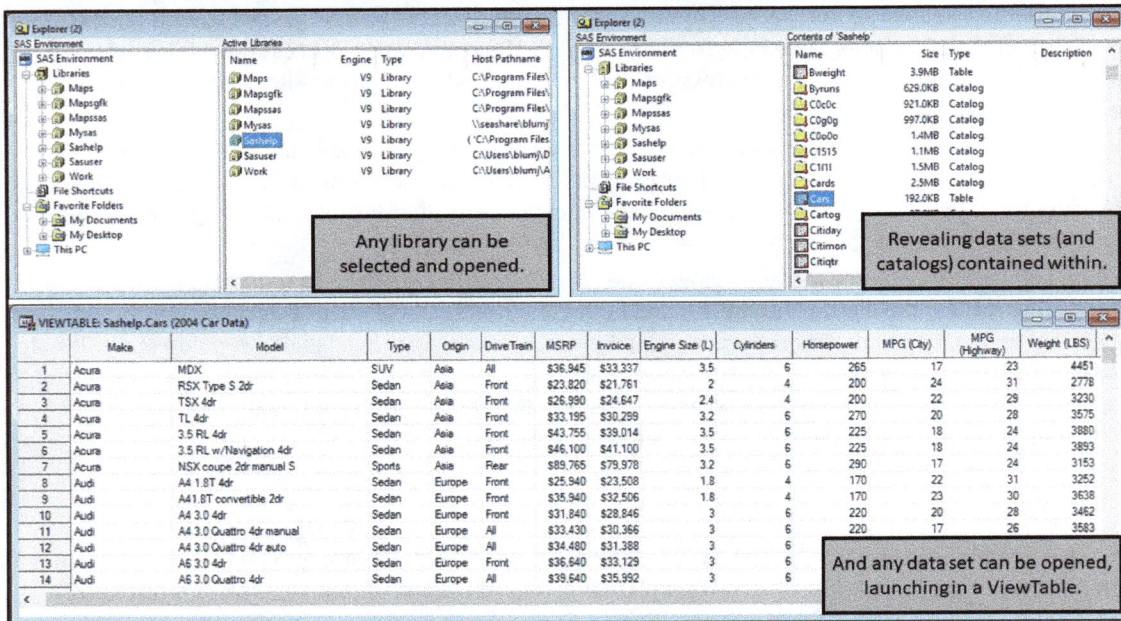

Figure 1.4.5 shows how to open the Cars data set in a SAS University Edition session, revealing several differences in the library navigation and the data view, which opens in a separate tab in the University Edition session.

Figure 1.4.5: Accessing the Cars Data Set in the Sashelp Library in University Edition

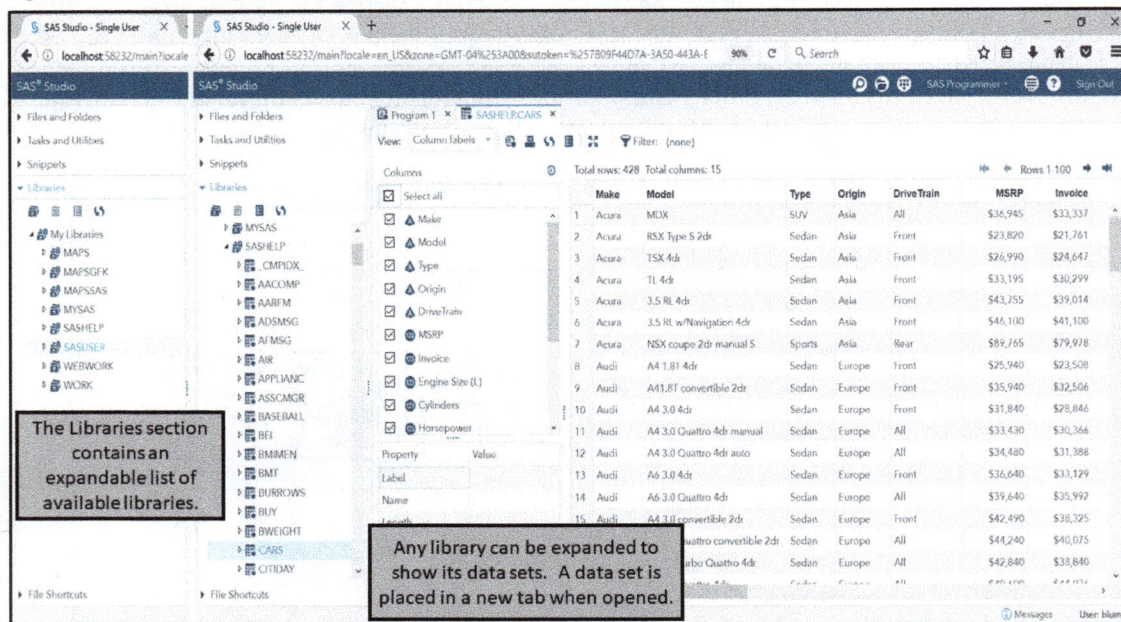

In the SAS windowing environment, options for the data view are driven by menus and toolbar buttons for the active window, while in SAS University Edition, each data set tab contains a set of buttons and menus in its toolbar. As part of this tab, SAS University Edition also offers boxes to select a subset of variables and gives properties for each variable as it is selected. Such changes are possible in the SAS windowing environment as well, but are menu-driven. For more information, see the SAS Documentation for the chosen environment.

Another major difference between the two data views is that the ViewTable in the SAS windowing environment has active control over the data set selected. The view in SAS University Edition is generated when the data set is opened, or re-generated if new options are selected, and control of the data set is released. For further detail on the implications of these differences, see Chapter Note 4 in Section 1.7.

Though there are ultimately several different forms of SAS libraries, the most basic simply assigns a library reference (or *libref*) to a folder which the SAS session can access. A library can be assigned in a program via the LIBNAME statement or through other tools available in the SAS windowing environment or SAS University Edition. In order to use this book, it is essential to assign library references to the data sets downloaded from the author web pages. Figures 1.4.6 and 1.4.7 show an assignment of a library named BookData to an assumed location. The path must be set to the actual location of the downloaded files, and the choice of library name must follow the naming conventions given previously in Program 1.4.3, with the additional restriction that the library reference is limited to 8 characters.

Figure 1.4.6: Assigning a Library in the SAS Windowing Environment

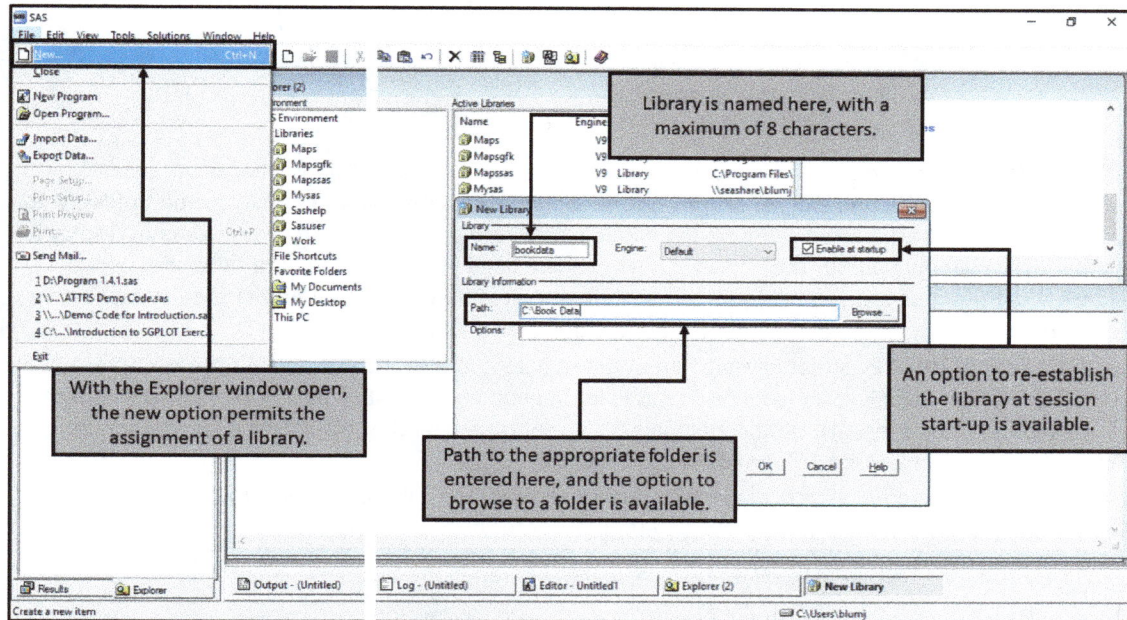

Figure 1.4.7: Assigning a Library in SAS University Edition

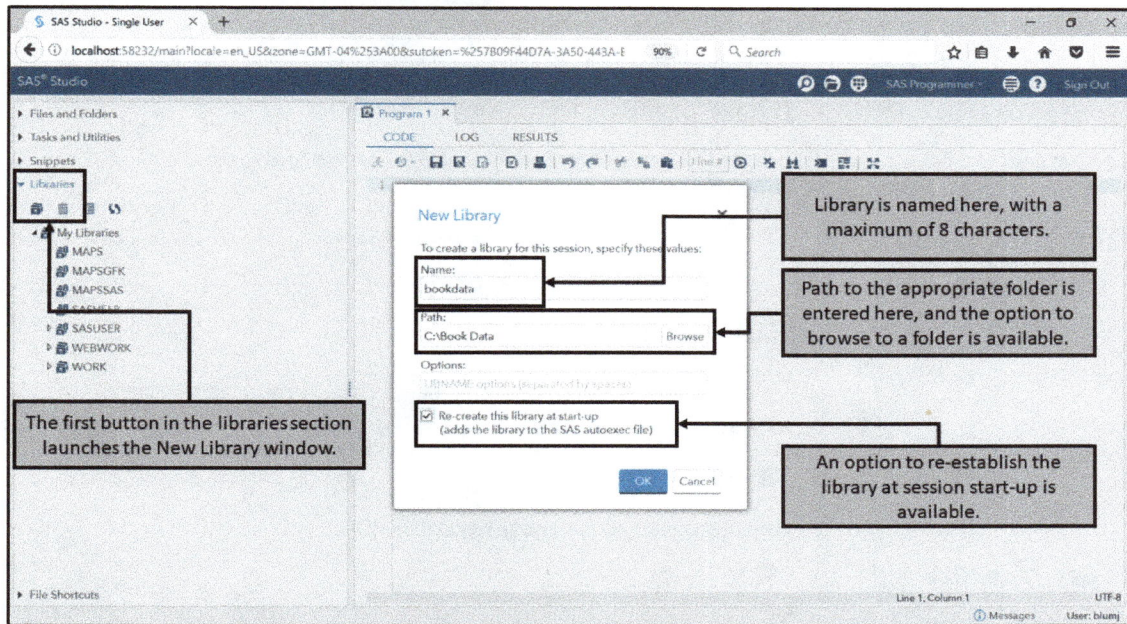

Submitting the following LIBNAME statement is equivalent to the assignments shown in the Figures 1.4.6 and 1.4.7, except for the fact that the assigned library is not re-created at start-up of the next session.

```
libname bookdata 'C:\Book Data';
```

Any of these assignments creates what is known as a permanent library, meaning that data sets and other files stored there remain in place until an explicit modification is made to them. Temporary libraries are expunged when the SAS session ends—in the SAS Windowing environment, Work is a temporary library; in SAS University Edition, Work and Webwork are temporary.

The PRINT and CONTENTS procedures provide information about data and metadata as program output. Program and Output 1.4.4 provide a demonstration of their use.

Program 1.4.4: Using the CONTENTS and PRINT Procedures to Display Metadata and Data

```
proc contents data=sashelp.cars; ❶
run;

proc print data=sashelp.cars(obs=10) label; ❷
  var make model msrp mpg_city mpg_highway; ❸
run;
```

❶ The CONTENTS procedure output shows a variety of metadata, including the number of variables, number of observations, and the full set of variables and their attributes. Adding the option VARNUM to the PROC CONTENTS statement reorders the variable attribute table in column order—default display is alphabetical by variable name. The keyword _ALL_ can be used in place of the data set name, in this instance, the output contains a full list of all library members followed by metadata for each data set in the library.

❷ The PRINT procedure directs the data portion of the selected data set to all active output destinations. By default, PROC PRINT displays variable names as column headers, the LABEL option changes these to the variable labels (when present). Display of labels is also controlled by the LABEL/NOLABEL system options, see Chapter Note 5 in Section 1.7 for additional details.

❸ Default behavior of the PRINT procedure is to output all rows and columns in the current data order. The VAR statement selects the set of columns and their order for display.

Output 1.4.4A: Output from PROC CONTENTS for Sashelp.Cars

Data Set Name	SASHELP.CARS	Observations	428
Member Type	DATA	Variables	15
Engine	V9	Indexes	0
Created	*Local Information Differs*	Observation Length	152
Last Modified	*Local Information Differs*	Deleted Observations	0
Protection		Compressed	NO
Data Set Type		Sorted	YES
Label	2004 Car Data		
Data Representation	WINDOWS_64		
Encoding	us-ascii ASCII (ANSI)		

Engine/Host Dependent Information	
Data Set Page Size	65536
Number of Data Set Pages	2
First Data Page	1
Max Obs per Page	430
Obs in First Data Page	413

Engine/Host Dependent Information	
Number of Data Set Repairs	0
ExtendObsCounter	YES
Filename	*Local Information Differs*
Release Created	9.0401M4
Host Created	X64_SR12R2
Owner Name	BUILTIN\Administrators
File Size	192KB
File Size (bytes)	196608

Alphabetic List of Variables and Attributes					
#	Variable	Type	Len	Format	Label
9	Cylinders	Num	8		
5	DriveTrain	Char	5		
8	EngineSize	Num	8		Engine Size (L)
10	Horsepower	Num	8		
7	Invoice	Num	8	DOLLAR8.	
15	Length	Num	8		Length (IN)
11	MPG_City	Num	8		MPG (City)
12	MPG_Highway	Num	8		MPG (Highway)
6	MSRP	Num	8	DOLLAR8.	
1	Make	Char	13		
2	Model	Char	40		
4	Origin	Char	6		
3	Type	Char	8		
13	Weight	Num	8		Weight (LBS)
14	Wheelbase	Num	8		Wheelbase (IN)

Sort Information	
Sortedby	Make Type
Validated	YES
Character Set	ANSI

Output 1.4.4B: Output from PROC PRINT (First 10 Rows) for Sashelp.Cars

Obs	Make	Model	MSRP	MPG (City)	MPG (Highway)
1	Acura	MDX	$36,945	17	23
2	Acura	RSX Type S 2dr	$23,820	24	31

Obs	Make	Model	MSRP	MPG (City)	MPG (Highway)
3	Acura	TSX 4dr	$26,990	22	29
4	Acura	TL 4dr	$33,195	20	28
5	Acura	3.5 RL 4dr	$43,755	18	24
6	Acura	3.5 RL w/Navigation 4dr	$46,100	18	24
7	Acura	NSX coupe 2dr manual S	$89,765	17	24
8	Audi	A4 1.8T 4dr	$25,940	22	31
9	Audi	A41.8T convertible 2dr	$35,940	23	30
10	Audi	A4 3.0 4dr	$31,840	20	28

1.4.4 The SAS Log

The SAS log tracks program submissions and generates information during compilation and execution to aid in the debugging process. Most of the information SAS displays in the log (besides repeating the code submission) falls into one of five categories:

- *Errors*: An error in the SAS log is an indication of a problem that has stopped the compilation or execution process. These may be generated by either syntax or logic errors, see the example in this section for a discussion of the differences in these two error types.

- *Warnings*: A warning in the SAS log is an indication of something unexpected during compilation or execution that was not sufficient to stop either from occurring. Most warnings are an indication of a logic error, but they can also reflect other events, such as an attempt by the compiler to correct a syntax error.

- *Notes*: Notes give various information about the submission process, including: process time, records and data set used, locations for file delivery, and other status information. However, some notes actually indicate potential problems during execution. Therefore, reviewing notes is important, and they should not be presumed to be benign. Program 1.4.5, along with others in later chapters, illustrates such an instance.

- *Additional Diagnostic Information*: Depending on the nature of the note, error, or warning, SAS may transmit additional information to the log to aid in diagnosing the problem.

- *Requested Information*: Based on various system options and other statements, a SAS program can request additional information be transmitted to the SAS log. The ODS TRACE statement is one such statement covered in Section 1.5. Other statements and options are included in later chapters.

Program 1.4.5 introduces errors into the code given in Program 1.4.1, with a review of the nature of the mistakes and the log entries corresponding to them shown in Figure 1.4.8. Errors can generally be split into two types: syntax and non-syntax errors. A syntax error occurs when the compiler is unable to recognize a portion of the code as a legal statement, option, or other language element; thus, it is a situation where programming statements do not conform to the rules of the SAS language. A non-syntax error occurs when correct syntax rules are used, but in a manner that leads to an incorrect result (including no result at all). In this book, non-syntax errors are also referred to as logic errors (an abbreviated phrase referring to errors in programming logic). Chapter Note 6 in Section 1.7 provides a further refinement of such error types.

Program 1.4.5: Program 1.4.1 Revised to Include Errors

```
options pagesize=100 linesize=90 number pageno=1 nodate;

data work.cars;
  set sashelp.cars;

  mpg_combo=0.6*mpg_city+0.4*mpg_highway;
```

```
    select(type);
      when('Sedan','Wagon') typeB='Sedan/Wagon';
      when('SUV','Truck') typeB='SUV/Truck';
      otherwise typeB=type;
    end;

    label mpg_combo='Combined MPG' type2❶='Simplified Type';
run;

Title 'Combined MPG Means';
proc sgplot daat❷=work.cars;
  hbar typeB / response=mpg_combo stat=mean limits=upper;
  where typeB ne 'Hybrid';
run;

Title 'MPG Five-Number Summary';
Titletwo❸ 'Across Types';
proc means data=car❹ min q1 median q3 max maxdec=1;
  class typeB;
  var mpg:;
run;
```

❶   This is a non-syntax error; the variable name Type2 is legal and is used correctly in the LABEL statement. However, no variable named Type2 has been defined in the data set.

❷   This is a syntax error, `daat` is not a legal option in this PROC statement.

❸   This is a syntax error, `titletwo` is not a legal statement name.

❹   This is a non-syntax error; the syntax is legal and directs the procedure to use a data set named Car in the Work library; however, no such data set exists.

Figure 1.4.8A: Checking the SAS Log for Program 1.4.4, First Page

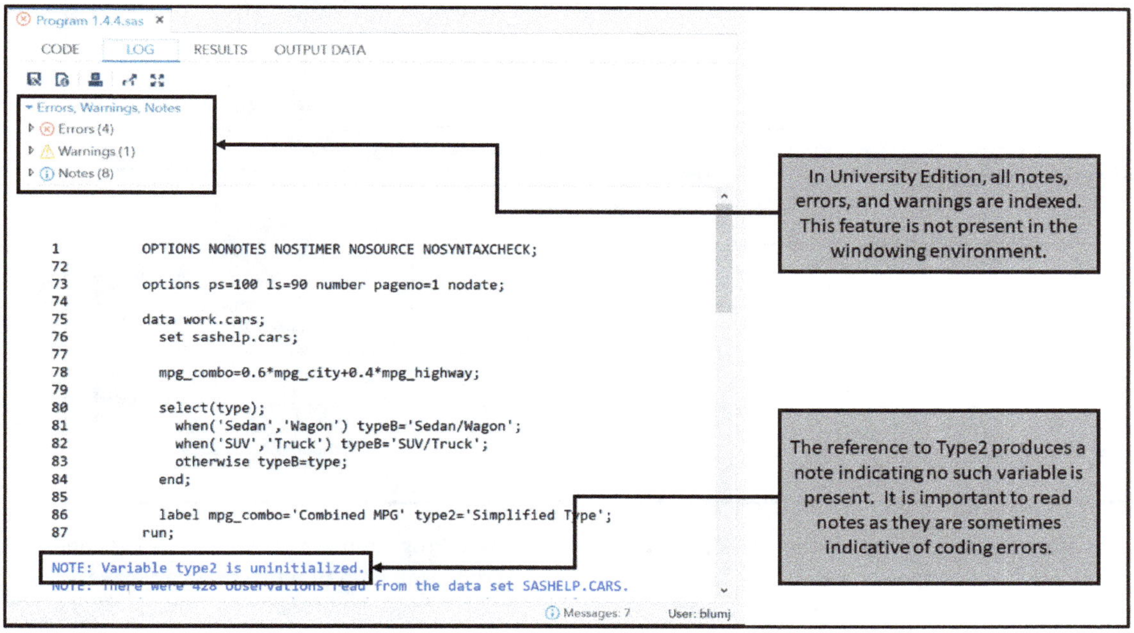

Figure 1.4.8B: Checking the SAS Log for Program 1.4.4, Second Page

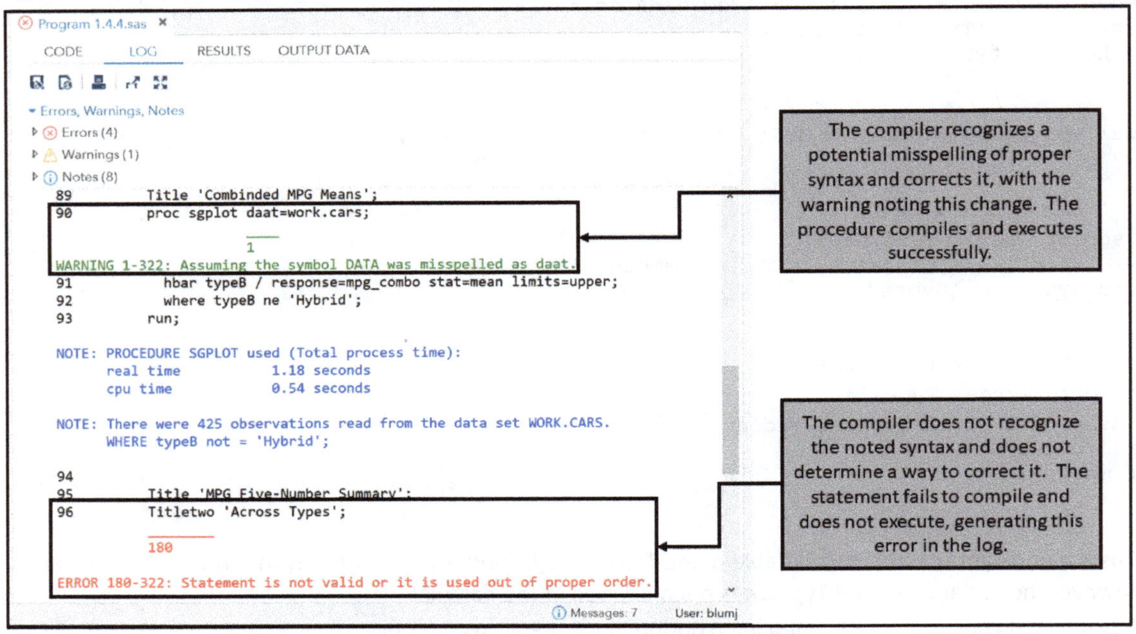

Figure 1.4.8C: Checking the SAS Log for Program 1.4.4, Third Page

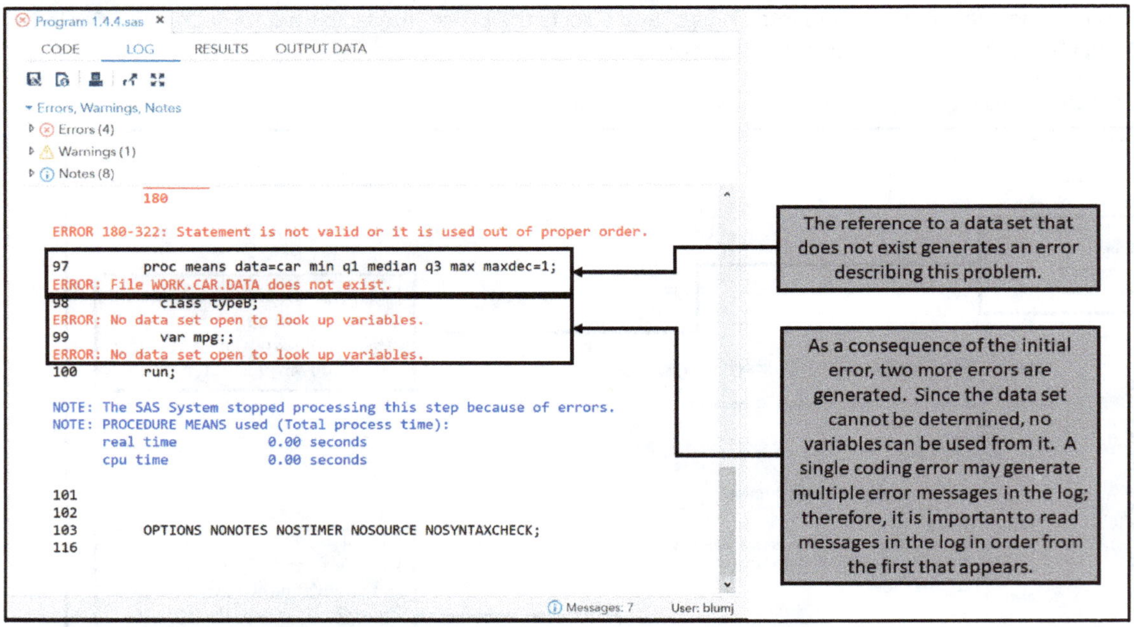

The value of a complete review of the SAS log cannot be overstated. Programmers often believe the code is correct if it produces output or if the log does not contain errors or warnings, a practice that can leave undetected problems in the code and the results.

Upon invocation of the SAS session, the log also displays notes, warnings, and errors as appropriate relating to the establishment of the SAS session. See the SAS Documentation for information about these messages.

1.5 Output Delivery System

The sample code presented in this section introduces SAS programming concepts that are important for working effectively in a SAS session and for re-creating samples shown in subsequent sections of this book. Delivery of output to various destinations, naming output files, and choosing the location where they are stored are included. Some differences in appearance that may arise between destinations are also discussed.

Program 1.5.1 revisits the CONTENTS procedure shown in Program 1.4.4, which generates output that is arranged and displayed in four tables. An Output Delivery System (ODS) statement, ODS TRACE ON, is supplied to deliver information to the log about all output objects generated.

Program 1.5.1: Using ODS TRACE to Track Output

```
ods trace on; ❶
proc contents data=sashelp.cars; ❷
run;

proc contents data=sashelp.cars varnum; ❷
run;
```

❶ There are many ODS statements available in SAS, some act globally—they remain in effect until another statement alters that effect—while others act locally—for the execution of the current or next procedure. The TRACE is a recording of all output objects generated by the code execution. ON delivers this information to the SAS log; OFF suppresses it. The effect of ODS TRACE is global, the ON or OFF condition only changes with a submission of a new ODS TRACE statement that makes the change. The typical default at the invocation of a SAS session is OFF.

❷ The VARNUM option in PROC CONTENTS rearranges the table showing the variable information from alphabetical order to position order. This also represents a change in the name of the table as indicated in the TRACE information shown in the log.

Log 1.5.1A: Using ODS TRACE to Track Output

```
74          ods trace on;
75             proc contents data=sashelp.cars;
76          run;


Output Added:
-------------
Name:       Attributes
Label:      Attributes
Template:   Base.Contents.Attributes
Path:       Contents.DataSet.Attributes
-------------

Output Added:
-------------
Name:       EngineHost
Label:      Engine/Host Information
Template:   Base.Contents.EngineHost
Path:       Contents.DataSet.EngineHost
-------------

Output Added:
-------------
Name:       Variables
Label:      Variables
Template:   Base.Contents.Variables
Path:       Contents.DataSet.Variables
-------------

Output Added:
-------------
Name:       Sortedby
Label:      Sortedby
Template:   Base.Contents.Sortedby
Path:       Contents.DataSet.Sortedby
-------------
NOTE: PROCEDURE CONTENTS used (Total process time):
      real time          0.29 seconds
      cpu time           0.20 seconds
```

Log 1.5.1B: Using ODS TRACE to Track Output

```
78            proc contents data=sashelp.cars varnum;
79            run;


Output Added:
-------------
Name:        Attributes
Label:       Attributes
Template:    Base.Contents.Attributes
Path:        Contents.DataSet.Attributes
-------------

Output Added:
-------------
Name:        EngineHost
Label:       Engine/Host Information
Template:    Base.Contents.EngineHost
Path:        Contents.DataSet.EngineHost
-------------

Output Added:
-------------
Name:        Position
Label:       Varnum
Template:    Base.Contents.Position
Path:        Contents.DataSet.Position
-------------

Output Added:
-------------
Name:        Sortedby
Label:       Sortedby
Template:    Base.Contents.Sortedby
Path:        Contents.DataSet.Sortedby
-------------
NOTE: PROCEDURE CONTENTS used (Total process time):
      real time            0.13 seconds
      cpu time             0.07 seconds
```

Each table generated by PROC CONTENTS has a name and a label; sometimes these are the same. Labels are free-form, while names follow the SAS naming conventions described earlier which are revisited in Section 1.6.2. The SAS Documentation also includes lists of ODS table names for each procedure, along with information about which are generated as default procedure output and which tables are generated as the result of including specific options. From the traces shown in Logs 1.5.1A and 1.5.1B, the rearrangement of the variable information when using the VARNUM option is actually a replacement of the Variables table with the Position table.

If ODS table names (and other output object names, such as graphs) are known, other forms of ODS statements are available to choose which output to include or not. Program 1.5.2 shows how to modify each of the CONTENTS procedures in Program 1.5.1 to only display the variable information.

Program 1.5.2: Using ODS SELECT to Subset Output

```
proc contents data=sashelp.cars;
  ods select Variables; ❶
run;

proc contents data=sashelp.cars varnum;
  ods select Position; ❷
run;
```

❶ ODS SELECT and ODS EXCLUDE each support a space-separated list of output object names. SELECT chooses output objects to be delivered; EXCLUDE chooses those that are not delivered. Only one should be used in any procedure, typically corresponding to whichever list of tables is shorter—those to be included or excluded. In place of the list of object names, one of the keywords of ALL or NONE can be used. ODS SELECT or ODS EXCLUDE can be placed directly before or within a procedure, and its effect is local if a list of objects is given, only applying to the execution of that procedure. If the ALL or NONE keywords are used, the effect is global, remaining in place until another statement alters it.

❷ The VARNUM option produces a table (Output 1.5.2B) with the same variable information as the first PROC CONTENTS but, as shown in the trace, it is a different table with a different name.

Output 1.5.2A: Using ODS SELECT to Subset Output

	Alphabetic List of Variables and Attributes				
#	Variable	Type	Len	Format	Label
9	Cylinders	Num	8		
5	DriveTrain	Char	5		
8	EngineSize	Num	8		Engine Size (L)
10	Horsepower	Num	8		
7	Invoice	Num	8	DOLLAR8.	
15	Length	Num	8		Length (IN)
11	MPG_City	Num	8		MPG (City)
12	MPG_Highway	Num	8		MPG (Highway)
6	MSRP	Num	8	DOLLAR8.	
1	Make	Char	13		
2	Model	Char	40		
4	Origin	Char	6		
3	Type	Char	8		
13	Weight	Num	8		Weight (LBS)
14	Wheelbase	Num	8		Wheelbase (IN)

Output 1.5.2B: Using ODS SELECT to Subset Output

#	Variable	Type	Len	Format	Label
	Variables in Creation Order				
1	Make	Char	13		
2	Model	Char	40		
3	Type	Char	8		
4	Origin	Char	6		
5	DriveTrain	Char	5		
6	MSRP	Num	8	DOLLAR8.	
7	Invoice	Num	8	DOLLAR8.	
8	EngineSize	Num	8		Engine Size (L)
9	Cylinders	Num	8		
10	Horsepower	Num	8		
11	MPG_City	Num	8		MPG (City)
12	MPG_Highway	Num	8		MPG (Highway)
13	Weight	Num	8		Weight (LBS)
14	Wheelbase	Num	8		Wheelbase (IN)
15	Length	Num	8		Length (IN)

ODS statements can be used to direct output to various destinations, including multiple destinations at any one time. Output styles can vary across destinations, as Program 1.5.3 demonstrates by delivering the same graph to a PDF and PNG file.

Program 1.5.3: Setting Output Destinations Using ODS Statements

```
x 'cd C:\Output'; ❶
ods _ALL_ CLOSE; ❷
ods listing; ❸
ods pdf file='Output 1-5-3.pdf'; ❹
proc sgplot data=sashelp.cars; ❺
        styleattrs datasymbols=(square circle triangle);
        scatter y=mpg_city x=horsepower/group=type;
        where type in ('Sedan','Wagon','Sports');
run;
ods pdf close; ❻
```

❶   The X command allows for submission of command line statements. CD is the change directory command in both Windows and Linux, here its effect is to change the SAS working directory. The SAS working directory is the default destination for any file reference that does not include a full path—one that starts with a drive letter or name. This directory must exist to successfully submit this code; therefore, either create the directory C:\Output or substitute another that the SAS session has write access to.

❷   The ODS _ALL_ CLOSE statement closes all output destinations.

❸   The ODS LISTING statement activates the listing destination, which is the destination for all graphics files created by the SGPLOT procedure. In the SAS windowing environment the ODS LISTING statement also activates the Output window, but graphics generated by PROC SGPLOT are not displayed there.

❹   The ODS PDF statement opens the PDF destination specified in the FILE= option (if this option is omitted the file is automatically named). Since the file name does not reference any path, it is placed in the location specified in ❶. A full-path reference, starting with a drive letter or name, can be given here. Commonly used destinations include PDF, RTF, HTML, and LISTING, but several others are available.

❺ Output 1.5.3A shows the graph generated by PROC SGPLOT and placed in the PDF file, while Output 1.5.3B shows the graphics file (a PNG file by default) generated. Note the difference in appearance between the two (and check the log)—different output destinations can have different options or styles in effect.

❻ The ODS PDF CLOSE statement closes the PDF destination opened in ❹ and completes writing of the file, which includes all output generated between the opening and closing ODS statements. In general, any ODS statement that opens a destination should have a complementary CLOSE statement.

Output 1.5.3A: Graph Delivered to PDF File

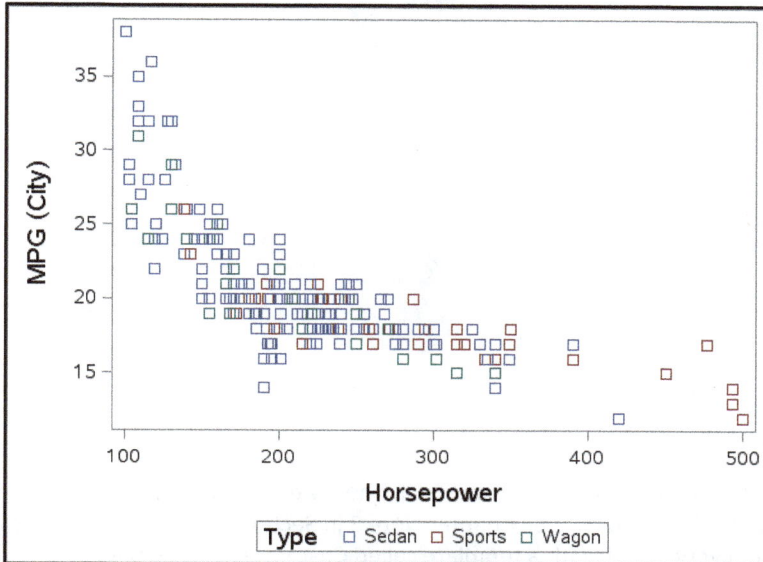

Output 1.5.3B: Graph Delivered to PNG File (Listing Destination)

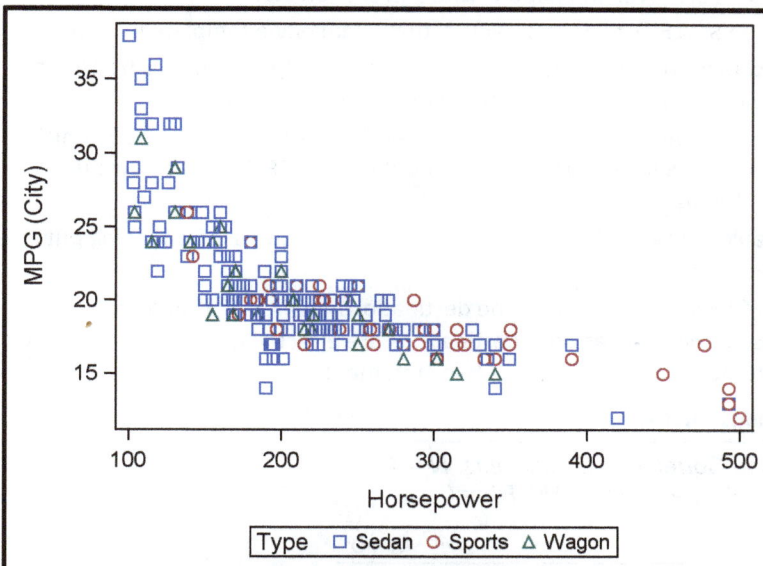

While the graph in the PDF file uses the same plotting shape and cycles the colors, the one delivered as an image file cycles through both different shapes and colors. The takeaway from this example, which applies in several instances, is that not all output destinations use the same styles. In this book, graphs are shown in the form generated by direct delivery to TIF files, see Chapter Note 7 in Section 1.7 for options used to generate these graphs.

Program 1.5.3 shows that ODS statements permit delivery of output to more than one destination at a time, and they also allow for different subsets of output to be delivered to each. While Program 1.5.3 shows that different destinations may have certain style elements that are different, it is also possible to specifically prescribe different styles to different output destinations. Program 1.5.4 opens a PDF and an RTF destination, sending different subsets of the output to each, and with different styles assigned to each.

Program 1.5.4: Setting Multiple Output Destinations and Styles Using ODS Statements

```
ods rtf file='RTF Output 1-5-4.rtf' style=journal; ❶
ods pdf file='PDF Output 1-5-4.pdf'; ❷

ods trace on; ❸
proc corr data=sashelp.cars;
  var mpg_city;
  with mpg_highway;
  ods select pearsoncorr;
run;

ods rtf exclude onewayfreqs; ❹
proc freq data=sashelp.cars;
  table type;
run;

ods pdf exclude summary; ❺
proc means data=sashelp.cars;
  class origin;
  var mpg_city;
run;
ods rtf close;
ods pdf close; ❻
```

❶ This opens an RTF destination and applies a style named Journal to it. The style templates available in a given SAS session can be viewed via PROC TEMPLATE, see Chapter Note 8 in Section 1.7 for details. Most tables in this book are the result of delivery to RTF with a template called CustomSapphire—the code required to create the CustomSapphire template is provided in the code that comes with the book, and details about how to set it up and use it are given in Chapter Note 9 in Section 1.7.

❷ This opens a PDF destination, since no STYLE= option is provided, the default style template is used.

❸ The ODS TRACE ON statement is included to ensure that information about output objects is transmitted to the SAS log. To understand the role of ❹ and ❺, review this information in the log.

❹ This PROC FREQ generates a single table named OneWayFreqs. Rather than using ODS EXCLUDE, which would leave it out of both destinations, ODS RTF EXCLUDE keeps it out of the RTF file, but it is included in the PDF file—see Output 1.5.4A and 1.5.4B.

❺ PROC MEANS generates only one table named SUMMARY, and this statement stops it from being put into the PDF file, but it does get delivered to the RTF file.

❻ The ODS RTF CLOSE and ODS PDF CLOSE statements close the destinations opened in ❶ and ❷, completing the writing of those files. In this case, as both destinations are effectively closed at the same time, a single ODS _ALL_ CLOSE statement can replace these two statements.

Output 1.5.4A: Multiple Output Destinations—RTF Results

Pearson Correlation Coefficients, N = 428	
Prob > |r| under H0: Rho=0	
	MPG_City
MPG_Highway	0.94102
MPG (Highway)	<.0001

Analysis Variable : MPG_City MPG (City)

Origin	N Obs	N	Mean	Std Dev	Minimum	Maximum
Asia	158	158	22.0126582	6.7333066	13.0000000	60.0000000
Europe	123	123	18.7317073	3.2895093	12.0000000	38.0000000
USA	147	147	19.0748299	3.9829920	10.0000000	29.0000000

Output 1.5.4B: Multiple Output Destinations—PDF Results

Pearson Correlation Coefficients, N = 428 Prob > |r| under H0: Rho=0	
	MPG_City
MPG_Highway MPG (Highway)	0.94102 <.0001

Type	Frequency	Percent	Cumulative Frequency	Cumulative Percent
Hybrid	3	0.70	3	0.70
SUV	60	14.02	63	14.72
Sedan	262	61.21	325	75.93
Sports	49	11.45	374	87.38
Truck	24	5.61	398	92.99
Wagon	30	7.01	428	100.00

1.6 SAS Language Basics

This chapter concludes with a review of the sample programs presented previously, highlighting language rules and variations in style and structure that are permitted.

1.6.1 SAS Language Structure

The following rules govern the structure of SAS programs:

- The SAS language is not case-sensitive. For example, `PROC SGPLOT`, `proc SGPLOT`, `Proc SGplot`, and other casing variations are all equivalent. This applies to all statements, names, functions, keywords, and other SAS language elements.

- In general, SAS statements end with a semicolon; otherwise, the SAS language is relatively free-form. The line breaks and indentations in Program 1.4.1, for example, are chosen to improve readability. In fact, as seen in later chapters, there are times when it is helpful to write one SAS statement across many lines with several levels of indentation. Good programming practice relating to code structure includes two fundamental rules:

  ○ Develop standards for easy readability of code

  ○ Follow these standards consistently

- Comments are available in SAS code, and good programming practice requires that code is commented to a level that makes its method and purpose clear. Two ways to write comments are shown in the following samples:

    ○ /*this is a comment*/

    ○ *this is also a comment;

1.6.2 SAS Naming Conventions

SAS language elements such as statements, names, functions, and keywords follow a standard set of naming conventions, as follows:

- Permitted characters include letters, numbers, and underscores.
- Names must begin with a letter or underscore
- Maximum length is 32 characters

Some methods are available to avoid these rules when it is deemed necessary, such as connections to other data sources that follow different rules, see Chapter Note 10 in Section 1.7 for more information. There are also some exceptions to these rules for certain SAS language elements. For example, the limitation to 8 characters for a library reference is discussed in Section 1.4.3. SAS formats, introduced in Chapter 2, are another example of a language element having some exceptions to these rules; however, most language elements, including data set and variable names, follow all three exactly.

Text literal values, such as title text or file paths, are encased in single or double quotation marks, as long as the opening quotation mark type matches the closing quotation mark. Be careful when copying text from other applications into SAS, various characters that fill the role of a quotation mark in other software are not interpreted in that manner by SAS—the color coding in the editors is often helpful in diagnosing this problem. In the role of a path, the first example of a text literal given below is generally interpreted as distinct from the second and third due to spacing. However, whether the second and third are taken as distinct due to casing is dependent on the operating system—for example, Microsoft Windows is not case-sensitive.

1. 'C:\MyFolder'
2. 'C:\My Folder'
3. 'C:\my folder'

1.7 Chapter Notes

1. *Managing Results in the SAS Windowing Environment.* In the SAS windowing environment, under default conditions, results of code submissions are cumulative in the HTML destination and in the Log window. If the listing destination is active, results are also cumulative in the Output window. For the Log and Output windows, the command **Clear All** from the **Edit** menu is used to clear either of these provided that window is active (take care not to use this command when an Editor window is active). Since the SAS windowing environment only allows for a single Log window and a single Output window during the session, the **New** command (from the **File** menu or using the toolbar button) also clears either window when active. Managing the **HTML** window is a bit more difficult; it too can be controlled by the ODS statements shown in this chapter. See the SAS Documentation for additional details. By default, SAS University Edition replaces the log and results on any code submission, so these steps are not necessary. As stated in Chapter Note 2, SAS University Edition also supports interactive mode and, when active, the Results and Log tabs are cumulative as they are in the SAS windowing environment.

2. *Interactive Mode.* The SAS windowing environment runs in interactive mode by default, while SAS University Edition runs in non-interactive mode by default, but can be set to run in interactive mode. There are two major differences between the two modes. First, results and logs from multiple code submissions are cumulative in interactive mode, while each code submission results in replacements for the log and results in non-interactive mode. (See Chapter Note 1 above.) Next, the final statement in any

code submission made in non-interactive mode is taken as a step boundary, which is not the case in interactive mode, so any submission in interactive mode must end with a step boundary.

3. *SAS Variable Lists.* To aid in simplifying references to sets of variables, SAS provides four types of variable lists:

   a. *Numbered Range Lists.* Numbered range lists are of the form VarM-VarN, where M and N are positive, whole numbers. This syntax selects all variables with the prefix given and numerical suffixes from M and N; for example, Name3-Name5 is equivalent to the list Name3 Name4 Name5. There is no restriction on the order of M and N, so Value6-Value3 is legal and is equivalent to the list Value6 Value5 Value4 Value3. All variables names corresponding to such a reference must be legal and, unless the variables are being created, all in the list must exist.

   b. *Name Range Lists.* Named range lists are of the form StartVar-EndVar, with no special restrictions on the variable names beyond their being legal. The set selected is the complete set of columns between the two variables in column order in the data set—PROC CONTENTS with the VARNUM options provides a method for checking column order. Referring to Output 1.5.2B and the Sashelp.Cars data set, the list Make--Origin is equivalent to Make Model Type Origin. For this list, the order of the two variable names is important, the first variable listed must precede the second variable listed in column order; for example, Origin--Make generates an error when used with Sashelp.Cars. It is possible to insert either of the keywords CHARACTER or NUMERIC between the two dashes, limiting the list to the variables of the chosen type.

   c. *Name Prefix Lists.* As used in example code in this chapter, a name prefix list is of the form var:, referencing all variables, in their column order, that start with the given prefix. For example, MPG: references MPG_City MPG_Highway in Sashelp.cars.

   d. *Special SAS Name Lists.* SAS also provides special lists for selection of variables without actually naming any variables. These are:

      i. _NUMERIC_ : All numeric variables in the data set, in column order

      ii. _CHARACTER_ : All character variables in the data set, in column order

      iii. _ALL_ : All variables in the data set, in column order

4. *Data View in SAS University Edition and SAS Windowing Environments.* Section 1.4.3 shows how to open a data set for viewing in each of the environments and the difference in appearance; however, there is another important difference in how these viewing utilities operate. The ViewTable in the SAS windowing environment maintains an active control over the data set in use, while the tab displayed in SAS University Edition does not. To see one problem that this can cause, submit the code from Program 1.4.1 in the SAS windowing environment, open the Cars data set from the Work library, then re-submit Program 1.4.1 (or, at least, the DATA step at the top of that program). An error message appears in the log, as shown in Figure 1.7.1, indicating the data set being open in a ViewTable has locked out any modifications to it.

Figure 1.7.1: Error Message for Updating a Data Set Open in a View Table

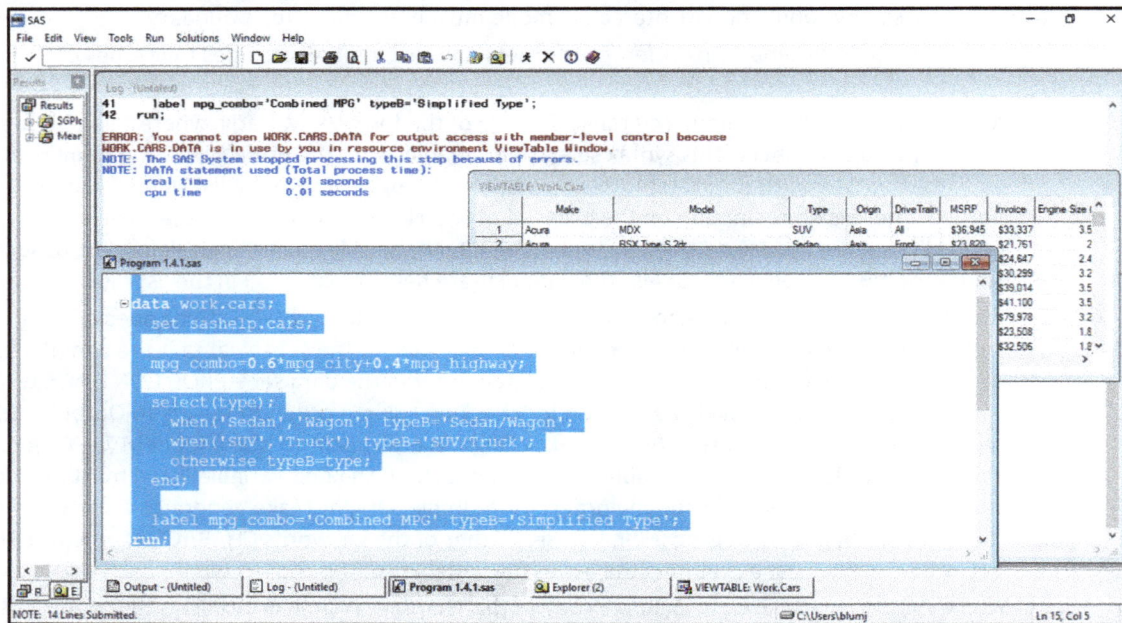

With this active control, the resource overhead in having large tables open in a ViewTable can be substantial. In contrast, the data views in SAS University Edition are based on results of a query of 100 records of the data set (and potentially a limited number of variables when many are present). Once the query is made, control of the data set is released and no resources beyond the current display are in use.

5. *LABEL/NOLABEL System Option.* By default, most procedures in SAS use variable labels (when present) in their output; however, this is in conjunction with the default system option LABEL. It is possible to suppress the use of most labels with the NOLABEL option in an OPTIONS statement, but some labels are still displayed (for example, labels defined in axis statements on a graph or chart). This not only affects variable labels, but also procedure labels—data sets can also have labels, which are unaffected by the NOLABEL option.

6. *Error Types.* The SAS Documentation separates errors into several types. Syntax errors are cases where programming statements do not conform to the rules of the SAS language—for example, misspellings of keywords or function names, missing semi-colons, or unbalanced parentheses. Semantic errors are those where the language element is correct, but the element is not valid for that usage—examples include using a character variable where a numeric variable is required or referencing a library or data set that does not exist. Execution-time errors are errors that occur when proper syntax leads to problems in processing—for example, invalid mathematical operations (such as division by zero) or incorrect sort orders when working with grouped data. Data errors are cases where data values are invalid, such as trying to store character values in numeric variables. Other error types involve SAS language elements that are beyond the scope of this book.

7. *Graphics File Setup.* All graphs shown in the book in Chapters 2 through 8 are generated as TIF files with a specific size and resolution. While running code copied directly from the book often produces similar results, as Output 1.5.3A and B show, it is not guaranteed to be the same. To match the specifications for the graphs in the book exactly, the following statements should precede any graph code (mostly generated with the SGPLOT and SGPANEL procedures):

```
ods listing image_dpi=300;
ods graphics / reset imagename='—give file name here--' width=4in imagefmt=tif;
```

The ODS LISTING statement directs the graphics output to a file, IMAGE_DPI= sets the resolution in dots per inch. The ODS GRAPHICS statement includes options after the slash (/): RESET resets all options to their default, including the sequence of file names. (Image files are not replaced by default, new files are given the same name with counting numbers attached as a suffix.) IMAGENAME= allows for a filename to be specified (a default name is given if none is specified). This can be given as a full-path reference or be built off the working directory. WIDTH= specifies the width with various units available, HEIGHT= can also be specified—when only one of height or width is specified, the image is produced in a 4:3 ratio. IMAGEFMT= allows for a file type to be chosen, most standard image types are available.

8. *Viewing Available Style Templates.* Lists of available style templates can be viewed using the TEMPLATE procedure with the LIST statement. The following code lists all style templates available in the default location—a template store named Tmplmst in the Sashelp library.

```
proc template;
  list styles;
run;
```

9. *Setting Up the CustomSapphire Template.* Nearly all output tables in the book are built as RTF tables using a custom template named CustomSapphire. The code to generate this template is provided as one of the files included with the text: CustomSapphire.sas. It is designed to store the CustomSapphire template in the BookData library in a template store named Template (which can be changed in the STORE= option in the DEFINE statement in the provided code). If the BookData library is assigned, submitting the CustomSapphire.sas code creates the template in that library. To use the template, two items are required. First, an ODS statement must be submitted to direct SAS to look for templates at this location, such as:

```
ods path (prepend) BookData.Template;
```

The ODS PATH includes a list of template stores that SAS searches whenever a request for a template has been made. Since multiple stores may have templates with the same name, the sequence matters, so the PREPEND option ensures the listed stores are at the start of the list (with the possible exception of the default template store in the WORK library). To see the current template stores listed in the path, submit the statement:

```
ods path show;
```

The template store(s) named in other ODS PATH statements appear in this list shown in the SAS log if they were correctly assigned. These two statements appear in the code given for every chapter of this book.

To see a list of templates available in any template store, a variation on the PROC TEMPLATE code given in Chapter Note 8 is given.

```
proc template;
  list / store=BookData.Template;
run;
```

Finally, to use the style template with any ODS file destination (assuming all previous steps are functional), use STYLE=CustomSapphire, similar to the use of STYLE=Journal in Program 1.5.4.

10. *Valid Variable and Other Names.* For most of the activities in this book, the naming conventions described in Section 1.6.2 apply. However, since SAS can connect to other data sources have different naming conventions, there are times when it is advantageous to alter these conventions. The VALIDVARNAME= system option allows for different rules to be enacted for variable names, while the VALIDMEMNAME= option allows for altering the naming conventions for data sets and data views. These are taken up in Section 7.6, which covers connections to Microsoft Excel workbooks and Access databases, which have different naming conventions than SAS.

1.8 Exercises

Concepts: Multiple Choice

1. Using the default rules for naming in SAS, which of the following is a valid library reference?
   a. ST445Data
   b. _LIB_
   c. My-Data
   d. 445Data

2. Which of the following is not a syntax error?
   a. Misspelling a keyword like PROC as PORC
   b. Misspelling a variable name like TypeB as TypB
   c. Omitting a semicolon at the end of a RUN statement
   d. Forgetting to close the quotation marks around the path in the LIBNAME statement

3. Which of the following is a temporary data set?
   a. Work.Employees
   b. Employees.Work
   c. Temp.Employees
   d. Employees.Temp

4. Using the default rules for naming data sets in SAS, which of the following cannot be used when naming a data set?
   a. Capital letters
   b. Digits
   c. Dashes
   d. Underscores

5. What statement is necessary to produce the following information?
   ```
   Output Added:
   -------------

   Name:        Report
   Label:
   Data Name:   ProcReportTable
   Path:        Report.Report.Report
   ```

   a. ODS RTF;
   b. ODS PDF;
   c. ODS TRACE ON;
   d. ODS LISTING;

Concepts: Short Answer

1. Under standard SAS naming conventions, decide whether each of the following names is legal syntax when used as a:
   i. Library reference
   ii. Data set name
   iii. Variable name

   Provide justification for each answer.
   a. mydata
   b. myvariable
   c. mylibrary
   d. left2right
   e. left-2-right

      f.   house2

      g.   2nd_house

      h.   _2nd

2.   Classify each of the following statements as: always true, sometimes true, or never true. Provide justification for each answer.

      a.   An error message in the log is an indication of a syntax error.

      b.   A warning message in the log is an indication of a logic error.

      c.   Notes in the log provide details about successful code execution.

      d.   Checking the log only for errors and warnings is considered a good programming practice.

3.   Classify each of the following statements as: always true, sometimes true, or never true. Provide justification for each answer.

      a.   SAS data sets can contain only numeric and character variables.

      b.   If the library is omitted from a SAS data set reference, the Work library is assumed.

      c.   Once a library is assigned in a SAS session, it is available automatically in subsequent SAS sessions on that machine.

      d.   A library and a data set can have the same name.

4.   Classify each of the following statements as: always true, sometimes true, or never true. Provide justification for each answer.

      a.   A PROC step can be nested with a DATA step.

      b.   The RUN statement must be used as a step boundary at the end of a DATA or PROC step.

      c.   It is a good programming practice to use an explicit step boundary, such as the RUN statement, at the end of any DATA or PROC step.

      d.   Global statements can be included inside DATA or PROC steps.

5.   Consider the program below.

```
options date label;
title 'Superhero Profile: Jennie Blockhus';
proc print data = superheroes label;
   where homeTown eq 'Redmond' and current = 'Themyscira';
   var Alias Powers FirstIssue Superfriend Nemesis;
run;

options nonumber nolabel;
proc freq data = superheroes;
   where homeTown eq "Redmond" and current = "Themyscira";
   table sightings*state;
run;
```

      a.   Determine the number of global statements.

      b.   Determine the number of steps.

      c.   What distinguishes global statements from other statements in the above program?

      d.   Is it a syntax error to enclose literals in both single and double quotation marks as done in the above program? Why or why not?

Programming Basics

1.   Complete the following steps, in either the SAS windowing environment or SAS University Edition:

      a.   Open Program 1.8.1 (shown below).

      b.   Download the data for the textbook and assign a library to its location.

      c.   Replace the comment with the library reference established in part (b).

      d.   Submit the code.

      e.   If the submission does not execute successfully, check the log and output for errors in the library assignment and/or reference.

      f.   Repeat the above steps as necessary until the code executes properly.

Program 1.8.1: Sample Program for Submission

```
proc means data=/*put library reference here*/.IPUMS2005Basic;
  class MortgageStatus;
  var HHIncome;
run;
```

Output 1.8.1: Expected Result from Program 1.8.1 (Colors and Fonts May Differ)

Analysis Variable : HHINCOME Total household income						
MortgageStatus	**N Obs**	**N**	**Mean**	**Std Dev**	**Minimum**	**Maximum**
N/A	303342	303342	37180.59	39475.13	-19998.00	1070000.00
No, owned free and clear	300349	300349	53569.08	63690.40	-22298.00	1739770.00
Yes, contract to purchase	9756	9756	51068.50	46069.11	-7599.00	834000.00
Yes, mortgaged/ deed of trust or similar debt	545615	545615	84203.70	72997.92	-29997.00	1407000.00

Case Study

For additional practice, multiple case studies are available in addition to the IPUMS CPS case study used in subsequent chapters. See Section 8.1 to apply the skills from this chapter to the Clinical Trials Case Study. For additional case studies, including extensions to the IPUMS CPS case study, see the author pages.

Chapter 2: Foundations for Analyzing Data and Reading Data from Other Sources

2.1 Learning Objectives

At the conclusion of this chapter, mastery of the concepts covered in the narrative includes the ability to:

- Apply the MEANS procedure to produce a variety of quantitative summaries, potentially grouped across several categories

- Apply the FREQ procedure to produce frequency and relative frequency tables, including cross-tabulations

- Categorize data for analyses in either the MEANS or FREQ procedures using internal SAS formats or user-defined formats

- Formulate a strategy for selecting only the necessary rows when processing a SAS data set

- Apply the DATA step to read data from delimited or fixed-position raw text files

- Describe the operations carried out during the compilation and execution phases of the DATA step

- Compare and contrast the input buffer and program data vector

- Apply DATA step statements to assist in debugging

- Apply the COMPARE procedure to compare and validate a data set against a standard

Use the concepts of this chapter to solve the problems in the wrap-up activity. Additional exercises and case-studies are also available to test these concepts.

2.2 Case Study Activity

This section introduces a case study that is used as a basis for most of the concepts and associated activities in this book. The data comes from the Current Population Survey by the Integrated Public Use Microdata Series (IPUMS CPS). IPUMS CPS contains a wide variety of information, only a subset of the data collected from 2001-2015 is included in the examples here. Further, the data used is introduced in various segments, starting with simple sets of variables and eventually adding more information that must be assembled to achieve the objectives of each section.

This chapter works with data that includes household-level information from the 2005 and 2010 IPUMS CPS data sets of over one million observations each. Included are variables on state, county, metropolitan area/city, household income, home value, mortgage status, ownership status, and mortgage payment. Outputs 2.2.1 through 2.2.4 show tabular summaries from the 2010 data, including quantitative statistics, frequencies, and/or percentages. Reproducing these tables in the wrap-up activity in Section 2.11 is the primary objective for this chapter.

The first sample output shown in Output 2.2.1 produces a set of six statistics on mortgage payments across metropolitan status for mortgages of $100 per month or more. In order to make this table, and the slightly more complicated Output 2.2.2, several components of the MEANS procedure must be understood.

Output 2.2.1: Basic Statistics on Mortgage Payments Grouped on Metropolitan Status

Analysis Variable : MortgagePayment Mortgage Payment						
Metro	N	Mean	Median	Std Dev	Minimum	Maximum
Not Identifiable	42927	970.2	800.0	668.5	100.0	7400.0
Not in Metro Area	97603	815.0	670.0	576.0	100.0	6800.0
Metro, Inside City	56039	1363.5	1100.0	974.8	100.0	7400.0
Metro, Outside City	185967	1480.8	1300.0	974.7	100.0	7400.0
Metro, City Status Unknown	163204	1233.2	1000.0	846.4	100.0	7400.0

Output 2.2.2: Minimum, Median, and Maximum on Mortgage Payments Across Multiple Categories

Metro	Household Income	Variable	Label	Minimum	Median	Maximum
Metro, Inside City	Negative	MortgagePayment HomeValue	Mortgage Payment Home Value	440 70000	1200 250000	4500 675000
	$0 to $45K	MortgagePayment HomeValue	Mortgage Payment Home Value	100 0	740 130000	6800 5303000
	$45K to $90K	MortgagePayment HomeValue	Mortgage Payment Home Value	100 0	1000 180000	7400 4915000
	Above $90K	MortgagePayment HomeValue	Mortgage Payment Home Value	100 0	1600 340000	7400 5303000
Metro, Outside City	Negative	MortgagePayment HomeValue	Mortgage Payment Home Value	100 10000	1450 250000	5400 4152000
	$0 to $45K	MortgagePayment HomeValue	Mortgage Payment Home Value	100 0	850 150000	7400 4304000
	$45K to $90K	MortgagePayment HomeValue	Mortgage Payment Home Value	100 0	1100 199000	6800 4915000
	Above $90K	MortgagePayment HomeValue	Mortgage Payment Home Value	100 0	1600 330000	7400 4915000
Metro, City Status Unknown	Negative	MortgagePayment HomeValue	Mortgage Payment Home Value	180 17000	1200 245000	5300 2948000
	$0 to $45K	MortgagePayment HomeValue	Mortgage Payment Home Value	100 0	720 125000	7400 4915000

Metro	Household Income	Variable	Label	Minimum	Median	Maximum
	$45K to $90K	MortgagePayment HomeValue	Mortgage Payment Home Value	100 0	960 160000	7400 4915000
	Above $90K	MortgagePayment HomeValue	Mortgage Payment Home Value	100 0	1400 270000	7400 4915000

In Outputs 2.2.3 and 2.2.4, frequencies and percentages are summarized across combinations of various categories, which requires mastery of the fundamentals of the FREQ procedure.

Output 2.2.3: Income Status Versus Mortgage Payment

Table of HHIncome by MortgagePayment					
HHIncome(Household Income)	**MortgagePayment(Mortgage Payment)**				
Frequency Row Pct	**$350 and Below**	**$351 to $1000**	**$1001 to $1600**	**Over $1600**	**Total**
Negative	30 9.93	97 32.12	92 30.46	83 27.48	302
$0 to $45K	22929 16.37	83125 59.33	22617 16.14	11436 8.16	140107
$45K to $90K	13877 6.96	103660 51.99	54778 27.48	27052 13.57	199367
Above $90K	5944 2.89	52679 25.58	62474 30.33	84867 41.20	205964
Total	42780	239561	139961	123438	545740

Output 2.2.4: Income Status Versus Mortgage Payment for Metropolitan Households (Table 1 of 3)

Table 1 of HHIncome by MortgagePayment					
Controlling for Metro=Metro, Inside City					
HHIncome(Household Income)	**MortgagePayment(Mortgage Payment)**				
Frequency Row Pct	**$350 and Below**	**$351 to $1000**	**$1001 to $1600**	**Over $1600**	**Total**
Negative	0 0.00	7 30.43	9 39.13	7 30.43	23
$0 to $45K	1596 10.75	8949 60.30	2597 17.50	1700 11.45	14842
$45K to $90K	910 4.75	9215 48.13	5571 29.10	3450 18.02	19146
Above $90K	504 2.29	4947 22.46	6321 28.70	10256 46.56	22028
Total	3010	23118	14498	15413	56039

2.3 Getting Started with Data Exploration in SAS

This section reviews and extends some fundamental SAS concepts demonstrated in code supplied for Chapter 1, with these examples built upon a simplified version of the case study data. First, Program 2.3.1 uses the CONTENTS and PRINT procedures to make an initial exploration of the Ipums2005Mini data set. To begin, make sure the BookData library is assigned as done in Chapter 1.

Program 2.3.1: Using the CONTENTS and PRINT Procedures to View Data and Attributes

```
proc contents data=bookdata.ipums2005mini❶;
  ods select variables;❷
run;

proc print data=bookdata.ipums2005mini(obs=5)❸;
  var state MortgageStatus MortgagePayment HomeValue Metro;❹
run;
```

❶ The BookData.Ipums2005Mini data set is a modification of a data set used later in this chapter, BookData.Ipums2005Basic. It subsets the original data set down to a few records and is used for illustration of these initial concepts.

❷ The ODS SELECT statement limits the output of a given procedure to the chosen tables, with the Variables table from PROC CONTENTS containing the names and attributes of the variables in the chosen data set. Look back to Program 1.4.4, paying attention to the ODS TRACE statement and its results, to review how this choice is made.

❸ The OBS= data set option limits the number of observations processed by the procedure. It is in place here simply to limit the size of the table shown in Output 2.3.1B. At various times in this text, the output shown may be limited in scope; however, the code given may not include this option for all such cases.

❹ The VAR statement is used in the PRINT procedure to select the variables to be shown and the column order in which they appear.

Output 2.3.1A: Using the CONTENTS Procedure to View Attributes

Alphabetic List of Variables and Attributes				
#	Variable	Type	Len	Format
4	CITYPOP	Num	8	
2	COUNTYFIPS	Num	8	
10	City	Char	43	
6	HHINCOME	Num	8	
7	HomeValue	Num	8	
3	METRO	Num	8	BEST12.
5	MortgagePayment	Num	8	
9	MortgageStatus	Char	45	
11	Ownership	Char	6	
1	SERIAL	Num	8	
8	state	Char	57	

Output 2.3.1B: Using the PRINT Procedure to View Data

Obs	state	MortgageStatus	MortgagePayment	HomeValue	METRO
1	South Carolina	Yes, mortgaged/ deed of trust or similar debt	200	32500	4
2	North Carolina	No, owned free and clear	0	5000	1
3	South Carolina	Yes, mortgaged/ deed of trust or similar debt	360	75000	4
4	South Carolina	Yes, contract to purchase	430	22500	3
5	North Carolina	Yes, mortgaged/ deed of trust or similar debt	450	65000	4

2.3.1 Assigning Labels and Using SAS Formats

As seen in Chapter 1, SAS variable names have a certain set of restrictions they must meet, including no special characters other than an underscore. This potentially limits the quality of the display for items such as the headers in PROC PRINT. SAS does permit the assignment of labels to variables, substituting more descriptive text into the output in place of the variable name, as demonstrated in Program 2.3.2.

Program 2.3.2: Assigning Labels

```
proc print data=bookdata.ipums2005mini(obs=5) noobs❶ label❷;
  var state MortgageStatus MortgagePayment HomeValue Metro;
  label HomeValue='Value of Home ($)' state='State'; ❸
run;
```

❶ By default, the output from PROC PRINT includes an Obs column, which is simply the row number for the record—the NOOBS option in the PROC PRINT statement suppresses this column.

❷ Most SAS procedures use labels when they are provided or assigned; however, PROC PRINT defaults to using variable names. To use labels, the LABEL option is provided in the PROC PRINT statement. See Chapter Note 1 in Section 2.12 for more details.

❸ The LABEL statement assigns labels to selected variables. The general syntax is: LABEL *variable1*='*label1*' *variable2*='*label2*' ...; where the labels are given as literal values in either single or double quotation marks, as long as the opening and closing quotation marks match.

Output 2.3.2: Assigning Labels

State	MortgageStatus	MortgagePayment	Value of Home ($)	METRO
South Carolina	Yes, mortgaged/ deed of trust or similar debt	200	32500	4
North Carolina	No, owned free and clear	0	5000	1
South Carolina	Yes, mortgaged/ deed of trust or similar debt	360	75000	4
South Carolina	Yes, contract to purchase	430	22500	3
North Carolina	Yes, mortgaged/ deed of trust or similar debt	450	65000	4

In addition to using labels to alter the display of variable names, altering the display of data values is possible with formats. The general form of a format reference is:

<$>*format*<w>.<d>

The <> symbols denote a portion of the syntax that is sometimes used/required—the <> characters are not part of the syntax. The dollar sign is required for any format that applies to a character variable (character formats) and is not permitted in formats used for numeric variables (numeric formats). The *w* value is the total number of characters (width) available for the formatted value, while *d* controls the number of values displayed after the decimal for numeric formats. The dot is required in all format assignments, and in many cases is the means by which the SAS compiler can distinguish between a variable name and a format name. The value of *format* is called the format name; however, standard numeric and character formats have a null name; for example, the 5.2 format assigns the standard numeric format with a total width of 5 and up to 2 digits displayed past the decimal. Program 2.3.3 uses the FORMAT statement to apply formats to the HomeValue, MortgagePayement, and MortgageStatus variables.

Program 2.3.3: Assigning Formats
```
proc print data=bookdata.ipums2005mini(obs=5) noobs label;
   var state MortgageStatus MortgagePayment HomeValue Metro;
   label HomeValue='Value of Home' state='State';
   format HomeValue MortgagePayment dollar9.❶ MortgageStatus $1.❷;
run;
```

❶ In the FORMAT statement, a list of one or more variables is followed by a format specification. Both HomeValue and MortgagePayment are assigned a dollar format with a total width of nine—any commas and dollar signs inserted by this format count toward the total width.

❷ The MortgageStatus variable is character and can only be assigned a character format. The $1. format is the standard character format with width one, which truncates the display of MortgageStatus to one letter, but does not alter the actual value. In general, formats assigned in procedures are temporary and only apply to the output for the procedure.

Output 2.3.3: Assigning Formats

State	MortgageStatus	MortgagePayment	Value of Home	METRO
South Carolina	Y	$200	$32,500	4
North Carolina	N	$0	$5,000	1
South Carolina	Y	$360	$75,000	4
South Carolina	Y	$430	$22,500	3
North Carolina	Y	$450	$65,000	4

2.3.2 PROC SORT and BY-Group Processing

Rows in a data set can be reordered using the SORT procedure to sort the data on the values of one or more variables in ascending or descending order. Program 2.3.4 sorts the BookData.Ipums2005Mini data set by the HomeValue variable.

Program 2.3.4: Sorting Data with the SORT Procedure

```
proc sort data=bookdata.ipums2005mini out=work.sorted❶;
  by HomeValue; ❷
run;

proc print data=work.sorted(obs=5) noobs label;
  var state MortgageStatus MortgagePayment HomeValue Metro;
  label HomeValue='Value of Home' state='State';
  format HomeValue MortgagePayment dollar9. MortgageStatus $1.;
run;
```

❶ The default behavior of the SORT procedure is to replace the input data set, specified in the DATA= option, with the sorted data set. To create a new data set from the sorted observations, use the OUT= option.

❷ The BY statement is required in PROC SORT and must name at least one variable. As shown in Output 2.3.4, the rows are now ordered in increasing levels of HomeValue.

Output 2.3.4: Sorting Data with the SORT Procedure

State	MortgageStatus	MortgagePayment	Value of Home	METRO
North Carolina	N	$0	$5,000	1
South Carolina	Y	$430	$22,500	3
North Carolina	Y	$300	$22,500	3
South Carolina	Y	$200	$32,500	4
North Carolina	N	$0	$45,000	1

Sorting on more than one variable gives a nested or hierarchical sorting. In those cases, values are ordered on the first variable, then for groups of records having the same value of the first variable those records are sorted on the second variable, and so forth. A specification of ascending (the default) or descending order is made for each variable. Program 2.3.5 sorts the BookData.Ipums2005Mini data set on three variables present in the data set.

Program 2.3.5: Sorting on Multiple Variables

```
proc sort data=bookdata.ipums2005mini out=work.sorted;
  by MortgagePayment❶ descending State❷ descending HomeValue❸;
run;
```

```
proc print data=work.sorted(obs=6) noobs label;
   var state MortgageStatus MortgagePayment HomeValue Metro;
   label HomeValue='Value of Home' state='State';
   format HomeValue MortgagePayment dollar9. MortgageStatus $1.;
run;
```

❶ The first sort is on MortgagePayment, in ascending order. Since 0 is the lowest value and that value occurs on six records in the data set, Output 2.3.5 shows one block of records with MortgagePayment 0.

❷ The next sort is on State in descending order—note that the DESCENDING option precedes the variable it applies to. For the six records shown in Output 2.3.5, the first three are South Carolina and the final three are North Carolina—descending alphabetical order. Note, when sorting character data, casing matters—uppercase values are before lowercase in such a sort. For more details about determining the sort order of character data, see Chapter Note 2 in Section 2.12.

❸ The final sort is on HomeValue, also in descending order—note that the DESCENDING option must precede each variable it applies to. So, within each State group in Output 2.3.5, values of the HomeValue variable are in descending order.

Output 2.3.5: Sorting on Multiple Variables

State	MortgageStatus	MortgagePayment	Value of Home	METRO
South Carolina	N	$0	$137,500	3
South Carolina	N	$0	$95,000	4
South Carolina	N	$0	$45,000	3
North Carolina	N	$0	$162,500	0
North Carolina	N	$0	$45,000	1
North Carolina	N	$0	$5,000	1

Most SAS procedures, including PROC PRINT, can take advantage of BY-group processing for data that is sorted into groups. The procedure must use a BY statement that corresponds to the sorting in the data set. If the data is sorted using PROC SORT, the BY statement in a subsequent procedure does not have to completely match the BY statement in PROC SORT; however, it must match the first level of sorting if only one variable is included, the first two levels if two variables are included, and so forth. It must also match ordering, ascending or descending, on each included variable. Program 2.3.6 groups output from the PRINT procedure based on BY grouping constructed with PROC SORT.

Program 2.3.6: BY-Group Processing in PROC PRINT

```
proc sort data=bookdata.ipums2005mini out= work.sorted;
   by MortgageStatus State descending HomeValue; ❶
run;

proc print data= work.sorted noobs label;
   by MortgageStatus State; ❷
   var MortgagePayment HomeValue Metro;
   label HomeValue='Value of Home' state='State';
   format HomeValue MortgagePayment dollar9. MortgageStatus $9.;
run;
```

❶ The original data is sorted first on MortgageStatus, then on State, and finally in descending order of HomeValue for each combination of MortgageStatus and State.

❷ PROC PRINT uses a BY statement matching on the MortgageStatus and State variables, which groups the output into sections based on each unique combination of values for these two variables, with the final sorting on HomeValue appearing in each table. Note that a BY statement with only MortgageStatus can be used as well, but a BY statement with only State cannot—the data is not sorted on State primarily.

Output 2.3.6: BY-Group Processing in PROC PRINT (First 2 of 6 Groups Shown)

MortgageStatus=No, owned State=North Carolina

MortgagePayment	Value of Home	METRO
$0	$162,500	0
$0	$45,000	1
$0	$5,000	1

MortgageStatus=No, owned State=South Carolina

MortgagePayment	Value of Home	METRO
$0	$137,500	3
$0	$95,000	4
$0	$45,000	3

The structure of BY groups in PROC PRINT can be altered slightly through use of an ID statement, as shown in Program 2.3.7. Assuming the variables listed in the ID statement match those in the BY statement, BY-group variables are placed as the left-most columns of each table, rather than between tables.

Program 2.3.7: Using BY and ID Statements Together in PROC PRINT

```
proc print data= work.sorted noobs label;
  by MortgageStatus State;
  id MortgageStatus State;
  var MortgagePayment HomeValue Metro;
  label HomeValue='Value of Home' state='State';
  format HomeValue MortgagePayment dollar9. MortgageStatus $9.;
run;
```

Output 2.3.7: Using BY and ID Statements Together in PROC PRINT (First 2 of 6 Groups Shown)

MortgageStatus	State	MortgagePayment	Value of Home	METRO
No, owned	North Carolina	$0	$162,500	0
		$0	$45,000	1
		$0	$5,000	1

MortgageStatus	State	MortgagePayment	Value of Home	METRO
No, owned	South Carolina	$0	$137,500	3
		$0	$95,000	4
		$0	$45,000	3

PROC PRINT is limited in its ability to do computations. (Later in this text, the REPORT procedure is used to create various summary tables.); however, it can do sums of numeric variables with the SUM statement, as shown in Program 2.3.8.

Program 2.3.8: Using the SUM Statement in PROC PRINT

```
proc print data= work.sorted noobs label;
  by MortgageStatus State;
  id MortgageStatus State;
  var MortgagePayment HomeValue Metro;
  sum MortgagePayment HomeValue;
  label HomeValue='Value of Home' state='State';
  format HomeValue MortgagePayment dollar9. MortgageStatus $9.;
run;
```

Output 2.3.8: Using the SUM Statement in PROC PRINT (Last of 6 Groups Shown)

MortgageStatus	State	MortgagePayment	Value of Home	METRO
Yes, mort	South Carolina	$360	$75,000	4
		$500	$65,000	3
		$200	$32,500	4
Yes, mort	South Carolina	$1,060	$172,500	
Yes, mort		$2,200	$315,000	
		$4,230	$1200000	

Sums are produced at the end of each BY group (and the SUMBY statement is available to modify this behavior), and at the end of the full table. Note that the format applied to the HomeValue column is not sufficient to display the grand total with the dollar sign and comma. If a format is of insufficient width, SAS removes what it determines to be the least important characters. However, it is considered good programming practice to determine the minimum format width needed for all values a format is applied to. If the format does not include sufficient width to display the value with full precision, then SAS may adjust the included format to a different format. See Chapter Note 3 in Section 2.12 for further discussion on format widths.

2.4 Using the MEANS Procedure for Quantitative Summaries

Producing tables of statistics like those shown for the case study in Outputs 2.2.1 and 2.2.2 uses MEANS procedure. This section covers the fundamentals of PROC MEANS, including how to select variables for analysis, choosing statistics, and separating analyses across categories.

2.4.1 Choosing Analysis Variables and Statistics in PROC MEANS

To begin, make sure the BookData library is assigned as done in Chapter 1, submit PROC CONTENTS on the IPUMS2005Basic SAS data set from the BookData library, and review the output. Also, to ensure familiarity with the data, open the data set for viewing or run the PRINT procedure to direct it to an output table. Once these steps are complete, enter and submit the code given in Program 2.4.1.

Program 2.4.1: Default Statistics and Behavior for PROC MEANS

```
options nolabel;
proc means data=BookData.IPUMS2005Basic;
run;
```

For variables that have labels, PROC MEANS includes them as a column in the output table; using NOLABEL in the OPTIONS statement suppresses their use. Here DATA= is technically an option; however, the default data set in any SAS session is the last data set created. If no data sets have been created during the session, which is the most likely scenario currently, PROC MEANS does not have a data set to process unless this option is provided. Beyond having a data set to work with, no other options or statements are required for PROC MEANS to compile and execute successfully. In this case, the default behavior, as shown in Output 2.4.1, is to summarize all numeric variables on a set of five statistics: number of nonmissing observations, mean, standard deviation, minimum, and maximum.

Output 2.4.1: Default Statistics and Behavior for PROC MEANS

Variable	N	Mean	Std Dev	Minimum	Maximum
SERIAL	1159062	621592.24	359865.41	2.0000000	1245246.00
COUNTYFIPS	1159062	42.2062901	78.9543285	0	810.0000000
METRO	1159062	2.5245354	1.3085302	0	4.0000000
CITYPOP	1159062	2916.66	12316.27	0	79561.00
MortgagePayment	1159062	500.2042634	737.9885592	0	7900.00
HHIncome	1159062	63679.84	66295.97	-29997.00	1739770.00
HomeValue	1159062	2793526.49	4294777.18	5000.00	9999999.00

SAS differentiates variable types as numeric and character only; therefore, variables stored as numeric that are not quantitative are summarized even if those summaries do not make sense. Here, the Serial, CountyFIPS, and Metro variables are stored as numbers, but means and standard deviations are of no utility on these since they are nominal. It is, of course, important to understand the true role and level of measurement (for instance, nominal versus ratio) for the variables in the data set being analyzed.

To select the variables for analysis, the MEANS procedure includes the VAR statement. Any variables listed in the VAR statement must be numeric, but should also be appropriate for quantitative summary statistics. As in the previous example, the summary for each variable is listed in its own row in the output table. (If only one variable is provided, it is named in the header above the table instead of in the first column.) Program 2.4.2 modifies Program 2.4.1 to summarize only the truly quantitative variables from BookData.IPUMS2005Basic, with the results shown in Output 2.4.2.

Program 2.4.2: Selecting Analysis Variables Using the VAR Statement in MEANS

```
proc means data=BookData.IPUMS2005Basic;
  var Citypop MortgagePayment HHIncome HomeValue;
run;
```

Output 2.4.2: Selecting Analysis Variables Using the VAR Statement in MEANS

Variable	N	Mean	Std Dev	Minimum	Maximum
CITYPOP	1159062	2916.66	12316.27	0	79561.00
MortgagePayment	1159062	500.2042634	737.9885592	0	7900.00
HHIncome	1159062	63679.84	66295.97	-29997.00	1739770.00
HomeValue	1159062	2793526.49	4294777.18	5000.00	9999999.00

The default summary statistics for PROC MEANS can be modified by including statistic keywords as options in the PROC MEANS statement. Several statistics are available, with the available set listed in the SAS Documentation, and any subset of those may be used. The listed order of the keywords corresponds to the order of the statistic columns in the table, and those replace the default statistic set. One common set of statistics is the five-number summary (minimum, first quartile, median, third quartile, and maximum), and Program 2.4.3 provides a way to generate these statistics for the four variables summarized in the previous example.

Program 2.4.3: Setting the Statistics to the Five-Number Summary in MEANS

```
proc means data=BookData.IPUMS2005Basic min q1 median q3 max;
  var Citypop MortgagePayment HHIncome HomeValue;
run;
```

Output 2.4.3: Setting the Statistics to the Five-Number Summary in MEANS

Variable	Minimum	Lower Quartile	Median	Upper Quartile	Maximum
CITYPOP	0	0	0	0	79561.00
MortgagePayment	0	0	0	830.0000000	7900.00
HHIncome	-29997.00	24000.00	47200.00	80900.00	1739770.00
HomeValue	5000.00	112500.00	225000.00	9999999.00	9999999.00

Confidence limits for the mean are included in the keyword set, both as a pair with the CLM keyword, and separately with LCLM and UCLM. The default confidence level is 95%, but is changeable by setting the error rate using the ALPHA= option. Consider Program 2.4.4, which constructs the 99% confidence intervals for the means, with the estimated mean between the lower and upper limits.

Program 2.4.4: Using the ALPHA= Option to Modify Confidence Levels

```
proc means data=BookData.IPUMS2005Basic lclm mean uclm alpha=0.01;
  var Citypop MortgagePayment HHIncome HomeValue;
run;
```

Output 2.4.4: Using the ALPHA= Option to Modify Confidence Levels

Variable	Lower 99% CL for Mean	Mean	Upper 99% CL for Mean
CITYPOP	2887.19	2916.66	2946.12
MortgagePayment	498.4385749	500.2042634	501.9699520
HHIncome	63521.22	63679.84	63838.46
HomeValue	2783250.94	2793526.49	2803802.04

There are also options for controlling the column display; rounding can be controlled by the MAXDEC= option (maximum number of decimal places). Program 2.4.5 modifies the previous example to report the statistics to a single decimal place.

Program 2.4.5: Using MAXDEC= to Control Precision of Results

```
proc means data=BookData.IPUMS2005Basic lclm mean uclm alpha=0.01 maxdec=1;
  var Citypop MortgagePayment HHIncome HomeValue;
run;
```

Output 2.4.5: Using MAXDEC= to Control Precision of Results

Variable	Lower 99% CL for Mean	Mean	Upper 99% CL for Mean
CITYPOP	2887.2	2916.7	2946.1
MortgagePayment	498.4	500.2	502.0
HHIncome	63521.2	63679.8	63838.5
HomeValue	2783250.9	2793526.5	2803802.0

MAXDEC= is limited in that it sets the precision for all columns. Also, no direct formatting of the statistics is available. The REPORT procedure, introduced in Chapter 4 and discussed in detail in Chapters 6 and 7, provides much more control over the displayed table at the cost of increased complexity of the syntax.

2.4.2 Using the CLASS Statement in PROC MEANS

In several instances, it is desirable to split an analysis across a set of categories and, if those categories are defined by a variable in the data set, PROC MEANS can separate those analyses using a CLASS statement. The

CLASS statement accepts either numeric or character variables; however, the role assigned to class variables by SAS is special. Any variable included in the CLASS statement (regardless of type) is taken as categorical, which results in each distinct value of the variable corresponding to a unique category. Therefore, variables used in the CLASS statement should provide useful groupings or, as shown in Section 2.5, be formatted into a set of desired groups. Two examples follow, the first (Program 2.4.6) providing an illustration of a reasonable class variable, the second (Program 2.4.7) showing a poor choice.

Program 2.4.6: Setting a Class Variable in PROC MEANS

```
proc means data=BookData.IPUMS2005Basic;
   class MortgageStatus;
   var HHIncome;
run;
```

Output 2.4.6: Setting a Class Variable in PROC MEANS

	Analysis Variable : HHIncome					
MortgageStatus	N Obs	N	Mean	Std Dev	Minimum	Maximum
N/A	303342	303342	37180.59	39475.13	-19998.00	1070000.00
No, owned free and clear	300349	300349	53569.08	63690.40	-22298.00	1739770.00
Yes, contract to purchase	9756	9756	51068.50	46069.11	-7599.00	834000.00
Yes, mortgaged/ deed of trust or similar debt	545615	545615	84203.70	72997.92	-29997.00	1407000.00

In this data, MortgageStatus provides a clear set of distinct categories and is potentially useful for subsetting the summarization of the data. In Program 2.4.7, Serial is used as an extreme example of a poor choice since Serial is unique to each household.

Program 2.4.7: A Poor Choice for a Class Variable

```
proc means data=BookData.IPUMS2005Basic;
   class Serial;
   var HHIncome;
run;
```

Output 2.4.7: A Poor Choice for a Class Variable (Partial Table Shown)

			Analysis Variable : HHIncome			
SERIAL	N Obs	N	Mean	Std Dev	Minimum	Maximum
2	1	1	12000.00	.	12000.00	12000.00
3	1	1	17800.00	.	17800.00	17800.00
4	1	1	185000.00	.	185000.00	185000.00
5	1	1	2000.00	.	2000.00	2000.00

Choosing Serial as a class variable results in each class being a single observation, making the mean, minimum, and maximum the same value and creating a situation where the standard deviation is undefined. Again, this would be an extreme case; however, class variables are best when structured to produce relatively few classes that represent a useful stratification of the data.

Of course, more than one variable can be used in a CLASS statement; the categories are then defined as all combinations of the categories from the individual variables. The order of the variables listed in the CLASS statement only alters the nesting order of the levels; therefore, the same information is produced in a different row order in the table. Consider the two MEANS procedures in Program 2.4.8.

Program 2.4.8: Using Multiple Class Variables and Effects of Order

```
proc means data=BookData.IPUMS2005Basic nonobs n mean std;
  class MortgageStatus Metro;
  var HHIncome;
run;

proc means data=BookData.IPUMS2005Basic nonobs n mean std;
  class Metro MortgageStatus;
  var HHIncome;
run;
```

Output 2.4.8A: Using Multiple Class Variables (Partial Listing)

Analysis Variable : HHIncome				
MortgageStatus	**METRO**	**N**	**Mean**	**Std Dev**
N/A	0	19009	31672.81	32122.89
	1	48618	29122.73	29160.23
	2	69201	38749.69	46226.50
	3	73234	43325.25	42072.78
	4	93280	36514.56	36974.63
No, owned free and clear	0	30370	46533.14	50232.50
	1	85696	42541.06	44664.64
	2	27286	60011.10	76580.75
	3	76727	63925.99	75404.62
	4	80270	55915.02	66293.39

Output 2.4.8B: Effects of Order (Partial Listing)

Analysis Variable : HHIncome				
METRO	**MortgageStatus**	**N**	**Mean**	**Std Dev**
0	N/A	19009	31672.81	32122.89
	No, owned free and clear	30370	46533.14	50232.50
	Yes, contract to purchase	1030	46069.26	36225.80
	Yes, mortgaged/ deed of trust or similar debt	41619	71611.01	55966.31
1	N/A	48618	29122.73	29160.23
	No, owned free and clear	85696	42541.06	44664.64
	Yes, contract to purchase	3034	42394.12	35590.14
	Yes, mortgaged/ deed of trust or similar debt	93427	62656.54	48808.66

The same statistics are present in both tables, but the primary ordering is on MortgageStatus in Output 2.4.8A as opposed to metropolitan status (Metro) in Output 2.4.8B. Two additional items of note from this example:

first, note the use of NONOBS in each. By default, using a CLASS statement always produces a column for the number of observations in each class level (NOBS), and this may be different from the statistic N due to missing data, but that is not an issue for this example. Second, the numeric values of Metro really have no clear meaning. Titles and footnotes, as shown in Chapter 1, are available to add information about the meaning of these numeric values. However, a better solution is to build a format and apply it to that variable, a concept covered in the next section.

2.5 User-Defined Formats

As seen in Section 2.3, SAS provides a variety of formats for altering the display of data values. It is also possible to define formats using the FORMAT procedure. These formats are used to assign replacements for individual data values or for groups or ranges of data, and they may be permanently stored in a library for subsequent use. Formats, both native SAS formats and user-defined formats, are an invaluable tool that are used in a variety of contexts throughout this book.

2.5.1 The FORMAT Procedure

The FORMAT procedure provides the ability to create custom formats, both for character and numeric variables. The principal tool used in writing formats is the VALUE statement, which defines the name of the format and its rules for converting data values to formatted values. Program 2.5.1 gives an example of a format written to improve the display of the Metro variable from the BookData.IPUMS2005Basic data set.

Program 2.5.1: Defining a Format for the Metro Variable
```
proc format;
   value❶ Metro❷
     0 = "Not Identifiable"
     1 = "Not in Metro Area"
     2 = "Metro, Inside City"
     3 = "Metro, Outside City"
     4 = "Metro, City Status Unknown"❸
   ;❹
run;
```

❶   The VALUE statement tends to be rather long given the number of items it defines. Remember, SAS code is generally free-form outside of required spaces and delimiters, along with the semicolon that ends every statement. Adopt a sound strategy for using indentation and line breaks to make code readable.

❷   The VALUE statement requires the format name, which follows the SAS naming conventions of up to 32 characters, but with some special restrictions. Format names must meet an additional restriction of being distinct from the names of any formats supplied by SAS. Also, given that numbers are used to define format widths, a number at the end of a format name would create an ambiguity in setting lengths; therefore, format names cannot end with a number. If the format is for character values, the name must begin with $, and that character counts toward the 32-character limit.

❸   In this format, individual values are set equal to their replacements (as literals) for all values intended to be formatted. Values other than 0, 1, 2, 3, and 4 may not appear as intended. For a discussion of displaying values other than those that appear in the VALUE statement, see Chapter Note 4 in Section 2.12.

❹   The semicolon that ends the value statement is set out on its own line here for readability—simply to make it easy to verify that it is present.

Submitting Program 2.5.1 makes a format named Metro in the format catalog in the Work library, it only takes effect when used, and it is used in effectively the same manner as a format supplied by SAS. Program 2.5.2 uses the Metro format for the class variable Metro to alter the appearance of its values in Output 2.5.2. Note that since the variable Metro and the format Metro have the same name, and since no width is required, the only syntax element that distinguishes these to the SAS compiler is the required dot (.) in the format name.

Program 2.5.2: Using the Metro Format

```
proc means data=BookData.IPUMS2005Basic nonobs maxdec=0;
  class Metro;
  var HHIncome;
  format Metro Metro.;
run;
```

Output 2.5.2: Using the Metro Format

METRO	N	Mean	Std Dev	Minimum	Maximum
Not Identifiable	92028	54800	52333	-19998	1076000
Not in Metro Area	230775	47856	45547	-29997	1050000
Metro, Inside City	154368	60328	70874	-19998	1391000
Metro, Outside City	340982	77648	75907	-29997	1739770
Metro, City Status Unknown	340909	64335	66110	-22298	1536000

Analysis Variable : HHIncome

For this case, a simplified format that distinguishes metro, non-metro, and non-identifiable observations may be desired. Program 2.5.3 contains two approaches to this, the first being clearly the most efficient.

Program 2.5.3: Assigning Multiple Values to the Same Formatted Value

```
proc format;
  value MetroB
    0 = "Not Identifiable"
    1 = "Not in Metro Area"
    2,3,4 ❶ = "In a Metro Area"
  ;
  value MetroC
    0 = "Not Identifiable"
    1 = "Not in Metro Area"
    2 = "In a Metro Area"❷
    3 = "In a Metro Area"❷
    4 = "In a Metro Area"❷
  ;
run;
```

❶ A comma-separated list of values is legal on the left side of each assignment, which assigns the formatted value to each listed data value.

❷ This format accomplishes the same result; however, it is important that the literal values on the right side of the assignment are exactly the same. Differences in even simple items like spacing or casing results in different formatted values.

Either format given in Program 2.5.3 can replace the Metro format in Program 2.5.2 to create the result in Output 2.5.3.

Output 2.5.3: Assigning Multiple Values to the Same Formatted Value

Analysis Variable : HHIncome					
METRO	N	Mean	Std Dev	Minimum	Maximum
Not Identifiable	92028	54800	52333	-19998	1076000
Not in Metro Area	230775	47856	45547	-29997	1050000
In a Metro Area	836259	69024	71495	-29997	1739770

It is also possible to use the dash character as an operator in the form of ValueA-ValueB to define a range on the left side of any assignment, which assigns the formatted value to every data value between ValueA and ValueB, inclusive. Program 2.5.4 gives an alternate strategy to constructing the formats given in Program 2.5.3 and that format can also be placed into Program 2.5.2 to produce Output 2.5.3.

Program 2.5.4: Assigning a Range of Values to a Single Formatted Value

```
proc format;
  value MetroD
    0 = "Not Identifiable"
    1 = "Not in Metro Area"
    2-4 = "In a Metro Area"
  ;
run;
```

Certain keywords are also available for use on the left side of an assignment, one of which is OTHER. OTHER applies the assigned format to any value not listed on the left side of an assignment elsewhere in the format definition. Program 2.5.5 uses OTHER to give another method for creating a format that can be used to generate Output 2.5.3. It is important to note that using OTHER often requires significant knowledge of exactly what values are present in the data set.

Program 2.5.5: Assigning a Range of Values to a Single Formatted Value

```
proc format;
  value MetroE
    0 = "Not Identifiable"
    1 = "Not in Metro Area"
    other = "In a Metro Area"
  ;
run;
```

In general, value ranges should be non-overlapping, and the < symbol—called an exclusion operator in this context—can be used at either end (or both ends) of the dash to indicate the value should not be included in the range. Overlapping ranges are discussed in Chapter Note 5 in Section 2.12. Using exclusion operators to create non-overlapping ranges allows for the categorization of a quantitative variable without having to know the precision of measurement. Program 2.5.6 gives two variations on creating bins for the MortgagePayment data and uses those bins as classes in PROC MEANS, with the results shown in Output 2.5.6A and Output 2.5.6B.

Program 2.5.6: Binning a Quantitative Variable Using a Format

```
proc format;
  value Mort
    0='None'
    1-350="$350 and Below"
    351-1000="$351 to $1000"
    1001-1600="$1001 to $1600"
    1601-high❶="Over $1600"
  ;❷
```

```
  value MortB
    0='None'
    1-350="$350 and Below"
    350<-1000="Over $350, up to $1000"
    1000<-1600="Over $1000, up to $1600"
    1600<-high="Over $1600"
  ; ❸
run;

proc means data=BookData.IPUMS2005Basic nonobs maxdec=0;
  class MortgagePayment;
  var HHIncome;
  format MortgagePayment Mort.; ❹
run;

proc means data=BookData.IPUMS2005Basic nonobs maxdec=0;
  class MortgagePayment;
  var HHIncome;
  format MortgagePayment MortB.;
run;
```

❶ The keywords LOW and HIGH are available so that the maximum and minimum values need not be known. When applied to character data, LOW and HIGH refer to the sorted alphanumeric values. Note that the LOW keyword excludes missing values for numeric variables but includes missing values for character variables.

❷ In these value ranges, the values used exploit the fact that the mortgage payments are reported to the nearest dollar.

❸ Using the < symbol to not include the starting ranges allows the bins to be mutually exclusive and exhaustive irrespective of the precision of the data values. The exclusion operator, <, omits the adjacent value from the range so that 350<-1000 omits only 350, 350-<1000 omits only 1000, and 350<-<1000 omits both 350 and 1000.

❹ When a format is present for a class variable, the format is used to construct the unique values for each category, and this behavior persists in most cases where SAS treats a variable as categorical.

Output 2.5.6A: Binning a Quantitative Variable Using the Mort Format

MortgagePayment	Analysis Variable : HHIncome				
	N	Mean	Std Dev	Minimum	Maximum
None	603691	45334	53557	-22298	1739770
$350 and Below	59856	47851	42062	-16897	841000
$351 to $1000	283111	64992	45107	-19998	1060000
$1001 to $1600	128801	96107	63008	-29997	1125000
Over $1600	83603	153085	117134	-29997	1407000

Output 2.5.6B: Binning a Quantitative Variable Using the MortB Format

Analysis Variable : HHIncome					
MortgagePayment	**N**	**Mean**	**Std Dev**	**Minimum**	**Maximum**
None	603691	45334	53557	-22298	1739770
$350 and Below	59856	47851	42062	-16897	841000
Over $350, up to $1000	283111	64992	45107	-19998	1060000
Over $1000, up to $1600	128801	96107	63008	-29997	1125000
Over $1600	83603	153085	117134	-29997	1407000

2.5.2 Permanent Storage and Inspection of Defined Formats

Formats can be permanently stored in a catalog (with the default name of Formats) in any assigned SAS library via the use of the LIBRARY= option in the PROC FORMAT statement. As an example, consider Program 2.5.7, which is a revision and extension of Program 2.5.6.

Program 2.5.7: Revisiting Program 2.5.6, Adding LIBRARY= and FMTLIB Options

```
proc format library=sasuser;
  value Mort
    0='None'
    1-350="$350 and Below"
    351-1000="$351 to $1000"
    1001-1600="$1001 to $1600"
    1601-high="Over $1600"
  ;
  value MortB
    0='None'
    1-350="$350 and Below"
    350<-1000="Over $350, up to $1000"
    1000<-1600="Over $1000, up to $1600"
    1600<-high="Over $1600"
  ;
run;

proc format fmtlib library=sasuser;
run;
```

Using the LIBRARY= option in this manner places the format definitions into the Formats catalog in the Sasuser library and accessing them in subsequent coding sessions requires the system option FMTSEARCH=(SASUSER) to be specified prior to their use. An alternate format catalog can also be used via two-level naming of the form *libref.catalog*, with the catalog being created if it does not already exist. Any catalog in any library that contains stored formats to be used in a given session can be listed as a set inside the parentheses following the FMTSEARCH= option. Those listed are searched in the given order, with WORK.FORMATS being defined implicitly as the first catalog to be searched unless it is included explicitly in the list.

The FMTLIB option shows information about the formats in the chosen library in the Output window, Output 2.5.7 shows the results for this case.

Output 2.5.7: Revisiting Program 2.5.6, Adding LIBRARY= and FMTLIB Options

```
        FORMAT NAME: MORT      LENGTH:    14    NUMBER OF VALUES:     5
    MIN LENGTH:    1   MAX LENGTH:   40   DEFAULT LENGTH:   14   FUZZ: STD

START                   END                 LABEL    (VER. V7|V8     17DEC2017:14:09:32)

                0                  0 None
                1                350 $350 and Below
              351               1000 $351 to $1000
             1001               1600 $1001 to $1600
             1601 HIGH               Over $1600
```

```
        FORMAT NAME: MORTB     LENGTH:    23    NUMBER OF VALUES:     5
    MIN LENGTH:    1   MAX LENGTH:   40   DEFAULT LENGTH:   23   FUZZ: STD

START                   END                 LABEL    (VER. V7|V8     17DEC2017:14:09:32)

                0                  0 None
                1                350 $350 and Below
             350<               1000 Over $350, up to $1000
            1000<               1600 Over $1000, up to $1600
            1600<HIGH               Over $1600
```

The top of the table includes general information about the format, including the name, various lengths, and number of format categories. The default length corresponds to the longest format label set in the VALUE statement. The rows below have columns for each format label and the start and end of each value range. Note that the first category in each of these formats is assigned to a range, even though it only contains a single value, with the start and end values being the same. The use of < as an exclusion operator is also shown in ranges where it is used, and the keyword HIGH is left-justified in the column where it is used. Note the exclusion operation is applied to the value of 1600 at the low end of the range, it is a syntax error to attempt to apply it to the keyword HIGH (or LOW).

2.6 Subsetting with the WHERE Statement

In many cases, only a subset of the data is used, with the subsetting criteria based on the values of variables in the data set. In these cases, using the WHERE statement allows conditions to be set which choose the records a SAS procedure processes while ignoring the others—no modification to the data set itself is required. If the OBS= data set option is in use, the number chosen corresponds to the number of observations meeting the WHERE condition.

In order to use the WHERE statement, it is important to understand the comparison and logical operators available. Basic comparisons like equality or various inequalities can be done with symbolic or mnemonic operators—Table 2.6.1 shows the set of comparison operators.

Table 2.6.1: Comparison Operators

Operation	Symbol	Mnemonic
Equal	=	EQ
Not Equal	^=	NE
Less Than	<	LT
Less Than or Equal	<=	LE

Operation	Symbol	Mnemonic
Greater Than	>	GT
Greater Than or Equal	>=	GE

In addition to comparison operators, Boolean operators for negation and compounding (along with some special operators) are also available—Table 2.6.2 summarizes these operators.

Table 2.6.2: Boolean and Associated Operators

Symbol	Mnemonic	Logic
&	AND	True result if both conditions are true
|	OR	True result if either, or both, conditions are true
	IN	True if matches any element in a list
	BETWEEN-AND	True if in a range of values (including endpoints)
~	NOT	Negates the condition that follows

Revisiting Program 2.5.2 and Output 2.5.2, subsetting the results to only include observations known to be in a metro area can be accomplished with any one of the following WHERE statements.

```
❶ where Metro eq 2 or Metro eq 3 or Metro eq 4;
❷ where Metro ge 2 and Metro le 4;
❸ where Metro in (2,3,4);
❹ where Metro between 2 and 4;
❺ where Metro not in (0,1);
```

❶ Each possible value can be checked by using the OR operator between equality comparisons for each possible value. When using OR, each comparison must be complete/specific. For example, it is not legal to say: `Metro eq 2 or eq 3 or eq 4`. It is legal, but unhelpful, to say `Metro eq 2 or 3 or 4`, as SAS uses numeric values for truth (since it does not include Boolean variables). The values 0 and missing are false, while any other value is true; hence, `Metro eq 2 or 3 or 4` is an immutably true condition.

❷ This conditioning takes advantage of the fact that the desired values fall into a range. As with OR, each condition joined by the AND must be complete; again, it is *not legal* to say: `Metro ge 2 and le 4`. Also, with knowledge of the values of Metro, this condition could have been simplified to `Metro ge 2`. However, good programming practice dictates that specificity is preferred to avoid incorrect assumptions about data values.

❸ IN allows for simplification of a set of conditions that might otherwise be written using the OR operator, as was done in ❶. The list is given as a set of values separated by commas or spaces and enclosed in parentheses.

❹ BETWEEN-AND allows for simplification of a value range that can otherwise be written using AND between appropriate comparisons, as was done in ❷.

❺ The NOT operator allows the truth condition to be made the opposite of what is specified. This is a slight improvement over ❸, as the list of values not desired is shorter than the list of those that are.

Adding any of these WHERE statements (or any other logically equivalent WHERE statement) to Program 2.5.2 produces the results shown in Table 2.6.3.

Table 2.6.3: Using WHERE to Subset Results to Specific Values of the Metro Variable

METRO	Analysis Variable : HHIncome				
	N	Mean	Std Dev	Minimum	Maximum
Metro, Inside City	154368	60328	70874	-19998	1391000
Metro, Outside City	340982	77648	75907	-29997	1739770
Metro, City Status Unknown	340909	64335	66110	-22298	1536000

The tools available allow for conditioning on more than one variable, and the variable(s) conditioned on need only be in the data set in use and do not have to be present in the output generated. In Program 2.6.1, the output is conditioned additionally on households known to have an outstanding mortgage.

Program 2.6.1: Conditioning on a Variable Not Used in the Analysis

```
proc means data=BookData.IPUMS2005Basic nonobs maxdec=0;
  class Metro;
  var HHIncome;
  format Metro Metro.;
  where Metro in (2,3,4)
        and
        MortgageStatus in
          ('Yes, contract to purchase',
           'Yes, mortgaged/ deed of trust or similar debt');
run;
```

Output 2.6.1: Conditioning on a Variable Not Used in the Analysis

METRO	Analysis Variable : HHIncome				
	N	Mean	Std Dev	Minimum	Maximum
Metro, Inside City	57881	86277	82749	-19998	1361000
Metro, Outside City	191021	96319	80292	-29997	1266000
Metro, City Status Unknown	167359	83879	72010	-19998	1407000

The condition on the MortgageStatus variable is a bit daunting, particularly noting that matching character values is a precise operation. Seemingly simple differences like casing or spacing lead to values that are non-matching. Therefore, the literals used in Program 2.6.1 are specified to be an exact match for the data. In Section 3.9, functions are introduced that are useful in creating consistency among character values, along with others that allow for extraction and use of relevant portions of a string. However, the WHERE statement provides some special operators, shown in Table 2.6.4, that allow for simplification in these types of cases without the need to intervene with a function.

Table 2.6.4: Operators for General Comparisons

Symbol	Mnemonic	Logic
?	CONTAINS	True result if the specified value is contained in the data value (character only).
	LIKE	True result if data value matches the specified value which may include wildcards. _ is any single character, % is any set of characters.

Program 2.6.2 offers two methods for simplifying the condition on MortgageStatus, one using CONTAINS, the other using LIKE. Either reproduces Output 2.6.1.

Program 2.6.2: Conditioning on a Variable Using General Comparison Operators
```
proc means data=BookData.IPUMS2005Basic nonobs maxdec=0;
  class Metro;
  var HHIncome;
  format Metro Metro.;
  where Metro in (2,3,4) and MortgageStatus contains ❶'Yes';
run;

proc means data=BookData.IPUMS2005Basic nonobs maxdec=0;
  class Metro;
  var HHIncome;
  format Metro Metro.;
  where Metro in (2,3,4) and MortgageStatus like ❷'%Yes%';
run;
```

❶ CONTAINS checks to see if the data value contains the string Yes; again, note that the casing must be correct to ensure a match. Also, ensure single or double quotation marks enclose the value to search for—in this case, without the quotation marks, Yes forms a legal variable name and is interpreted by the compiler as a reference to a variable.

❷ LIKE allows for the use of wildcards as substitutes for non-essential character values. Here the % wildcard before and after Yes results in a true condition if Yes appears anywhere in the string and is thus logically equivalent to the CONTAINS conditioning above.

2.7 Using the FREQ Procedure for Categorical Summaries

To produce tables of frequencies and relative frequencies (percentages) like those shown for the case study in Outputs 2.2.3 and 2.2.4, the FREQ procedure is the tool of choice, and this section covers its fundamentals.

2.7.1 Choosing Analysis Variables in PROC FREQ

As in previous sections, the examples here use the IPUMS2005Basic SAS data set, so make sure the BookData library is assigned. As a first step, enter and submit Program 2.7.1. (Note that the use of labels has been re-established in the OPTIONS statement.)

Program 2.7.1: PROC FREQ with Variables Listed Individually in the TABLE Statement
```
options label;
proc freq data=BookData.IPUMS2005Basic;
  table metro mortgageStatus;
run;
```

The TABLE statement allows for specification of the variables to summarize, and a space-delimited list of variables produces a one-way frequency table for each, as shown in Output 2.7.1.

Output 2.7.1: PROC FREQ with Variables Listed Individually in the TABLE Statement

	Metropolitan status			
METRO	Frequency	Percent	Cumulative Frequency	Cumulative Percent
0	92028	7.94	92028	7.94
1	230775	19.91	322803	27.85
2	154368	13.32	477171	41.17
3	340982	29.42	818153	70.59
4	340909	29.41	1159062	100.00

MortgageStatus	Frequency	Percent	Cumulative Frequency	Cumulative Percent
N/A	303342	26.17	303342	26.17
No, owned free and clear	300349	25.91	603691	52.08
Yes, contract to purchase	9756	0.84	613447	52.93
Yes, mortgaged/ deed of trust or similar debt	545615	47.07	1159062	100.00

The TABLE statement is not required; however, in that case, the default behavior produces a one-way frequency table for every variable in the data set. Therefore, both types of SAS variables, character or numeric, are legal in the TABLE statement. Given that variables listed in the TABLE statement are treated as categorical (in the same manner as variables listed in the CLASS statement in PROC MEANS), it is best to have the summary variables be categorical or be formatted into a set of categories.

The default summaries in a one-way frequency table are: frequency (count), percent, cumulative frequency, and cumulative percent. Of course, the cumulative statistics only make sense if the categories are ordinal, which these are not. Many options are available in the table statement to control what is displayed, and one is given in Program 2.7.2 to remove the cumulative statistics.

Program 2.7.2: PROC FREQ Option for Removing Cumulative Statistics

```
proc freq data=BookData.IPUMS2005Basic;
  table metro mortgageStatus / nocum;
run;
```

As with the CLASS statement in the MEANS procedure, variables listed in the TABLE statement in PROC FREQ use the format provided with the variable to construct the categories. Program 2.7.3 uses a format defined in Program 2.5.6 to bin the MortgagePayment variable into categories and, as this is an ordinal set, the cumulative statistics are appropriate.

Program 2.7.3: Using a Format to Control Categories for a Variable in the TABLE Statement

```
proc format;
  value Mort
    0='None'
    1-350="$350 and Below"
    351-1000="$351 to $1000"
    1001-1600="$1001 to $1600"
    1601-high="Over $1600"
  ;
run;
```

```
proc freq data=BookData.IPUMS2005Basic;
  table MortgagePayment;
  format MortgagePayment Mort.;
run;
```

Output 2.7.3: Using a Format to Control Categories for a Variable in the TABLE Statement

First mortgage monthly payment				
MortgagePayment	Frequency	Percent	Cumulative Frequency	Cumulative Percent
None	603691	52.08	603691	52.08
$350 and Below	59856	5.16	663547	57.25
$351 to $1000	283111	24.43	946658	81.67
$1001 to $1600	128801	11.11	1075459	92.79
Over $1600	83603	7.21	1159062	100.00

The FREQ procedure is not limited to one-way frequencies—special operators between variables in the TABLE statement allow for construction of multi-way tables.

2.7.2 Multi-Way Tables in PROC FREQ

The * operator constructs cross-tabular summaries for two categorical variables, which includes the following statistics:

- cross-tabular and marginal frequencies
- cross-tabular and marginal percentages
- conditional percentages within each row and column

Program 2.7.4 summarizes all combinations of Metro and MortgagePayment, with Metro formatted to add detail and MortgagePayment formatted into the bins used in the previous example.

Program 2.7.4: Using the * Operator to Create a Cross-Tabular Summary with PROC FREQ

```
proc format;
  value METRO
    0 = "Not Identifiable"
    1 = "Not in Metro Area"
    2 = "Metro, Inside City"
    3 = "Metro, Outside City"
    4 = "Metro, City Status Unknown"
  ;
  value Mort
    0='None'
    1-350="$350 and Below"
    351-1000="$351 to $1000"
    1001-1600="$1001 to $1600"
    1601-high="Over $1600"
  ;
run;

proc freq data=BookData.IPUMS2005Basic;
  table Metro*MortgagePayment; ❶
  format Metro Metro. MortgagePayment Mort.; ❷
run;
```

❶ The first variable listed in any request of the form A*B is placed on the rows in the table. Requesting MortgagePayment*Metro transposes the table and the included summary statistics.

❷ The format applied to the Metro variable is merely a change in display and has no effect on the structure of the table—it is five rows with or without the format. The format on MortgagePayment is essential to the column structure—allowing each unique value of MortgagePayment to form a column does not produce a useful summary table.

Output 2.7.4: Using the * Operator to Create a Cross-Tabular Summary with PROC FREQ

Table of METRO by MortgagePayment						
METRO(Metropolitan status)	**MortgagePayment(First mortgage monthly payment)**					
Frequency Percent Row Pct Col Pct	None	$350 and Below	$351 to $1000	$1001 to $1600	Over $1600	Total
Not Identifiable	49379 4.26 53.66 8.18	6979 0.60 7.58 11.66	25488 2.20 27.70 9.00	7307 0.63 7.94 5.67	2875 0.25 3.12 3.44	92028 7.94
Not in Metro Area	134314 11.59 58.20 22.25	21698 1.87 9.40 36.25	60948 5.26 26.41 21.53	10464 0.90 4.53 8.12	3351 0.29 1.45 4.01	230775 19.91
Metro, Inside City	96487 8.32 62.50 15.98	4410 0.38 2.86 7.37	28866 2.49 18.70 10.20	14049 1.21 9.10 10.91	10556 0.91 6.84 12.63	154368 13.32
Metro, Outside City	149961 12.94 43.98 24.84	12148 1.05 3.56 20.30	79388 6.85 23.28 28.04	56330 4.86 16.52 43.73	43155 3.72 12.66 51.62	340982 29.42
Metro, City Status Unknown	173550 14.97 50.91 28.75	14621 1.26 4.29 24.43	88421 7.63 25.94 31.23	40651 3.51 11.92 31.56	23666 2.04 6.94 28.31	340909 29.41
Total	603691 52.08	59856 5.16	283111 24.43	128801 11.11	83603 7.21	1159062 100.00

Various options are available to control the displayed statistics. Program 2.7.5 illustrates some of these with the result shown in Output 2.7.5.

Program 2.7.5: Using Options in the TABLE Statement.

```
proc freq data=BookData.IPUMS2005Basic;
  table Metro*MortgagePayment / nocol nopercent❶ format=comma10.❷;
  format Metro Metro. MortgagePayment Mort.;
run;
```

❶ NOCOL and NOPERCENT suppress the column and overall percentages, respectively, with NOPERCENT also applying to the marginal totals. NOROW and NOFREQ are also available, with NOFREQ also applying to the marginal totals.

❷ A format can be applied to the frequency statistic; however, this only applies to cross-tabular frequency tables and has no effect in one-way tables.

Output 2.7.5: Using Options in the TABLE Statement

METRO(Metropolitan status)	MortgagePayment(First mortgage monthly payment)					
Frequency Row Pct	None	$350 and Below	$351 to $1000	$1001 to $1600	Over $1600	Total
Not Identifiable	49,379 53.66	6,979 7.58	25,488 27.70	7,307 7.94	2,875 3.12	92,028
Not in Metro Area	134,314 58.20	21,698 9.40	60,948 26.41	10,464 4.53	3,351 1.45	230,775
Metro, Inside City	96,487 62.50	4,410 2.86	28,866 18.70	14,049 9.10	10,556 6.84	154,368
Metro, Outside City	149,961 43.98	12,148 3.56	79,388 23.28	56,330 16.52	43,155 12.66	340,982
Metro, City Status Unknown	173,550 50.91	14,621 4.29	88,421 25.94	40,651 11.92	23,666 6.94	340,909
Total	603,691	59,856	283,111	128,801	83,603	1,159,062

Table of METRO by MortgagePayment

Higher dimensional requests can be made; however, they are constructed as a series of two-dimensional tables. Therefore, a request of A*B*C in the TABLE statement creates the B*C table for each level of A, while a request of A*B*C*D makes the C*D table for each combination of A and B, and so forth. Program 2.7.6 generates a three-way table, where a cross-tabulation of Metro and HomeValue is built for each level of Mortgage Status as shown in Output 2.7.6. The VALUE statement that defines the character format $MortStatus takes advantage of the fact that value ranges are legal for character variables. Be sure to understand the difference between uppercase and lowercase letters when ordering the values of a character variable.

Program 2.7.6: A Three-Way Table in PROC FREQ

```
proc format;
  value MetroB
    0 = "Not Identifiable"
    1 = "Not in Metro Area"
    other = "In a Metro Area"
  ;
  value $MortStatus
    'No'-'Nz'='No'
    'Yes'-'Yz'='Yes'
  ;
  value Hvalue
    0-65000='$65,000 and Below'
    65000<-110000='$65,001 to $110,000'
    110000<-225000='$110,001 to $225,000'
    225000<-500000='$225,001 to $500,000'
    500000-high='Above $500,000'
  ;
run;

proc freq data=BookData.IPUMS2005Basic;
  table MortgageStatus*Metro*HomeValue/nocol nopercent format=comma10.;
  format MortgageStatus $MortStatus. Metro MetroB. HomeValue Hvalue.;
  where MortgageStatus ne 'N/A';
run;
```

Output 2.7.6: A Three-Way Table in PROC FREQ

METRO(Metropolitan status)	HomeValue(House value)					
	Table 1 of METRO by HomeValue					
	Controlling for MortgageStatus=No					
Frequency Row Pct	$65,000 and Below	$65,001 to $110,000	$110,001 to $225,000	$225,001 to $500,000	Above $500,000	Total
Not Identifiable	10,777 35.49	5,460 17.98	10,415 34.29	2,584 8.51	1,134 3.73	30,370
Not in Metro Area	34,766 40.57	16,261 18.98	26,889 31.38	5,553 6.48	2,227 2.60	85,696
In a Metro Area	34,176 18.55	23,706 12.86	71,133 38.60	33,590 18.23	21,678 11.76	184,283
Total	79,719	45,427	108,437	41,727	25,039	300,349

METRO(Metropolitan status)	HomeValue(House value)					
	Table 2 of METRO by HomeValue					
	Controlling for MortgageStatus=Yes					
Frequency Row Pct	$65,000 and Below	$65,001 to $110,000	$110,001 to $225,000	$225,001 to $500,000	Above $500,000	Total
Not Identifiable	7,486 17.55	7,142 16.75	19,453 45.61	6,468 15.17	2,100 4.92	42,649
Not in Metro Area	24,443 25.34	19,396 20.11	40,668 42.16	9,164 9.50	2,790 2.89	96,461
In a Metro Area	26,351 6.33	37,345 8.97	175,482 42.16	110,412 26.52	66,671 16.02	416,261
Total	58,280	63,883	235,603	126,044	71,561	555,371

It is also possible for the FREQ procedure to count based on a quantitative variable using the WEIGHT statement, effectively tabulating the sum of the weights. Program 2.7.7 uses the weight statement to summarize total HomeValue for combinations of Metro and MortgagePayment.

Program 2.7.7: Using the WEIGHT Statement to Summarize a Quantitative Value.

```
proc freq data=BookData.IPUMS2005Basic;
  table Metro*MortgagePayment /nocol nopercent format=dollar14.;
  weight HomeValue;
  format Metro MetroB. MortgagePayment Mort.;
run;
```

Output 2.7.7: Using the WEIGHT Statement to Summarize a Quantitative Value

Table of HomeValue by METRO				
HomeValue(House value)	METRO(Metropolitan status)			
Frequency Row Pct	Not Identifiable	Not in Metro Area	In a Metro Area	Total
$65,000 and Below	$2,737,530 12.20	$8,736,986 38.93	$10,969,600 48.88	$22,444,116
$65,001 to $110,000	$3,770,840 10.74	$9,887,454 28.16	$21,448,052 61.09	$35,106,346
$110,001 to $225,000	$15,896,854 7.82	$30,632,556 15.07	$156,700,074 77.10	$203,229,484
$225,001 to $500,000	$8,192,908 4.80	$10,741,258 6.30	$151,601,294 88.90	$170,535,460
Above $500,000	$3,854,280 2.60	$4,735,288 3.19	$139,862,780 94.21	$148,452,348
Total	$34,452,412	$64,733,542	$480,581,800	$579,767,754

2.8 Reading Raw Data

Often data is not available as a SAS data set; in practice, data often comes from external sources including raw files such as text files, spreadsheets such as Microsoft Excel ®, or relational databases such as Oracle ®. In this section, work with external data sources begins by exploring how to read raw data files.

Raw data refers to certain files that contain unprocessed data that is not in a SAS data set. (Certain other structures also qualify as raw data. See Chapter Note 6 in Section 2.12 for additional details.) Generally, these are plain-text files and some common file types are:

- tab-delimited text (.txt or .tsv)
- comma-separated values (.csv)
- fixed-position files (.dat)

Choices for file extensions are not fixed; therefore, the extension does not dictate the type of data the file contains, and many other file types exist. Therefore, it is always important to explore the raw data before importing it to SAS. While SAS provides multiple ways to read raw data; this chapter focuses on using the DATA step due to its flexibility and ubiquity—understanding the DATA step is a necessity for a successful SAS programmer.

To assist in finding the column numbers when displaying raw files, a ruler is included in the first line when presenting raw data in the book, but the ruler is not present in the actual data file. Input Data 2.8.1 provides an example of such a ruler. Each dash in the ruler represents a column, while a plus represents multiples of five, and a digit represents multiples of ten. For example, the 1 in the ruler represents column 10 in the raw file and the plus sign between the 1 and the 2 represents column 15 in the raw file.

Input Data 2.8.1: Space Delimited Raw File (Partial Listing)

```
----+----1----+----2----+
1 1800 9998 9998 9998
2 480 1440 9998 9998
3 2040 360 100 9998
4 3000 9998 360 9998
5 840 1320 90 9998
```

2.8.1 Introduction to Reading Delimited Files

Delimiters, often used in raw files, are a single character such as a tab, space, comma, or pipe (vertical bar) used to indicate the break between one value and the next in a single record. Input Data 2.8.1 includes a partial representation of the first five records from a space-delimited file (Utility 2001.prn). Reading in this file, or any raw file, requires determining whether the file is delimited and, if so, what delimiters are present. If a file is delimited, it is important to note whether the delimiters also appear as part of the values for one or more variables. The data presented in Input Data 2.8.1 follows a basic structure and uses spaces to separate each record into five distinct values or fields. SAS can read this file correctly using simple list input without the need for additional options or statements using the following rules:

1. At least one blank/space must separate the input values and SAS treats multiple, sequential blanks as a single blank.
2. Character values cannot contain embedded blanks.
3. Character variables are given a length of eight bytes by default.
4. Data must be in standard numeric or character format. Standard numeric values must only contain digits, decimal point, +/-, and E for scientific notation.

Input Data 2.8.1 satisfies these rules using the default delimiter (space). Options and statements are available to help control the behavior associated with rules 1 through 3, which are covered in subsequent sections of this chapter. Violating rule 4 precludes the use of simple list input but is easily addressed with modified list input, as shown in Chapter 3. However, no such options or modifications are required to read Input Data 2.8.1, which is done using Program 2.8.1.

Program 2.8.1: Reading the Utility 2001 Data

```
data Utility2001; ❶
   infile "--insert path here--\Utility 2001.prn"; ❷
   input Serial$ Electric Gas Water Fuel; ❸
run; ❹

proc print data = Utility2001 (obs=5 ❺);
run;
```

❶ The DATA statement begins the DATA step and here names the data set as Utility2001, placing it in the Work library given the single-level naming. Explicit specification of the library is available with two-level naming, for example, Sasuser.Utility2001 or Work.Utility2001—see Program 1.4.3. If no data set name appears, SAS provides the name as DATA*n*, where *n* is the smallest whole number (1, 2, 3, ...) that makes the data set name unique.

❷ The INFILE statement specifies the location of the file via a full path specification to the file—this path must be completed to reflect the location of the raw file for the code to execute successfully.

❸ The INPUT statement sets the names of each variable from the raw file in the INFILE statement with those names following the conventions outlined in Section 1.6.2. By default, SAS assumes the incoming variables are numeric. One way to indicate character data is shown here – place a dollar sign after each character variable.

❹ Good programming practice dictates that all steps end with an explicit step boundary, including the DATA step.

❺ The OBS= option selects the last observation for processing. Because procedures start with the first observation by default, this step uses the first five observations from the Utility2001 data set, as shown in Output 2.8.1.

Output 2.8.1: Reading the Utility 2001 Data (Partial Listing)

Obs	Serial	Electric	Gas	Water	Fuel
1	1	1800	9998	9998	9998
2	2	480	1440	9998	9998
3	3	2040	360	100	9998
4	4	3000	9998	360	9998
5	5	840	1320	90	9998

In Program 2.8.1, Serial is read as a character variable; however, it contains only digits and therefore can be stored as numeric. The major advantage in storing Serial as character is size—its maximum value is six digits long and therefore requires six bytes of storage as character, while all numeric variables have a default size of eight bytes. The major disadvantage to storing Serial as character is ordering—for example, as a character value, 11 comes before 2. While the other four variables can be read as character as well, it is a very poor choice as no mathematical or statistical operations can be done on those values. For examples in subsequent sections, Serial is read as numeric.

In Program 2.8.1, the INFILE statement is used to specify the raw data file that the DATA step reads. In general, the INFILE statement may include references to a single file or to multiple files, with each reference provided one of the following ways:

- A physical path to the files. Physical paths can be either relative or absolute.
- A file reference created via the FILENAME statement.

Program 2.8.1 is set up to use the first method, with either an absolute or relative path chosen. An absolute path starts with a drive letter or name, while any other specification is a relative path. All relative paths are built from the current working directory. (Refer to Section 1.5 for a discussion of the working directory and setting its value.) It is often more efficient to use a FILENAME statement to build references to external files or folders. Programs 2.8.2 and 2.8.3 demonstrate these uses of the FILENAME statement, producing the same data set as Program 2.8.1.

Program 2.8.2: Using the FILENAME Statement to Point to an Individual File

```
filename Util2001 ❶ "--insert path here--\Utility 2001.prn"❷;

data work.Utility2001A;
  infile Util2001; ❸
  input Serial$ Electric Gas Water Fuel;
run;
```

❶ The FILENAME statement creates a file reference, called a *fileref*, named Util2001. Naming conventions for a *fileref* are the same as those for a *libref*.

❷ The path specified, which can be relative or absolute as in Program 2.8.1, includes the file name. SAS assigns the *fileref* Util2001 to this file.

❸ The INFILE statement now references the *fileref* Util2001 rather than the path or file name. Note, quotation marks are not used on Util2001 since it is to be interpreted as a *fileref* and not a file name or path.

Program 2.8.3: Associating the FILENAME Statement with a Folder

```
filename RawData '--insert path to folder here--'; ❶

data work.Utility2001B;
  infile RawData("Utility 2001.prn"); ❷
  input Serial$ Electric Gas Water Fuel;
run;
```

❶ It is assumed here that the path, either relative or absolute, points to a folder and not a specific file. In that case, the FILENAME statement associates a folder with the *fileref* RawData. The path specified should be to the folder containing the raw files downloaded from the author page, much like the BookData library was assigned to the folder containing the SAS data sets.

❷ The INFILE statement references both the *fileref* and the file name. Although the file reference can be made without the quotation marks in certain cases, good programming practice includes the quotation marks.

Since each of Programs 2.8.2 and 2.8.3 generate the same result as Program 2.8.1 but actually require slightly more code, the benefits of using the FILENAME statement may not be obvious. The form of the FILENAME in Program 2.8.3 is useful if a single file needs to be read repeatedly under different conditions, allowing the multiple references to that file to be shortened. More commonly, the form used in Program 2.8.4 is more efficient when reading multiple files from a common location. Again, if the path specified is to the folder containing the raw files downloaded from the author page, the *fileref* RawData refers to the location for all non-SAS data sets used in examples for Chapters 2 through 7.

2.8.2 More with List Input

Input Data 2.8.4 includes a partial representation of the first five records from a comma-delimited file (IPUMS2005Basic.csv). Due to the width of the file, Input Data 2.8.4 truncates the third and fifth records.

Input Data 2.8.4: Comma Delimited Raw File (Partial Listing)

```
----+----1----+----2----+----3----+----4----+----5----+----6----+----7----+----8----+
2,Alabama,Not in identifiable city (or size group),0,4,73,Rented,N/A,0,12000,9999999
3,Alabama,Not in identifiable city (or size group),0,1,0,Rented,N/A,0,17800,9999999
4,Alabama,Not in identifiable city (or size group),0,4,73,Owned,"Yes, mortgaged/ deed
5,Alabama,Not in identifiable city (or size group),0,1,0,Rented,N/A,0,2000,9999999
6,Alabama,Not in identifiable city (or size group),0,3,97,Owned,"No, owned free and
```

Not only is this file delimited by commas, but the eighth field on the third and fifth rows also includes data values containing a comma, with those values embedded in quotation marks. (Recall these records are truncated in the text due to their length so the final quote is not shown for these two records.) To successfully read this file, the DATA step must recognize the delimiter as a comma, but also that commas embedded in quoted values are not delimiters. The DSD option is introduced in Program 2.8.4 to read such a file.

Program 2.8.4: Reading the 2005 Basic IPUMS CPS Data
```
data work.Ipums2005Basic;
  infile RawData("IPUMS2005basic.csv") dsd; ❶
  input Serial State $ City $ CityPop Metro
        CountyFIPS Ownership $ MortgageStatus $
        MortgagePayment HHIncome HomeValue; ❷
run;

proc print data = work.Ipums2005Basic (obs=5);
run;
```

❶ The DSD option included in the INFILE statement modifies the delimiter and some additional default behavior as listed below.

❷ Again, the INPUT statement names each of the variables read from the raw file in the INFILE statement and sets their types. By default, SAS assumes the incoming variables are numeric; however, State, City, Ownership, and MortgageStatus must be read as character values.

Output 2.8.4 shows that, while Program 2.8.4 executes successfully, the resulting data set does not correctly represent the values from Input Data 2.8.4—the City and MortgageStatus variables are truncated. This truncation occurs due to the default length of 8 assigned to character variables; therefore, SAS did not allocate enough memory to store the values in their entirety. Only the first five records are shown; however, further investigation reveals this truncation occurs for the variable State as well.

Output 2.8.4: Reading the 2005 Basic IPUMS CPS Data (Partial Listing).

Obs	Serial	State	City	CityPop	Metro	CountyFIPS	Ownership
1	2	Alabama	Not in i	0	4	73	Rented
2	3	Alabama	Not in i	0	1	0	Rented
3	4	Alabama	Not in i	0	4	73	Owned
4	5	Alabama	Not in i	0	1	0	Rented
5	6	Alabama	Not in i	0	3	97	Owned

Obs	MortgageStatus	MortgagePayment	HHIncome	HomeValue
1	N/A	0	12000	9999999
2	N/A	0	17800	9999999
3	Yes, mor	900	185000	137500
4	N/A	0	2000	9999999
5	No, owne	0	72600	95000

Program 2.8.4 uses the DSD option in the INFILE statement to change three default behaviors:

1. Change the delimiter to comma
2. Treat two consecutive delimiters as a missing value
3. Treat delimiters inside quoted strings as part of a character value and strip off the quotation marks

For Input Data 2.8.4, the first and third actions are necessary to successfully match the structure of the delimiters in the data since (a) the file uses commas as delimiters and (b) commas are included in the quoted strings in the data for the MortgageStatus variable. Because the file does not contain consecutive delimiters, the second modification has no effect.

Of course, it might be necessary to produce the second and third effects while using blanks—or any other character—as the delimiter. It is also often necessary to change the delimiter without making the other modifications included with the DSD option. In those cases, use the DLM= option to specify one or more delimiters by placing them in a single set of quotation marks, as shown in the following examples.

1. DLM = '/' causes SAS to move to a new field when it encounters a forward slash
2. DLM = ', ' causes SAS to move to a new field when it encounters a comma
3. DLM = ',/' causes SAS to move to a new field when it encounters either a comma or forward slash

Introduction to Variable Attributes

In SAS, the amount of memory allocated to a variable is called the variable's length; length is one of several attributes that each variable possesses. Other attributes include the name of the variable, its position in the data set (1st column, 2nd column, ...), and its type (character or numeric). As with all the variable attributes, the length is set either by use of a default value or by explicitly setting a value.

By default, both numeric and character variables have a length of eight bytes. For character variables, one byte of memory can hold one character in the English language. Thus, the DATA step truncates several values of State, City, and MortgageStatus from Input Data 2.8.4 since they exceed the default length of eight bytes. For numeric variables, the default length of eight bytes is sufficient to store up to 16 decimal digits (commonly known as double-precision). When using the Microsoft Windows® operating system, numeric variables have a minimum allowable length of three bytes and a maximum length of eight bytes. Character variables may have a minimum length of 1 byte and a maximum length of 32,767 bytes. While there are many options and statements that affect the length of a variable implicitly, the LENGTH statement allows for explicit declaration of the length and type attributes for any variables. Program 2.8.5 demonstrates the usage of the LENGTH statement.

Program 2.8.5: Using the LENGTH Statement
```
data work.Ipums2005Basic;
  length state $ 20 City$ 25 MortgageStatus$50; ❶
  infile RawData("IPUMS2005basic.csv") dsd;
  input Serial State City ❷ CityPop Metro
        CountyFIPS Ownership $ ❷ MortgageStatus$ ❸
        MortgagePayment HHIncome HomeValue;
run;

proc print data = work.Ipums2005Basic(obs = 5);
run;
```

❶ The LENGTH statement sets the lengths of State, City, and MortgageStatus to 20, 25, and 50 characters, respectively, with the dollar sign indicating these are character variables. Separating the dollar sign from the variable name or length value is optional, though good programming practices dictate using a consistent style to improve readability.

❷ Type (character or numeric) is an attribute that cannot be changed in the DATA step once it has been established. Because the LENGTH statement sets these variables as character, the dollar sign is optional in the INPUT statement. However, good programming practices generally dictate including it for readability and so that removal of the LENGTH statement does not lead to a data type mismatch. (This would be an execution-time error.)

❸ As in ❶, the spacing between the dollar sign and variable name is optional in the INPUT statement as well. Good programming practices still dictate selecting a consistent spacing style.

Output 2.8.5 shows the results of explicitly setting the length of the State, City, and MortgageStatus variables. In addition to the lengths of these three variables changing, their column position in the SAS data set has changed as well. Variables are added to the data set based on the order they are encountered during compilation of the DATA step, so since the LENGTH statement precedes the INPUT statement, it has actually changed two attributes—length and position—for these three variables (while also defining the type attribute as character).

Output 2.8.5: Using the LENGTH Statement (Partial Listing)

Obs	state	City	MortgageStatus	Serial	CityPop	Metro	CountyFIPS
1	Alabama	Not in identifiable city	N/A	2	0	4	73
2	Alabama	Not in identifiable city	N/A	3	0	1	0
3	Alabama	Not in identifiable city	Yes, mortgaged/ deed of trust or similar debt	4	0	4	73
4	Alabama	Not in identifiable city	N/A	5	0	1	0
5	Alabama	Not in identifiable city	No, owned free and clear	6	0	3	97

Obs	Ownership	MortgagePayment	HHIncome	HomeValue
1	Rented	0	12000	9999999
2	Rented	0	17800	9999999
3	Owned	900	185000	137500
4	Rented	0	2000	9999999
5	Owned	0	72600	95000

Like the type attribute, SAS does not allow the position and length attributes to change after their initial values are set. Attempting to change the length attribute after the INPUT statement, as shown in Program 2.8.6, results in a warning in the Log.

Program 2.8.6: Using the LENGTH Statement After the INPUT Statement

```
data work.Ipums2005Basic;
   infile RawData("IPUMS2005basic.csv") dsd;
   input Serial State $ City $ CityPop Metro
         CountyFIPS Ownership $ MortgageStatus $
         MortgagePayment HHIncome HomeValue;
   length state $20 City $25 MortgageStatus $50;
run;
```

Log 2.8.6: Warning Generated by Attempting to Reset Length

```
WARNING: Length of character variable State has already been set. Use the LENGTH
statement as the very first statement in the DATA STEP to declare the length of a
character variable.
```

Tab-Delimited Files

If the delimiter is not a standard keyboard character, such as the tab used in tab-delimited files, an alternate method is used to specify the delimiter via its hexadecimal code. While the correct hexadecimal representation depends on the operating system, Microsoft Windows and Unix/Linux machines typically use ASCII codes. The ASCII hexadecimal code for a tab is 09 and is written in the SAS language as '09'x; the x appended to the literal value of 09 instructs the compiler to make the conversion from hexadecimal. Program 2.8.7 uses hexadecimal encoding in the DLM= option to correctly set the delimiter to a tab. The results of Program 2.8.7 are identical to those of Program 2.8.5.

Program 2.8.7: Reading Tab-Delimited Data

```
data work.Ipums2005Basic;
   length state $ 20 City $ 25 MortgageStatus $ 50;
   infile RawData ('ipums2005basic.txt') dlm = '09'x;
   input Serial State $ City $ CityPop Metro
         CountyFIPS Ownership $ MortgageStatus $
         MortgagePayment HHIncome HomeValue;
run;
```

Because there are no missing values denoted by sequential tabs, nor any tabs included in data values, the DSD option is no longer needed in the INFILE statement for this program.

To specify multiple delimiters that include the tab, each must use a hexadecimal representation—for example, DLM= '2C09'x selects commas and tabs as delimiters since 2C is the hexadecimal value for a comma. For records with different delimiters within the same DATA step, see Chapter Note 7 in Section 2.12.

2.8.3 Introduction to Reading Fixed-Position Data

While delimited data takes advantage of delimiting characters in the data, other files depend on the starting and stopping position of the values being read. These types of files are referred to by several names: fixed-width, fixed-position, and fixed-field, among others. The first five records from a fixed-position file

(IPUMS2005Basic.dat) are shown in Input Data 2.8.8. As with Input Data 2.8.4, truncation of this display occurs due to the length of the record—now occurring in each of the five records.

Input Data 2.8.8: Excerpt from a Fixed-Position Data File

```
----+----1----+----2----+----3----+----4----+----5----+----6----+----7----+----8----+
       2 Alabama              Not in identifiable city (or size group)      0    4   73
       3 Alabama              Not in identifiable city (or size group)      0    1    0
       4 Alabama              Not in identifiable city (or size group)      0    4   73
       5 Alabama              Not in identifiable city (or size group)      0    1    0
       6 Alabama              Not in identifiable city (or size group)      0    3   97
```

Since fixed-position files do not use delimiters, reading a fixed-position file requires knowledge of the starting position of each data value. In addition, either the length or stopping position of the data value must be known. Using the ruler, the first displayed field, Serial, appears to begin and end in column 8. However, inspection of the complete raw file reveals that is only the case for the single-digit values of Serial. The longest value is eight-digits wide, so the variable Serial truly starts in column 1 and ends in column 8. Similarly, the next field, State, begins in column 10 and ends in column 29. Some text editors, such as Notepad++ and Visual Studio Code, show the column number in a status bar as the cursor moves across lines in the file.

The DATA step for reading fixed-position data looks similar to the DATA step for reading delimited data, but there are several important modifications. For fixed-position files, the syntax of the INPUT statement provides information about column positions of the variable values in the raw file, as it cannot rely on delimiters for separating values. Therefore, delimiter-modifying INFILE options such as DSD and DLM= have no utility with fixed-position data. Two different forms of input are commonly used for fixed-position data: column input or formatted input. This section focuses on column input while Chapter 4 discusses formatted input.

Column Input

Column input takes advantage of the fixed positions in which variable values are found by directly placing the starting and ending column positions into the INPUT statement. Program 2.8.8 shows how to use column input to read the IPUMS CPS 2005 basic data. The results of Program 2.8.8 are identical to Output 2.8.5.

Program 2.8.8: Reading Data Using Column Input
```
data work.ipums2005basicFPa;
  infile RawData ('ipums2005basic.dat');
  input serial 1-8 state $ 10-29 ❶ city $ 31-70 ❷ cityPop 72-76 ❸
        metro 78-80 countyFips 82-84 ownership $ 86-91
        mortgageStatus $ 93-137 mortgagePayment 139-142
        HHIncome 144-150 homeValue 152-158;
run;
```

❶ The LENGTH statement is no longer needed—when using column input, SAS assigns the length based on the number of columns read if the length attribute is not previously defined. Here, SAS assigns State a length of 20 bytes, just as was done in the LENGTH statement in Program 2.8.5.

❷ The first value indicates the column position—31—from which SAS should start reading for the current variable, City. The second number—70—indicates the last column SAS reads to determine the value of City.

❸ The default length of eight bytes is still used for numeric variables, regardless of the number of columns.

Beyond the differences between column input and list input shown in Program 2.8.8, since column input uses the column positions, the INPUT statement can read variables in any order, and can even reread columns if necessary. Furthermore, the INPUT statement can skip unwanted variables. Program 2.8.9 reads Input Data 2.8.8 and demonstrates the ability to reorder and reread columns.

Program 2.8.9: Reading the Input Variables Differently than Column Order
```
data work.ipums2005basicFPb;
  infile RawData('ipums2005basic.dat');
  input serial 1-8 hhIncome 144-150 homeValue 152-158 ❶
        ownership $ 86-91 ownershipCoded $ 86 ❷
        state $ 10-29 city $ 31-70 cityPop 72-76
```

```
          metro 78-80 countyFips 82-84
          mortgageStatus $ 93-137 mortgagePayment 139-142;
run;

proc print data = work.ipums2005basicFPb(obs = 5);
  var serial --❸ state;
run;
```

❶   Output 2.8.9 shows that HHIncome and HomeValue are now earlier in the data set. Column input allows
     for reading variables in a user-specified order.

❷   Column 86 is read twice: first as part of a full value for Ownership, and second as a simplified version
     using only the first character as the value of a new variable, OwnershipCoded.

❸   As discussed in Chapter Note 3 in Section 1.7, the double-dash selects all variables between Serial and
     State, inclusive.

Output 2.8.9: Reading the Input Variables Differently than Column Order

Obs	serial	hhIncome	homeValue	ownership	ownershipCoded	state
1	2	12000	9999999	Rented	R	Alabama
2	3	17800	9999999	Rented	R	Alabama
3	4	185000	137500	Owned	O	Alabama
4	5	2000	9999999	Rented	R	Alabama
5	6	72600	95000	Owned	O	Alabama

Mixed Input

Programs 2.8.1 through 2.8.7 make use of simple list input for every variable, and Programs 2.8.8 and 2.8.9
use column input for every variable. However, it may not always be the case of making a choice between one
or the other. If files contain some delimited fields while other fields have fixed positions, it is necessary to use
multiple input styles simultaneously. This process, called mixed input, requires mastery of two other input
methods covered in Chapter 3, modified list input and formatted input, along with a substantial
understanding of how the DATA step processes raw data. For a discussion of the fifth and final input style,
named input, see the SAS Documentation.

2.9 Details of the DATA Step Process

This section provides further details about how the DATA step functions. While this material can initially be
considered optional for many readers, understanding it makes writing high-quality code easier by providing a
foundation for how certain coding decisions lead to particular outcomes. This material is also essential for
successful completion of the base certification exam.

2.9.1 Introduction to the Compilation and Execution Phases

SAS processes every step in Base SAS, including the DATA step, in two phases: compilation and execution.
Each of the DATA steps seen so far in this text have several elements in common: they each read data from
one or more sources (for example, a SAS data set or a raw data file), and they each create a data set as a
result of the DATA step. For DATA steps such as these, the flowchart in Figure 2.9.1 provides a high-level
overview of the actions taken by SAS upon submission of a DATA step. Details about the individual actions are
included in this section, in the Chapter Notes in Section 2.12, and in subsequent chapters.

Figure 2.9.1: Flowchart of Default DATA Step Actions

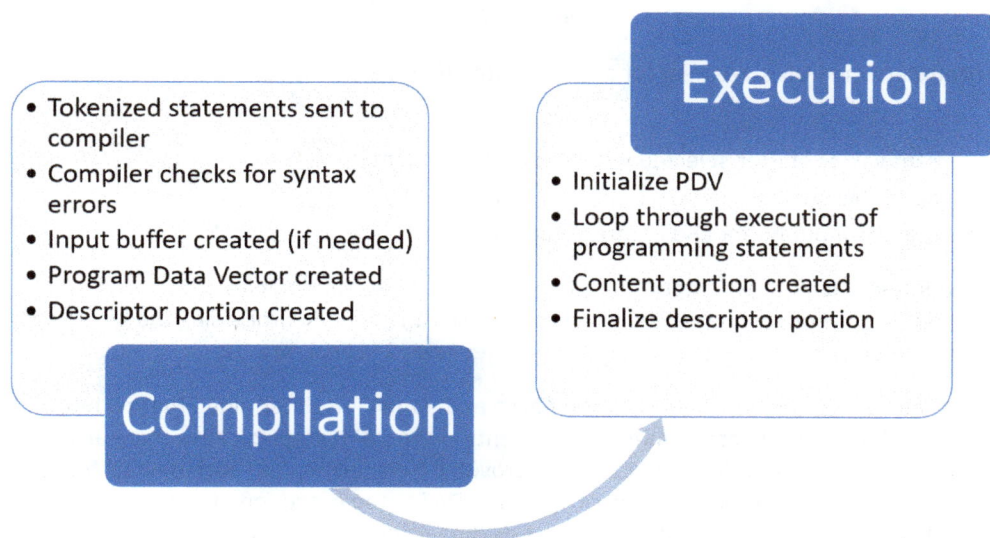

Compilation

- Tokenized statements sent to compiler
- Compiler checks for syntax errors
- Input buffer created (if needed)
- Program Data Vector created
- Descriptor portion created

Execution

- Initialize PDV
- Loop through execution of programming statements
- Content portion created
- Finalize descriptor portion

Compilation Phase

During the compilation phase, SAS begins by tokenizing the submitted code and sending complete statements to the compiler. (For more details, see Chapter Note 8 in Section 2.12.) Once a complete statement is sent to the compiler, the compiler checks the statement for syntax errors. If there is a syntax error, SAS attempts to make a substitution that creates legal syntax and prints a warning to the SAS log indicating the substitution made. For example, misspelling the keyword DATA as DAAT produces the following warning.

```
WARNING 14-169: Assuming the symbol DATA was misspelled as daat.
```

Be sure to review these warnings and correct the syntax even if SAS makes an appropriate substitution. If there is a syntax error and SAS cannot make a substitution, then an error message is printed to the log, and the current step is not executed. For example, misspelling the keyword DATA as DSTS results in the following error.

```
ERROR 180-322: Statement is not valid or it is used out of proper order.
```

If there is not a syntax error, or if SAS can make a substitution to correct a syntax error, then the compilation phase continues to the next statement, tokenizes it, and checks it for syntax errors. This process continues until SAS compiles all statements in the current DATA step.

When reading raw data, SAS creates an input buffer to load individual records from the raw data and creates the program data vector to assign the parsed values to variables for later delivery to the SAS data set. During this process, SAS also creates the shell for the descriptor portion, or metadata, for the data set, which is accessible via procedures such as the CONTENTS procedure from Chapter 1. Of course, not all elements of the descriptor portion, such as the number of observations, are known during the compilation phase. Once the compilation phase ends, SAS enters the execution phase where the compiled code is executed. At the conclusion of the execution phase, SAS populates any such remaining elements of the descriptor portion.

Execution Phase

The compilation phase creates the input buffer (when reading from a raw data source) and creates the program data vector; however, it is the execution phase that populates them. SAS begins by initializing the variables in the program data vector based on data type (character or numeric) and variable origin (for example, a raw data file or a SAS data set). SAS then executes the programming statements included in the DATA step. Certain statements, such as the LENGTH or FORMAT statements shown earlier in this chapter, are considered compile-time statements because SAS completes their actions during the compilation phase. Compile-time statements take effect during the compilation phase, and their effects cannot be altered during the execution phase. Statements active during the execution phase are referred to as execution-time statements.

Finally, when SAS encounters the RUN statement (or any other step boundary) the default actions are as follows:

1. output the current values of user-selected variables to the data set
2. return to the top of the DATA step
3. reset the values in the input buffer and program data vector

At this point, the input buffer (if it exists) is empty, and the program data vector variables are incremented/reinitialized as appropriate so that the execution phase can continue processing the incoming data set. For more information about step boundaries, see Chapter Note 9 in Section 2.12.

When reading in data from various sources, the execution phase ends when it is determined that no more data can or should be read, based on the programming statements in the DATA step. Because there are multiple factors that affect this, an in-depth discussion is not provided here. Instead, as each new technique for reading and combining data is presented, a review of when the DATA step execution phase ends is included. This chapter includes examples on reading a single raw data using an INFILE statement and, in this case, the execution phase ends when SAS encounters an end-of-file (EOF) marker in the incoming data source. For plain text files, the EOF marker is a non-printable character that lets software reading the file know that the file contains no further information. At the conclusion of the execution phase, SAS completes the content portion of the data set, which contains the data values, and finalizes the descriptor portion.

2.9.2 Building blocks of a Data Set: Input Buffers and Program Data Vectors

Input Buffer

When reading raw data, SAS needs to parse the characters from the plain text in order to determine the values to place in the data set. Parsing involves dividing the text into groups of characters and interpreting each group as a value for a variable. To facilitate this, the default is for SAS to read a single line of text from the raw file and place it into the input buffer—a section of logical memory. In the input buffer, SAS places each character into its own column and uses a column pointer to keep track of the column the INPUT statement is currently reading.

Program Data Vector

Regardless of the data source used in a DATA step (raw data files or SAS data sets), a program data vector (PDV) is created. Like the input buffer, the PDV is a section of logical memory; but, instead of storing raw, unstructured data, the PDV is where SAS stores variable values. SAS determines these values in potentially many ways: by parsing information in the input buffer, by reading values from structured sources such as Excel spreadsheets or SAS data sets, or by executing programming statements in the DATA step. Just as the input buffer holds a single line of raw text, the PDV holds only the values of each variable for a single record.

In addition to user-defined variables, SAS places automatic variables into the PDV. Two automatic variables, _N_ and _ERROR_, are present in every PDV. By default, the DATA step acts as a loop that repeatedly processes any executable statements and builds the final data set one record at a time. These loops are referred to as iterations and are tracked by the automatic variable, _N_. _N_ is a counter that keeps track of the number of DATA step iterations—how many times the DATA statement has executed—and is initialized to one at invocation of the DATA step. _N_ is not necessarily the same as the number of observations in the data set since programming elements are available to selectively output records to the final data set. Similarly, certain statements and options are available to only select a subset of the variables in the final data set.

The second automatic variable, _ERROR_, is an indicator that SAS initializes to zero and sets to one at the first instance of certain non-syntax errors. Details about the errors it tracks are discussed in Chapter Note 10 in Section 2.12. Automatic variables are not written to the resulting data set, though their values can be assigned to new variables or used in other DATA step programming statements.

Example

Some aspects of the compilation and execution phases are demonstrated below using a raw data set having the five variables shown in Input Data 2.9.1: flight number (FlightNum), flight date (Date), destination city (Destination), number of first-class passengers (FirstClass), and number of economy passengers (EconClass). The first line contains a ruler to help locate the values; it is not included in the raw file. Program 2.9.1 reads in this data set.

Input Data 2.9.1: Flights.prn data set

```
----+----1----+----2----+----3
439 12/11/2000   LAX 20 137
921 12/11/2000 DFW 20 131
114   12/12/2000 LAX 15 170
982 12/12/2000   dfw 5   85
439 12/13/2000 LAX 14 196
982   12/13/2000 DFW 15 116
431 12/14/2000 LaX 17 166
982 12/14/2000   DFW 7   88
114   12/15/2000 LAX   0   187
982 12/15/2000   DFW 14 31
```

Program 2.9.1: Demonstrating the Input Buffer and Program Data Vector (PDV)

```
data work.flights;
  infile RawData('flights.prn');
  input FlightNum Date $ Destination $ FirstClass EconClass;
run;
```

During the compilation phase, SAS scans each statement for syntax errors, and finding none in this code, it creates various elements as each statement compiles. The DATA statement triggers the initial creation of the PDV with the two automatic variables: _N_ and _ERROR_. SAS then determines an input buffer is necessary when it encounters the INFILE statement. SAS automatically allocates the maximum amount of memory, 32,767 bytes, when it creates the input buffer. If explicit control is needed, the INFILE option LRECL= allows specification of a value. Compilation of the input statement completes the PDV with the five variables in the input statement established, in the same order they are encountered, along with their attributes.

A visual representation of the input buffer and PDV at various points in the compilation phase is given in Tables 2.9.1 through 2.9.10, with the input buffer showing 26 columns here for the sake of brevity—the actual size is 32,767 columns.

Table 2.9.1: Representation of Input Buffer During Compilation Phase

01	02	03	04	05	06	07	08	09	10	11	12	13	14	15	16	17	18	19	20	21	22	23	24	25	26

Table 2.9.2: Representation of the PDV at the Completion of the Compilation Phase

_N_	_ERROR_	FlightNum	Date	Destination	FirstClass	EconClass

Note that while SAS is not case-sensitive when referencing variables, variable names are stored as they are first referenced.

Once the compilation phase has completed, the execution phase begins by initializing the variables in the PDV. Each of the user-defined variables in this example comes from a raw data file, so they are all initialized to missing. Recall missing numeric data is represented with a single period, and missing character values are represented as a null string. The automatic variables _N_ and _ERROR_ are initialized to one and zero, respectively, since this is the first iteration through the DATA step, and no errors tracked by _ERROR_ have been encountered.

Table 2.9.3: Representation of the PDV at the Beginning of the Execution Phase

_N_	_ERROR_	FlightNum	Date	Destination	FirstClass	EconClass
1	0	.			.	.

When SAS encounters the INPUT statement on the first iteration of the DATA step, it reads the first line of data and places it in the input buffer with each character in its own column. Table 2.9.4 illustrates this for the first record of Flights.prn.

Table 2.9.4: Illustration of the Input Buffer After Reaching the INPUT Statement on the First Iteration

01	02	03	04	05	06	07	08	09	10	11	12	13	14	15	16	17	18	19	20	21	22	23	24	25	26
4	3	9		1	2	/	1	1	/	2	0	0	0			L	A	X		2	0		1	3	7

To move raw data from the input buffer to the PDV, SAS must parse the character string from the input buffer to determine which characters should be grouped together and whether any of these character groupings must be converted to numeric values. The parsing process uses information found in the INFILE statement (for example, DSD and DLM=), the INPUT statement (such as the $ for character data), and other sources (like the LENGTH statement) to determine the values it places in the PDV.

No delimiter options are present in the INFILE statement in Program 2.9.1; thus, a space is used as the default delimiter, and the first variable, FlightNum, is read using simple list input. SAS uses column pointers to keep track of where the parsing begins and ends for each variable and, with simple list input, SAS begins in the first column and scans until the first non-delimiter character is found. In this record, the first column is non-blank, so the starting pointer is placed there, indicated by the blue triangle below Table 2.9.5. Next, SAS scans until it finds another delimiter (a blank in this case), which is indicated below Table 2.9.5 with the red octagon. Thus, when reading the input buffer to create FlightNum, SAS has read from column 1 up to column 4.

Table 2.9.5: Column Pointers at the Starting and Ending Positions for Parsing FlightNum

01	02	03	04	05	06	07	08	09	10	11	12	13	14	15	16	17	18	19	20	21	22	23	24	25	26
4	3	9		1	2	/	1	1	/	2	0	0	0			L	A	X		2	0		1	3	7

Based on information defined in the descriptor portion of the data during compilation of the INPUT statement, FlightNum is a numeric variable with a default length of eight bytes. There are no additional instructions on how this value should be handled, so SAS converts the extracted character string "439" to the number 439 and sends it to the PDV. Note that the blank found in column 4 is not part of the parsed value— only non-delimiter columns are included. Table 2.9.6 shows the results of parsing FlightNum from the first record.

Table 2.9.6: Representation of the PDV After FlightNum is Read During the First Iteration

_N_	_ERROR_	FlightNum	Date	Destination	FirstClass	EconClass
1	0	439			.	.

Before parsing begins for the next value, SAS advances the column pointer one position, in this case advancing to column 5. This prevents SAS from beginning the next value in a column that was used to create a previous value. This automatic advancement of a single column occurs regardless of the input style, even though it is only demonstrated here for simple list input.

Since Date is also read in using simple list input, the parsing process is similar to how FlightNum was read. SAS begins at column 5 and reads until it encounters the next delimiter, which is in column 15. Table 2.9.7 shows this with, as before, the blue triangle indicating the starting column and the red octagon the ending column.

Table 2.9.7: Column Pointers at the Starting and Ending Positions for Parsing FlightNum

01	02	03	04	05	06	07	08	09	10	11	12	13	14	15	16	17	18	19	20	21	22	23	24	25	26
4	3	9		1	2	/	1	1	/	2	0	0	0		L	A	X		2	0		1	3	7	

▲ ⬢

The characters read for Date are "12/11/2000", and SAS must parse this string using the provided instructions, which are given by the dollar sign used in the INPUT statement for this variable. This declares its type as character and, since there are no instructions about the length, the default length of eight is used. Unlike numeric variables, where the length attribute does not control the displayed width, the length of a character variable is typically equal to the number of characters that can be stored. (Check the SAS Documentation to determine how length is related to printed width for character values in various languages and encodings.) The resulting value for date, shown in Table 2.9.8, has been truncated to the first 8 characters. This highlights that while list input reads from delimiter to delimiter, it only stores the value parsed from the input buffer subject to any attributes, such as type and length, previously established.

Table 2.9.8: Representation of the PDV after Date is Read During the First Iteration

_N_	_ERROR_	FlightNum	Date	Destination	FirstClass	EconClass
1	0	439	12/11/20		.	.

As demonstrated in Program 2.8.5, one way to prevent the truncation of Date values is to use a LENGTH statement to set the length of Date to at least 10 bytes. Another means of avoiding this issue with Date is to read it with an informat, a concept covered in Chapter 3. This has the added benefits of converting the Date values to numeric; allowing for easier sorting, computations, and changes in display formats.

SAS continues through the input buffer and, since all variables in this example are read using simple list input, the reading and parsing follows the same process as before. Table 2.9.9 shows the starting and stopping position of each of the remaining variables. Note that because SAS automatically advances the column pointer by one column after every variable and because list input scans for the next non-delimiter, the starting position for Destination in this record is column 17 rather than column 16.

Table 2.9.9: Column Pointers at the Starting and Ending Positions for Parsing the Remaining Values

0 1	0 2	0 3	0 4	0 5	0 6	0 7	0 8	0 9	1 0	1 1	1 2	1 3	1 4	1 5	1 6	17	1 8	1 9	20	21	2 2	23	24	2 5	26
4	3	9		1	2	/	1	1	/	2	0	0	0			L	A	X		2	0		1	3	7

Table 2.9.10 contains the final PDV for the first record—the values that are sent to the data set at the end of the first iteration of the DATA step which corresponds to the location of the RUN statement (or other step boundary).

Table 2.9.10: Representation of the PDV After Reading All Values

_N_	_ERROR_	FlightNum	Date	Destination	FirstClass	EconClass
1	0	439	12/11/20	LAX	20	137

After this record is sent to the data set, the execution phase returns to the top of the DATA step and the variables are reinitialized as needed. _N_ is incremented to 2, _ERROR_ remains at zero since no tracked errors have been encountered, and the remaining variables are set back to missing in this case. Further iterations continue the process: the next row is loaded into the input buffer, values are parsed to the PDV, and those values are sent to the data set at the bottom of the DATA step. This implicit iteration terminates when the end-of-file marker is encountered.

2.9.3 Debugging the DATA Step

Following the process from Section 2.9.1 may seem tedious at first, but understanding how SAS parses data when moving it from the raw file through the input buffer then to the PDV (and ultimately to the data set) is crucial for success in more complex cases. It is often inefficient to develop a program using a trial-and-error approach; instead, knowledge of the data-handling process ensures a smoother, more reliable process for developing programs. This section discusses several statements that SAS provides to help follow aspects of the parsing process through iterations of the DATA step. Program 2.9.2 demonstrates the LIST statement using Input Data 2.9.1.

Program 2.9.2: Demonstrating the LIST Statements

```
data work.flights;
   infile RawData('flights.prn');
   input FlightNum Date $ Destination $ FirstClass EconClass;
   list;
run;
```

The LIST statement writes the contents of the input buffer to the log at the end of each iteration of the DATA step, placing a ruler before the first observation is printed. Log 2.9.2 shows the results of including the LIST statement in Program 2.9.2 for the first five records. The complete input buffer, including the delimiters, appear for each record.

Log 2.9.2: Demonstrating the LIST Statements

```
RULE:       ----+----1----+----2----+----3----+----4 ❶
1           439 12/11/2000   LAX 20 137 ❷ 26 ❸
2           921 12/11/2000 DFW 20 131 25
3           114  12/12/2000 LAX 15 170 26
4           982 12/12/2000   dfw 5   85 25
5           439 12/13/2000 LAX 14 196 25
```

❶   Before writing the input buffer contents for the first time, the INPUT statement prints a ruler to the log.

❷   The LIST statement writes the complete input buffer for the record.

❸   If the records have variable length, then the LIST statement also prints the number of characters in the input buffer.

Since the log is a plain-text environment, SAS cannot display non-printable characters such as tabs. However, in these cases, SAS prints additional information to the log to ensure an unambiguous representation of the input buffer. Program 2.9.3 demonstrates the results of using the LIST statement with a tab-delimited file.

Program 2.9.3: LIST Statement Results with Non-Printable Characters

```
data work.flights;
   infile RawData('flights.txt')❶ dlm = '09'x;❷
   input FlightNum Date $ Destination $ FirstClass EconClass;
   list;❸
run;
```

❶   Note the different file extension. This data is similar to the data in Program 2.9.2, but in a tab-delimited file.

❷   The DLM= option uses the hexadecimal representation of a tab, 09, along with hexadecimal literal modifier, x.

❸   No change is necessary in the LIST statement, regardless of what, if any, delimiter is used in the file.

Because of the increase in information present in the log, Log 2.9.3 only shows the results for the first two records. For each record, the contents from the input buffer now occupy three lines in the log.

Log 2.9.3: LIST Statement Results with Non-Printable Characters

```
RULE:        ----+----1----+----2----+----3----+----4

1     ❶CHAR   439.12/11/2000.LAX.20.137 25
      ❷ZONE   33303323323333304450330333
      ❷NUMR   439912F11F20009C189209137
                 ❸
2      CHAR   921.12/11/2000.DFW.20.131 25
       ZONE   33303323323333304450330333
       NUMR   921912F11F200094679209131
```

❶   The CHAR line represents the printable data from the input buffer. It displays non-printable characters as periods.

❷   The ZONE and NUMR rows represent the two digits in the hexadecimal representation.

❸   Note the fourth column in the input buffer appears to be a period. However, combining the information from the ZONE and NUMR lines indicates the hexadecimal value is 09—a tab.

Because SAS converts all non-printable characters to a period when writing the input buffer to the log, the ZONE and NUMR lines provide crucial information to determine the actual value stored in the input buffer. In particular, they provide a way to differentiate a period that was in the original data (hexadecimal code 2E) from a period that appears as a result of a non-printable character (for example, a tab with the hexadecimal code 09).

When debugging, two other useful statements are the PUT and PUTLOG statements. Both PUT and PUTLOG statements provide a way for SAS to write out information from the PDV along with other user-defined messages. The statements differ only in their destination—the PUTLOG statement can only write to the SAS log, while the PUT statement can write to the log or any file destination specified in the DATA step. The PUT statement is covered in more detail in Chapter 7; this section focuses on the PUTLOG statement. Program 2.9.4 uses Input Data 2.9.1 to demonstrate various uses of the PUTLOG statement, with Log 2.9.4 showing the results for the first record.

Program 2.9.4: Demonstrating the PUTLOG Statement

```
data work.flights;
   infile RawData('flights.prn');
```

```
   input FlightNum Date $ Destination $ FirstClass EconClass;
   putlog 'NOTE: It is easy to write the PDV to the log';❶
   putlog _all_;❷
   putlog 'NOTE: Selecting individual variables is also easy';
   putlog 'NOTE: ' ❸ FlightNum= ❹ Date ❺;
   putlog 'WARNING: Without the equals sign variable names are omitted';❻
run;
```

❶ This PUTLOG statement writes the quoted string to the log once for every record. If the string starts with the string 'NOTE:', then SAS color-codes the statement just like a system-generated note, and it is indexed in SAS University Edition with other system notes.

❷ The _ALL_ keyword selects every variable from the PDV, including the automatic variables _N_ and _ERROR_.

❸ The PUTLOG statements accept a mix of quoted strings and variable names or keywords.

❹ A variable name followed by the equals sign prints both the name and current value of the variable.

❺ Omitting the equals sign only prints the value of the variable, with a potentially adverse effect on readability.

❻ Beginning the string with 'WARNING:' or 'ERROR:' also ensures SAS color-codes the messages to match the formatting of the system-generated warnings and errors, and indexes them in SAS University Edition. If this behavior is not desired, use alternate terms such as QC_NOTE, QC_WARNING, and QC_ERROR to differentiate these user-generated quality control messages from automatically generated messages.

Log 2.9.4 Demonstrating the PUTLOG Statement

```
NOTE: It is easy to write the PDV to the log
FlightNum=439 Date=12/11/20 Destination=LAX FirstClass=20 EconClass=137 _ERROR_=0
_N_=1
NOTE: Selecting individual variables is also easy
FlightNum=439 12/11/20
WARNING: Without the equals sign variable names are omitted
```

When used in conjunction in a single DATA step, the PUTLOG and LIST statements allow for easy access to both the PDV and input buffer contents, providing a simple way to track down the source of data errors. In these examples, the PUTLOG results shown in Log 2.9.4 reveal the truncation in the Date values in the PDV, while the LIST results from Log 2.9.2 or 2.9.3 show the full date is present in the input buffer. Using them together, it is clear the issue with Date values is not present in the raw data and must relate to how the INPUT statement has parsed those values from the input buffer.

One additional debugging tool, the ERROR statement, acts exactly as the PUTLOG statement does, while also setting the automatic variable _ERROR_ to one.

2.10 Validation

The term validation has many definitions in the context of programming. It can refer to ensuring software is correctly performing the intended actions—for example, confirming that PROC MEANS is accurately calculating statistics on the provided data set. Validation can also refer to the processes of self-validation, independent validation, or both. Self-validation occurs when a programmer performs checks to ensure any data sets read in accurately reflect the source data, data set manipulations are correct, or that any analyses represent the input data sets. Independent validation occurs when two programmers develop code separately to achieve the same results. Once both programmers have produced their intended results, those results (such as listings of data sets, tables of summary statistics, or graphics) are compared and are considered validated when no substantive differences appear between the two sets of results. SAS provides a variety of tools to use for validation, including PROC COMPARE, which is an essential tool for independent validation of data sets.

The COMPARE procedure compares the contents of two SAS data sets, selected variables across different data sets, or selected variables within a single data set. When two data sets are used, one is specified as the base data set and the other as the comparison data set. Program 2.10.1 shows the options used to specify these data sets, BASE= and COMPARE=, respectively.

Program 2.10.1: Comparing the Contents of Two Data Sets

```
proc compare base = sashelp.fish compare = sashelp.heart;
run;
```

If no statements beyond those shown in Program 2.10.1 are included, the complete contents portions of the two data sets are compared, along with meta-information such as types, lengths, labels, and formats. PROC COMPARE only compares attributes and values for variables with common names across the two data sets. Thus, even though the full data sets are specified for comparison, it is possible for individual variables in one data set to not be compared against any variable in the other set. Program 2.10.1 compares two data sets from the Sashelp library: Fish and Heart. These data sets are not intended to match, so submitting Program 2.10.1 produces a summary of the mismatches to demonstrate the types of output available from the COMPARE procedure.

Output 2.10.1: Comparing the Contents of Two Data Sets

```
                                  Data Set Summary

Dataset              Created         Modified    NVar   NObs  Label

SASHELP.FISH   25JUN15:01:23:43  25JUN15:01:23:43    7    159  Measurements of 159 Fish Caught in Lake Laengelmavesi, Finland
SASHELP.HEART  25JUN15:01:15:36  25JUN15:01:15:36   17   5209  Framingham Heart Study


                                  Variables Summary

              Number of Variables in Common: 2.
              Number of Variables in SASHELP.FISH but not in SASHELP.HEART: 5.
              Number of Variables in SASHELP.HEART but not in SASHELP.FISH: 15.


                                 Observation Summary

                   Observation       Base  Compare

                   First Obs            1        1
                   First Unequal        1        1
                   Last  Unequal      159      159
                   Last  Match        159      159
                   Last  Obs            .     5209

         Number of Observations in Common: 159.
         Number of Observations in SASHELP.HEART but not in SASHELP.FISH: 5050.
         Total Number of Observations Read from SASHELP.FISH: 159.
         Total Number of Observations Read from SASHELP.HEART: 5209.

         Number of Observations with Some Compared Variables Unequal: 159.
         Number of Observations with All Compared Variables Equal: 0.
```

While the output in this text is normally delivered in an RTF format using the Output Delivery System, the output from PROC COMPARE is not well-suited to such an environment, so Output 2.10.1 shows the results as they appear in the Output window in the SAS Windowing Environment. Regardless of the destination, the output from this example of the COMPARE procedure includes sections for the following:

- *Data set summary*—data set names, number of variables, number of observations

- Variables summary—number of variables in common, along with the number in each data set which are not found in the other set

- Observation summary—location of first/last unequal record and number of matching/nonmatching observations

- Values comparison summary—number of variables compared with matches/mismatches, listing of mismatched variables and their differences

A review of the output provided by PROC COMPARE shows, in this case, only two variables are compared, despite the Fish and Heart data sets containing 7 and 17 variables, respectively. This is because only two variables (Weight and Height) have names in common. As such, even if the results indicate the base and comparison data sets have no mismatches, it is important to confirm that all variables were compared before declaring the data sets are identical. Similarly, the number of records compared is the minimum of the number of records in the two data sets, so the number of records must be compared as well. Several options and statements exist to alter how comparisons are done and to direct some comparison information to data sets.

Since the Heart and Fish data sets are not expected to be similar, applying PROC COMPARE to them is a simplistic demonstration of the procedure. A more typical comparison is given in Program 2.10.2, which applies the COMPARE procedure to the data set read in by Program 2.8.8 (using fixed-position data) and the IPUMS2005Basic data set in the BookData library.

Program 2.10.2: Comparing IPUMS 2005 Basic Data Generated from Different Sources

```
data work.ipums2005basicSubset;
  set work.ipums2005basicFPa;
  where homeValue ne 9999999; ❶
run;

proc compare base = BookData.ipums2005basic compare = work.ipums2005basicSubset
            out = work.diff ❷ outbase ❸ outcompare ❹ outdif ❺ outnoequal ❻
            method = absolute ❼ criterion = 1E-9 ❽;
run;

proc print data = work.diff(obs=6);
  var _type_ _obs_ serial countyfips metro citypop homevalue;
run;

proc print data = work.diff(obs=6);
  var _type_ _obs_ city ownership;
run;
```

❶ To create a data set which differs from the provided BookData.IPUMS2005Basic data set, a WHERE statement is used to remove any homes with a home value of $9,999,999.

❷ OUT= produces a data set containing information about the differences for each pair of compared observations for all matching variables. SAS includes all compared variables and two automatic variables, _TYPE_ and _OBS_.

❸ OUTBASE copies the record being compared in the BASE= data set into the OUT= data set.

❹ Like OUTBASE, OUTCOMPARE copies the record being compared in the COMPARE= data set into the OUT= data set.

❺ OUTDIF produces a record that contains the difference between the OUTBASE and OUTCOMPARE records.

❻ OUTNOEQUAL outputs the requested records (base, compare, and difference) only if a difference is found in at least one of the compared variables for the two records.

❼ METHOD= selects the way in which two numeric values are compared. ABSOLUTE considers two numbers equal if their absolute difference is smaller than a fixed precision level.

❽ CRITERION= specifies the precision used when comparing two numeric values.

Output 2.10.2A: Comparing IPUMS 2005 Basic Data Generated from Different Sources

Obs	_TYPE_	_OBS_	SERIAL	COUNTYFIPS	METRO	CITYPOP	HomeValue
1	BASE	1	2	73	4	0	9999999
2	COMPARE	1	4	73	4	0	137500
3	DIF	1	2	E	E	E	-9862499
4	BASE	2	3	0	1	0	9999999
5	COMPARE	2	6	97	3	0	95000
6	DIF	2	3	97	2	E	-9904999

Output 2.10.2A shows the first six records and nine columns of the data set created by the OUT= option in PROC COMPARE in Program 2.10.2, including _TYPE_ and _OBS_, which are both automatic variables. The automatic variable _TYPE_ is a character variable (with the default length of 8) that indicates the source of the observation—BASE, COMPARE, or DIF—based on the options in use. The second automatic variable is _OBS_, which uniquely identifies the observations from the source data sets that were used for comparison. In Output 2.10.2A, the first three records are: the first record from the base data set, the first record from the compare data set, and the difference between them. The second set of three records repeats this pattern for the second observation from the BASE= and COMPARE= data sets.

The next seven columns include the values compared and their differences. The first compared column, Serial, has different values in the BASE (Serial = 2) and COMPARE (Serial = 4) data sets. Since the absolute value of the difference ($|$BASE-COMPARE$|$ = $|4 - 2|$ = 2) is larger than the specified criterion, 1×10^{-9}, these two values are evaluated as unequal and the difference is placed in this column on the DIF record. For the next three columns, CountyFips, Metro, and CityPop, the values are evaluated as equal. In these cases, numeric variables display a single E as the value in the difference record when using the OUTNOEQUAL option, unless a format overrides this behavior. All comparisons are made on unformatted values; the comparison of formats is done as part of the metadata comparison and is shown in the procedure output.

Output 2.10.2B: Results of Additional Options in Program 2.10.2

Obs	_TYPE_	_OBS_	City	Ownership
1	BASE	1	Not in identifiable city (or size group)	Rented
2	COMPARE	1	Not in identifiable city (or size group)	Owned
3	DIF	1	....	XX.XXX
4	BASE	2	Not in identifiable city (or size group)	Rented
5	COMPARE	2	Not in identifiable city (or size group)	Owned
6	DIF	2	....	XX.XXX

Output 2.10.2B shows the comparison between two character variables, City and Ownership, for the same records. SAS compares strings character by character, and if the strings are not of the same length, the end of the shorter string is padded with spaces until the strings are the same length. In any position where identical characters are encountered, a single period is used to represent the match in the _TYPE_ = DIF record. Since the two values for the City variable are identical at every character, the difference records (the third and sixth lines) only contain periods. However, when SAS encounters non-matching characters, that position in the difference record is indicated with a single X. Thus, for the variable Ownership, the first two characters do not match, the third matches, and the last three do not match. This is of particular use when comparing long character strings, as the difference record not only indicates that a mismatch is present but also exactly where in the string mismatches occur.

Finally, since Program 2.10.2 uses the OUTNOEQUAL option, only records corresponding to differences appear in the output data set. Thus, a critical indicator of a successful comparison is an empty OUT= data set when OUTNOEQUAL option is active. However, because the OUT= data set contains only value comparisons on the default matching of variables on common names, care must still be taken to ensure that the comparison is complete and accurate.

The data set, variables, observations, and values comparison summaries produced in the PROC COMPARE output are invaluable for ensuring that all variables and records were compared and, if necessary, that their attributes matched as well. PROC COMPARE cannot be used to directly create an output data set for attribute comparisons in the same way as it does for the content portion of a data set.

Earlier, this section mentions that PROC COMPARE can implement three distinct types of comparisons: comparing the complete content portions of two different data sets, comparing user-selected variables across data sets, or comparing user-selected variables from a single data set. The options shown in Program 2.10.2 are useful for all three scenarios; however, the second and third comparison types require the use of additional statements in PROC COMPARE. Specifically, VAR and WITH statements can be used to facilitate those comparison types. By default, observations are compared in all three scenarios based on their row order; that is, the first observations from each data set (base and comparison) are compared. To compare based on key variables instead of by position, the ID statement can be added. Details of using the ID, VAR, and WITH statements can be found in the SAS Documentation.

2.11 Wrap-Up Activity

Use the lessons and examples contained in this chapter to complete the activity shown in Section 2.2.

Data

Use the basic IPUMS CPS data from 2010 to complete the activity. This data set is similar in structure and contents to the 2005 data used for many of the in-chapter examples. Completing this activity requires the following files:

- Ipums2010basic.sas7bdat
- Ipums2010basic.txt
- Ipums2010basic.csv
- Ipums2010basic.dat

Scenario

Read in the raw files and validate them to ensure the SAS data sets created match the provided SAS data set. Use any of these SAS data sets to generate the tables shown in Outputs 2.2.1 through 2.2.4.

2.12 Chapter Notes

1. *LABEL/NOLABEL.* In addition to variable labels, data sets may have temporary or permanent labels and each procedure—such as FREQ and MEANS—is labeled in SAS output destinations if that destination includes a table of contents. By default, the system option LABEL is in effect, which allows procedures to display variable labels (if they exist) instead of variable names when producing output. Similarly, this option allows procedures to appear with custom labels in the table of contents if a custom label is provided in the ODS PROCLABEL statement. For procedures that do not use labels by default, such as PROC PRINT, local LABEL options are typically available to allow the procedure access to variable labels. However, when the NOLABEL option is in effect, no procedure can display variable labels—even if its local LABEL option is in effect—and procedure labels given via ODS PROCLABEL do not appear in the table of contents for output destinations. Data set labels are unaffected by the NOLABEL system option.

2. *Determining Sort Order.* PROC SORT determines the order of the resulting data set based on the collating sequence option given, or the default sequence if no option is given. The SORT procedure code in this text always uses the default collating sequence option, which depends on information provided to SAS by the operating system. The examples in this text use ASCII, for which the default options sort character values so that, for example, capital letters are given higher sort priority than lowercase letters. Similarly, it sorts other characters (digits, underscores, spaces, and other special characters) before all letters. It is important to understand the sort method used since the default is machine-dependent.

   In English, the two primary collating sequences are ASCII and EBCDIC, which are two different methods for encoding characters as numbers. EBCDIC is primarily used in IBM systems, while most other systems use ASCII—either of these being available as collating sequence options in PROC SORT. Other options provide foreign language support (for example, DANISH and SWEDISH) or even customized methods via the SORTSEQ= option. PROC SORT also supplies options for controlling many other aspects of sorting. For programs where code portability across operating systems is needed, it is a good programming practice to use sequencing options explicitly to prevent machine-dependent results. For more information about the available options in PROC SORT, see the SAS Documentation.

3. *Widths of Formats.* Every named format includes a default width that SAS applies when no other width is specified—for example, the DOLLAR format has a default width of 6, the $REVERSJ format has a default width of 1, and some other formats have default lengths that only apply if the length of the variable has not been previously defined. When values exceed the width—either default or user-provided—SAS applies various rules depending on several contributing factors.

- *Character.* For character formats provided by SAS, SAS formats values longer than the stated width by simply truncating the value after the specified width has been reached. For example, applying the format $9. to the value 'This is only a test.' yields a value of 'This is o' because SAS truncates after reaching the ninth character.

- *Numeric.* For numeric formats provided by SAS, SAS applies a variety of rules to limit the impact on the representation of the number. For a complete description, see the SAS Documentation. In general, SAS begins by removing characters such as dollar signs, percent signs, and commas that do not affect the precision of the displayed value. If this is still not sufficient, SAS may apply a different format such as BEST, represent the value using scientific notation, or simply display asterisks in place of the value. For example, applying the format DOLLAR6. to the value 12345.67 yields $12346, omitting the comma. Applying DOLLAR4. to that same value results in 12E3, making use of scientific notation. Using DOLLAR2. does not provide enough width to display a usable value and is displayed as **. In certain cases, SAS also produces the following note in the log.

```
NOTE: At least one W.D format was too small for the number to be printed. The
      decimal may be shifted by the "BEST" format.
```

It is important to recognize the source of this note since the note always refers to the W.D format even when other formats are used.

4. *Displaying Values Not Defined in a User-Defined Format.* When creating a custom format, it is not necessary to format every possible value. However, values that do not appear in the format definition may not be formatted as intended when the custom format is used. Values that do not correspond to a formatted value are displayed using their internal, unformatted, value. However, the width SAS uses to display the value is inherited from the custom format applied, even though the value did not appear in that format definition. To provide an example, consider the following format intended to classify a numeric value into two categories.

```
proc format;
   value Audit 70-high = 'AU'
                  0-69 = 'NR'
   ;
run;
```

For the value 69.82, which is not in either value range defined in the Audit format, the result is an unexpected display value. This occurs due to AUDIT having a default length of two (the default length of a custom format is the length of the longest formatted value), which SAS applies to the internal value of 69.82. Thus, SAS rounds the value to 70 to accommodate the width of two and displays that value—a result which is confusing since a value of 70 is defined by the format to appear as AU. Of course, either using AUDIT5. or including all values in the format definition alleviates the issue. As shown in Program 2.5.7, the default width of a format is part of the information provided by the FMTLIB option in PROC FORMAT. Good programming practice dictates that format definitions cover all values the format is applied to.

5. *Overlapping Ranges in PROC FORMAT.* As stated in Section 2.5.1, ranges used in the VALUE statement should be non-overlapping in most cases. If two ranges overlap only at the boundary (for example, 1-3 and 3-5), then no error occurs and SAS assigns a formatted value based on the range that it encountered first in the VALUE statement. Use the MULTILABEL option if overlapping formatted ranges are necessary. Using this option allows assignment of overlapping ranges to unique formatted values or assignment of a single range to multiple formatted values. However, not all procedures support the multilabel formats. The exclusion operator, <, can be used to prevent overlapping ranges, as shown in Section 2.5.1. Another useful option is FUZZ=, which allows specification of an allowable margin when matching values to a range. For example, using FUZZ=0.25 with the assignment 1='Low' would be equivalent to assigning 0.75-1.25='Low' in the VALUE statement. For more details, see the SAS Documentation.

6. *Raw Data.* Not all external data qualifies as raw data. Recall raw data was defined as data that had not been processed into a SAS data set; however, Database Management System (DBMS) files are not considered raw data, despite not being SAS data sets. This includes files from Microsoft Access and Excel, Oracle, SPSS, JMP, Stata, and many other non-SAS sources. To retrieve data from DBMS sources and process it in SAS, the LIBNAME statement provided by SAS/ACCESS can potentially be used if the SAS/ACCESS product is licensed. To determine what SAS products are licensed in a given session (including SAS/ACCESS), run Program 2.12.1. This program creates a list of all licensed products and prints the list in the Log window.

Program 2.12.1: Determining Licensed SAS Products

```
proc setinit;
run;
```

7. *Variable Delimiters.* Occasionally, such as when reading data from multiple raw files in a single DATA step, the delimiters are not consistent across records. In cases such as these, it is possible to use a variable, rather than a single string, to identify the delimiter. For example, setting DLM = MyDLM in the INFILE statement uses the current value of MyDLM as the delimiter(s) for that record. Note, the value of MyDLM must already be populated in the PDV before SAS can read the delimited fields.

8. *Tokens.* When submitting a program, SAS sends the code to a location called the input stack to parse the characters submitted. To facilitate this, SAS uses a word scanner that reads the program character by character and groups the characters into tokens, a process called tokenization. There are four classes of tokens that SAS recognizes:

   1. *Numbers*—Any numeric value including real numbers; date, time, and datetime constants; and hexadecimal constants. The decimal point, sign (+ or -), and E for scientific notation are allowed. For example, -6.2, 7E-2, '27OCT1977'd, '13:44't, and 1C99.

   2. *Name*—A group of characters not beginning with a digit, but which contain only letters, underscores, and digits. For example, PROC, HHIncome, COMPARE, _OBS_.

   3. *Literal*—A string of characters contained in single or double quotation marks. For example, '$350 and Below' or 'Metro, Inside City'.

   4. *Special Character*—Any character other than letters, numbers, underscores, and blanks. For example, = ( ) @ ^ &.

   Tokens end when a new token begins, a blank is encountered (after name and number tokens), or when a literal token is ended by the same type of quotation mark (single or double) that began the literal string.

9. *Step Boundaries.* In this chapter, each DATA and PROC step included a RUN statement as the final statement in the step. For example, Program 2.10.2 uses a RUN statement at the end of each of the four steps. Since the nesting of steps is not permitted, invocation of any step is seen as a step boundary, ending the previous step and forcing its compilation and execution prior to those processes occurring for the next step. For an example of this behavior, see Program 1.4.2. In order to explicitly indicate a step or procedure is complete, include a statement or action that creates a step boundary; this is the most common purpose for the RUN statement (though other statements, such as QUIT, may sometimes fit this role). Even though the additional RUN statements are not required syntax, they are considered a good programming practice. RUN statements do not impact the run time of a program and can prevent ambiguity when submitting global statements, such as TITLE or OPTIONS. They also allow for easy submissions of sections of code, creating a much simpler debugging process.

10. *_ERROR_.* This error flag does not track all errors and is not a counter. If _ERROR_ is zero, it is not an indication that the code has no errors, only that no tracked errors are present; and if it is one, there is at least one tracked error. The _ERROR_ automatic variable only tracks data errors and some semantic and execution-time errors. To review the error types, see the SAS Documentation.

2.13 Exercises

Concepts: Multiple Choice

1. In the following FORMAT procedure, the dollar sign preceding the format name Codes is:

   ```
   proc format;
      value $codes
         '1'='East'
         '2'='West'
         '3'='North'
         '4'='South'
      ;
   run;
   ```

   a. A syntax error

   b. A logic error

   c. Both a syntax and a logic error

   d. None of the above

2. Which of the following answer choices correctly identifies the variables included in the results of PROC FREQ if only a data set is specified?

   a. All numeric variables in the data set.

   b. All character variables in the data set.

   c. All variables in the data set.

   d. No variables—a TABLE statement is required.

3. Which of the following answer choices correctly identifies the variables included in the results of PROC MEANS if only a data set is specified?

   a. All numeric variables in the data set.

   b. All character variables in the data set.

   c. All variables in the data set.

   d. No variables—a VAR statement is required.

4. Which of the MEANS procedures given below generates the table shown?

Analysis Variable : MPG_Highway MPG (Highway)					
Origin	Type	N	Mean	Median	Std Dev
Asia	Sedan	94	29.9680851	29.0000000	4.8845865
	Wagon	11	28.1818182	28.0000000	5.3817875
Europe	Sedan	78	27.1153846	27.0000000	3.8069682
	Wagon	12	26.5833333	26.5000000	2.8431204
USA	Sedan	90	28.5444444	28.0000000	4.1412182
	Wagon	7	29.7142857	30.0000000	4.8550416

   a.
   ```
   proc means data=sashelp.cars nonobs n mean median std;
      class origin type;
      var mpg_Highway;
      where type eq ('Sedan','Wagon');
   run;
   ```

   b.
   ```
   proc means data=sashelp.cars n mean median std;
      class origin type;
      var mpg_Highway;
      where type in ('Sedan','Wagon');
   run;
   ```

c.

```
proc means data=sashelp.cars nonobs n mean median std;
  class type origin;
  var mpg_Highway;
  where type in ('Sedan','Wagon');
run;
```

d.

```
proc means data=sashelp.cars nonobs n mean median std;
  class origin type;
  var mpg_Highway;
  where type in ('Sedan','Wagon');
run;
```

5. Which of the answer choices contains the FREQ procedure that generates the given table?

Table of Cause by Day						
Cause(Cause of Failure)	**Day**					
Frequency Row Pct	**Monday**	**Tuesday**	**Wednesday**	**Thursday**	**Friday**	**Total**
Contamination	23 20.91	25 22.73	24 21.82	14 12.73	24 21.82	110
Corrosion	4 25.00	3 18.75	2 12.50	3 18.75	4 25.00	16
Total	27	28	26	17	28	126

a.

```
proc freq data=sashelp.failure;
  table day*cause / nocol nopercent;
  weight count;
  where cause eq 'Corrosion' or cause eq 'Contamination';
run;
```

b.

```
proc freq data=sashelp.failure;
  table cause*day / norow nopercent;
  weight count;
  where cause eq 'Corrosion' or cause eq 'Contamination';
run;
```

c.

```
proc freq data=sashelp.failure;
  table cause*day / nocol nopercent;
  weight count;
  where cause eq 'Corrosion' or cause eq 'Contamination';
run;
```

d.

```
proc freq data=sashelp.failure;
  table cause*day / nocol nopercent;
  weight count;
  where cause eq 'Corrosion' and cause eq 'Contamination';
run;
```

6. Which of the following format definitions could be used to alter the PROC MEANS output in the first table to match that shown in the second table?

Analysis Variable : Systolic				
Smoking Status	**N Obs**	**Lower Quartile**	**Median**	**Upper Quartile**
Heavy (16-25)	1046	120.0000000	130.0000000	142.0000000
Light (1-5)	579	120.0000000	130.0000000	144.0000000
Moderate (6-15)	576	118.0000000	126.0000000	140.0000000
Non-smoker	2501	122.0000000	136.0000000	152.0000000
Very Heavy (> 25)	471	122.0000000	134.0000000	145.0000000

Analysis Variable : Systolic				
Smoking Status	**N Obs**	**Lower Quartile**	**Median**	**Upper Quartile**
Non-Smoker	2501	122.0000000	136.0000000	152.0000000
Smoker	2672	120.0000000	130.0000000	144.0000000

a.
```
proc format;
   value $smk
      low - < 'N','O' - high='Smoker'
      other = 'Non-Smoker';
run;
```

b.
```
proc format;
   value $smk
      'Non-smoker' = 'Non-Smoker'
      Other = 'Smoker';
run;
```

c. Both a and b

d. Neither a nor b

7. Based on the following program, what is the correct order of the user-defined variables in the PDV?
```
data Athletes;
   format Salary dollar12.;
   infile Teams;
   input Division $ League $ FirstName $ LastName $ Salary;
run;
```
a. Salary Division League FirstName LastName

b. Division League FirstName LastName Salary

c. Salary Teams Division League FirstName LastName

d. Teams Division League FirstName LastName Salary

8. Given the raw data file shown below, which of the following answer choices contains an INPUT statement that produces the following table?

```
----+----1----+
Bill30E8567411
Joe 32D7344211
Sue 29C6652327
Jane33C9359121
```

Name	Age	IDNum	JobCode	Salary
Bill	30	E85	E	67411
Joe	32	D73	D	44211
Sue	29	C66	C	52327
Jane	33	C93	C	59121

a. `input name $ age IDNum $ jobcode $ salary;`

b. `input name $ 1-4 age 5-6 IDNum $ 7-9 jobcode $ 7 salary 10-14;`

c. `input name $ 1-4 age 5-6 IDNum 8-9 jobcode $ 7 salary 10-14;`

d. `input name $ low-4 age 5-6 IDNum $ 8-9 jobcode $ 7 salary 10-high;`

9. Given the data table created with the OUT= option in PROC COMPARE, which Compare procedure would create this table?

one	two	three	four	five
1999	Children Sports	14	288.80	580.40
0	....	0	0.00	0.00
1999	Children Sports	19	344.80	693.40
0	....	0	0.00	0.00
1999	Clothes	25	3767.90	7192.10
0	.XXXXXXX.XXXXXX..........	4	-3271.50	-6193.50

a.
```
proc compare base=b compare=c out=diff outbase outdiff;
run;
```

b.
```
proc compare base=b compare=c out=diff outbase outdiff outnoequal;
run;
```

c.
```
proc compare base=b compare=c out=diff outnoequal;
run;
```

d.
```
proc compare base=b compare=c out=diff outall;
run;
```

10. Which of the following initial objectives should be verified first (via PROC COMPARE or another method) before going on to use PROC COMPARE as a validation tool for variable values?

    a. Ensure the number of records in each data set is the same.

    b. Ensure the variables in each data set have matching names and types.

    c. Ensure the data sets are both sorted in the same order.

    d. All of the above.

Concepts: Short-Answer

1. For each of the following, find the syntax error in the provided code.

    a.

```
proc format;
  value bins
    low-<10='Less than 10'
    10-20='10 to 20'
    20<-high='More than 20';
  value bins2
    low-<5='Less than 5'
    5-15='5 to 15'
    15<-high='More than 15';
run;
```

    b.

```
 proc format;
  value $codes;
    '1'='New'
    '2'='Old'
    '3'='Unknown';
  run;
```

    c.

```
proc means data=stuff;
  class category;
  var sales;
  where store ge 1 and le 4;
run;
```

2. If a variable is used in the CLASS statement in PROC MEANS or the TABLE statement in PROC FREQ, how does SAS determine the levels of the CLASS variables? To alter the set of classes without altering the data, what approach can be taken?

3. When the DSD option is used in the INFILE statement, what three characteristics for handling delimiters does it alter and how do they change? Which of these three can be further modified with an additional option in the INFILE statement?

4. True or false: When reading fixed-position data, it is possible to list the variables in any order in the input statement. Based on the concepts from this chapter, is the answer the same for list input?

5. Even when using OUT= in PROC COMPARE to create a data set containing the information requested on differences, why is it also important to carefully consider the tables PROC COMPARE places in the Output window?

6. Suppose the SORT procedure shown below is successfully executed.

```
proc sort data = example out = SortedExample;
  by w x descending y z;
run;
```

Explain whether the following BY statements are valid in a subsequent step that uses the SortedExample data set.

   a. `by w;`

   b. `by x;`

   c. `by w x y;`

   d. `by w x descending y;`

   e. `by w x descending y descending z;`

7. Consider the comma-delimited data set shown below and answer the following items. Note that this is a variation on the data in Input Data 2.9.1 where the value of 0 for the FirstClass in the ninth observation is now missing.

```
----+----1----+----2----+----3
439,12/11/2000,LAX,20,137
921,12/11/2000,DFW,20,131
114,12/12/2000,LAX,15,170
982,12/12/2000,dfw,5,85
439,12/13/2000,LAX,14,196
982,12/13/2000,DFW,15,116
431,12/14/2000,LaX,17,166
982,12/14/2000,DFW,7,88
114,12/15/2000,LAX,,187
982,12/15/2000,DFW,14,31
```

   a. Without submitting any code, give the input buffer and program data vector for the ninth observation if the following DATA step is used.

```
data PartA;
  infile myData dsd;
  input FlightNum $ Date $ Destination $ FirstClass EconClass;
run;
```

   b. Without submitting any code, give the input buffer and value of FlightNum in the PDV for the first observation if the following DATA step is used.

```
data PartB;
  infile myData dsd dlm=' ';
  input FlightNum $ Date $ Destination $ FirstClass EconClass;
run;
```

Programming Basics

1. Using the Sashelp.Cars data set, create the tables given below.

Analysis Variable : MPG_City MPG (City)						
Origin	N Obs	Minimum	Lower Quartile	Median	Upper Quartile	Maximum
Asia	158	13.0	18.0	20.5	24.0	60.0
Europe	123	12.0	17.0	19.0	20.0	38.0
USA	147	10.0	17.0	18.0	21.0	29.0

Table of Type by Origin

Type	Origin			
Frequency **Row Pct**	**Asia**	**Europe**	**USA**	**Total**
SUV	25 41.67	10 16.67	25 41.67	60
Sedan	94 35.88	78 29.77	90 34.35	262
Sports	17 34.69	23 46.94	9 18.37	49
Wagon	11 36.67	12 40.00	7 23.33	30
Total	147	123	131	401

2.   Using the Sashelp.Heart data set, create the tables given below:

Systolic	N Obs	Variable	Mean	Median	Minimum	Maximum
Normal (Below 120)	1029	Cholesterol Weight	214.2 140.6	209.0 138.0	118.0 71.0	479.0 226.0
Pre-Hypertension (120-139)	2157	Cholesterol Weight	224.9 151.3	221.0 149.0	96.0 67.0	435.0 276.0
Stage 1 (140-159)	1226	Cholesterol Weight	232.8 160.4	228.0 159.0	124.0 89.0	568.0 281.0
Stage 2 (160 or More)	797	Cholesterol Weight	242.9 162.8	238.0 160.0	142.0 92.0	492.0 300.0

Table of Systolic by Sex

Systolic	Sex		
Percent **Row Pct** **Col Pct**	**Female**	**Male**	**Total**
Normal (Below 120)	12.50 63.27 22.66	7.26 36.73 16.18	19.75
Pre-Hypertension (120-139)	21.65 52.29 39.26	19.75 47.71 44.05	41.41
Stage 1 (140-159)	11.63 49.43 21.09	11.90 50.57 26.54	23.54
Stage 2 (160 or More)	9.37 61.23 16.99	5.93 38.77 13.23	15.30
Total	2873 55.15	2336 44.85	5209 100.00

3. Write a DATA step that reads the data in the file Heart.csv and uses PROC COMPARE to demonstrate that the result is equivalent to the Sashelp.Heart data set.

4. Write a DATA step that reads the data in the file Heart.dat and uses PROC COMPARE to demonstrate that the result is equivalent to the Sashelp.Heart data set.

5. Program 2.3.8 introduces the SUM statement for computing cumulative sums in PROC PRINT and shows that, in addition to the overall sum, the SUM statement computes certain cumulative sums associated with the BY groups. The SUMBY statement is available to provide additional control over which cumulative sums PROC PRINT displays. Use the SUMBY statement to produce the following report from the Sashelp.Mdv data set.

TYPE	CODE	COUNTRY	SALES95	_4CAST96
MD11	THIRD DAY	UNITED STATES	$1,672.00	$1,880.10
MD11	THIRD DAY		$10,563.00	$11,278.39
MD11			$32,720.19	$35,865.77
			$240,958.47	$267,273.82

6. Write a program that does the following:

   ○ Reads in the Flights.dat data set

   ○ Uses the PUTLOG statement to write the FlightNum and FirstClass values along with the associated variable names

   ○ Uses the LIST statement

   ○ Compares the data set against the Flights data set created in Program 2.9.1

7. Write a program that does the following:

   ○ Reads in the Flights.txt data set

   ○ Uses the PUTLOG statement to write the FlightNum and FirstClass values along with the associated variable names

   ○ Uses the LIST statement

   ○ Compares the data set against the Flights data set created in Program 2.9.1

Case Studies

For additional practice, multiple case studies are available in addition to the IPUMS CPS case study used in the chapters. See Section 8.2 to apply the skills from this chapter to the Clinical Trials Case Study. For additional case studies including extensions to the IPUMS CPS case study, see the author pages.

Chapter 3: Bar Chart Basics, Data Diagnostics and Cleaning, and More on Reading Data from Other Sources

3.1 Learning Objectives

At the conclusion of this chapter, mastery of the concepts covered in the narrative includes the ability to:

- Differentiate between the appropriate use of formatted input and modified list input and apply the correct method in a given scenario

- Apply column pointer controls to establish where SAS begins parsing a variable

- Apply informat widths to control column positions and set character variable lengths

- Differentiate between the appropriate use of FLOWOVER, MISSOVER, or TRUNCOVER to handle certain types of missing data and apply the correct method in a given scenario

- Apply the MEANS and FREQ procedures to detect anomalous or inconsistent data values

- Apply functions to manipulate and correct anomalous or inconsistent data values

- Create data sets from computations using the MEANS and FREQ procedures for subsequent use

- Create bar charts, including charts created from pre-summarized data

- Modify bar charts with styling elements available in the HBAR and VBAR statements in PROC SGPLOT
- Modify charts with axis and legend styling elements to enhance their readability
- Apply PROC IMPORT to read raw data files
- Compare and contrast PROC IMPORT and the DATA step for reading raw files

Use the concepts of this chapter to solve the problems in the wrap-up activity. Additional exercises and case-studies are also available to test these concepts.

3.2 Case Study Activity

Continuing the case study introduced in Chapter 2, Output 3.2.1 shows a summary of the 2010 IPUMS CPS Basic data set. For the purposes of the case study in this chapter, raw data sets are provided which deviate from the data in its original form, having data integrity errors introduced intentionally. The objective, as described in the Wrap-Up Activity, is to read in the dirty data, clean it, and produce summaries that demonstrate the successful cleaning of the data. Suggested summaries are shown below, which are prepared from a cleaned version of the dirty raw data.

Output 3.2.1: Listing of Ownership Categories and Frequencies

Ownership	Frequency	Percent	Cumulative Frequency	Cumulative Percent
N/A	79899	6.22	79899	6.22
Owned	856700	66.74	936599	72.96
Rented	347077	27.04	1283676	100.00

Output 3.2.2: Listing of Mortgage Status Categories and Frequencies

MortgageStatus	Frequency	Percent	Cumulative Frequency	Cumulative Percent
N/A	426976	33.26	426976	33.26
No, owned free and clear	307864	23.98	734840	57.24
Yes, contract to purchase	7127	0.56	741967	57.80
Yes, mortgaged/ deed of trust or similar debt	541709	42.20	1283676	100.00

Output 3.2.3: Partial Listing of Cities

City	Frequency	Percent	CumFrequency	CumPercent
Akron, OH	824	0.06	824	0.06
Albany, NY	470	0.04	1294	0.10
Alexandria, VA	681	0.05	1975	0.15
Allentown, PA	371	0.03	2346	0.18
Anaheim, CA	1025	0.08	3371	0.26
Anchorage, AK	806	0.06	4177	0.33
Ann Arbor, MI	484	0.04	4661	0.36
Arlington, TX	1267	0.10	5928	0.46

Output 3.2.4: Partial Listing of State Names

state	Frequency	Percent	Cumulative Frequency	Cumulative Percent
Alabama	20808	1.62	20808	1.62
Alaska	2711	0.21	23519	1.83
Arizona	26288	2.05	49807	3.88
Arkansas	12649	0.99	62456	4.87
California	136838	10.66	199294	15.53
Colorado	21546	1.68	220840	17.20
Connecticut	15281	1.19	236121	18.39

Output 3.2.5: Quantiles on Mortgage Payments

			Analysis Variable : MortgagePayment			
N	N Miss	Minimum	25th Pctl	50th Pctl	75th Pctl	Maximum
1283676	0	0.0	0.0	0.0	880.0	7400.0

Output 3.2.6: More Percentiles on Mortgage Payments

				Analysis Variable : MortgagePayment					
N	50th Pctl	60th Pctl	70th Pctl	75th Pctl	80th Pctl	90th Pctl	95th Pctl	99th Pctl	Maximum
1283676	0.0	310.0	700.0	880.0	1100.0	1600.0	2200.0	3700.0	7400.0

Output 3.2.7: Charting Mortgage Payments Versus Metro Status

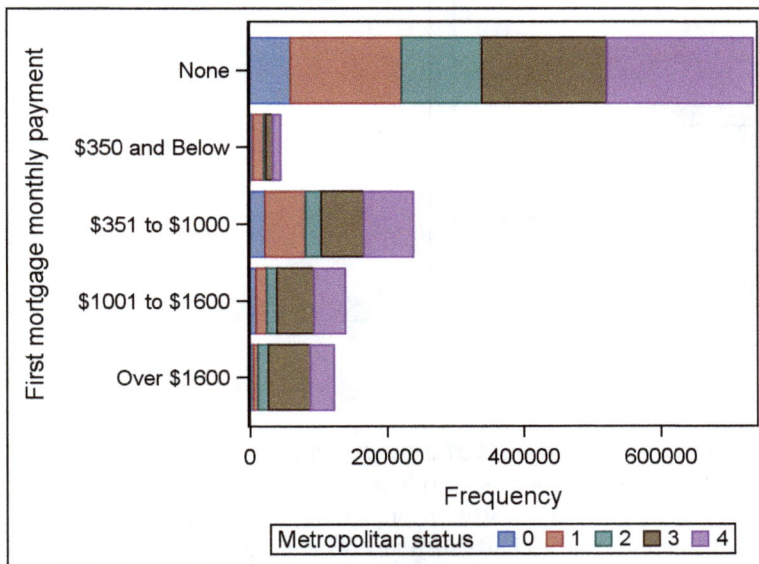

Output 3.2.8: Cumulative Distribution of Mortgage Payments

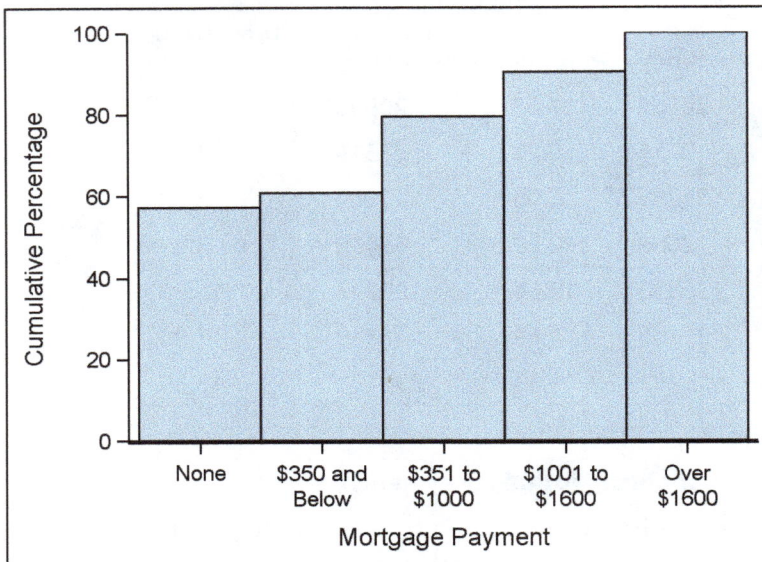

Output 3.2.9: Household Income Levels and Mortgage Payments

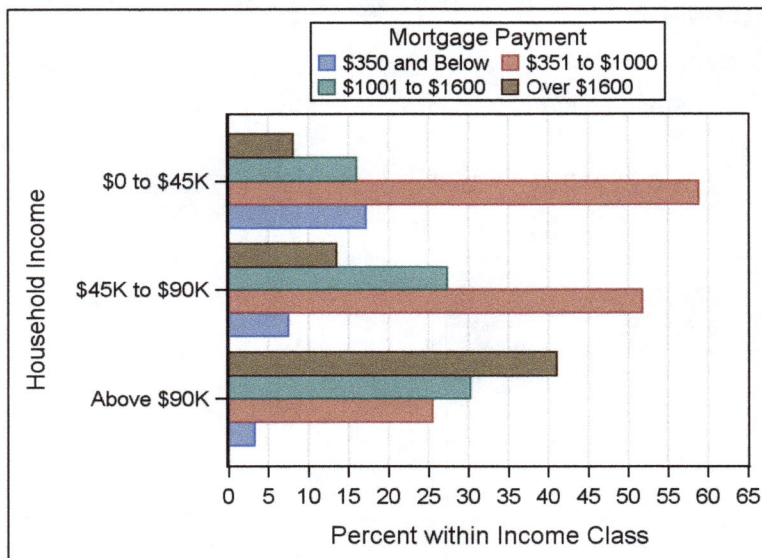

3.3 Bar Charts

To produce bar charts for the case-study in Section 3.2, PROC SGPLOT is used with the HBAR and VBAR statements producing horizontal and vertical bar charts, respectively. This section covers the fundamental structure and options for the HBAR and VBAR statements in the SGPLOT procedure, along with additional statements that control the styling of certain elements in the charts such as axes and legends.

3.3.1 Bar Charts on a Single Variable for Frequency and Relative Frequency

As in previous chapters, make sure the BookData library is assigned in the current SAS session, and review the Ipums2005Basic SAS data set used for the examples in Chapter 2. As a first step, enter and submit the code given in Program 3.3.1.

Program 3.3.1: A Simple Vertical Bar Chart

```
proc sgplot data=BookData.Ipums2005Basic;
  vbar metro;
run;
```

The SGPLOT procedure provides a variety of plot and chart types, one of which is a vertical bar chart, generated with the VBAR statement (other statements in PROC SGPLOT can also generate vertical bar charts). The variable provided is treated as categorical, as seen in Chapter 2 for variables listed in the CLASS statement in PROC MEANS or the TABLE statement in PROC FREQ. The default summary is the frequency in each category, as shown in Output 3.3.1.

Output 3.3.1: Vertical Bar Chart for Counts on Each Level of Metro

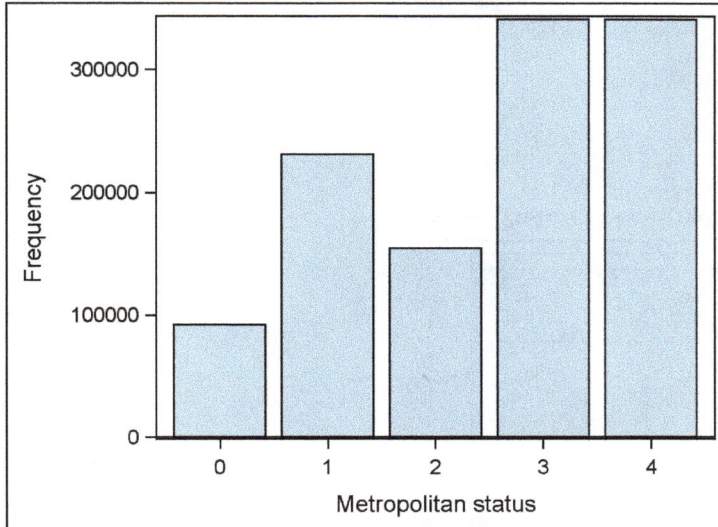

Horizontal bar charts are also possible, HBAR is one statement available in PROC SGPLOT to create one—consider Program 3.3.2 and its corresponding output.

Program 3.3.2: A Simple Horizontal Bar Chart

```
proc sgplot data=BookData.Ipums2005Basic;
  hbar metro;
run;
```

Output 3.3.2: Horizontal Bar Chart for Counts on Each Level of Metro

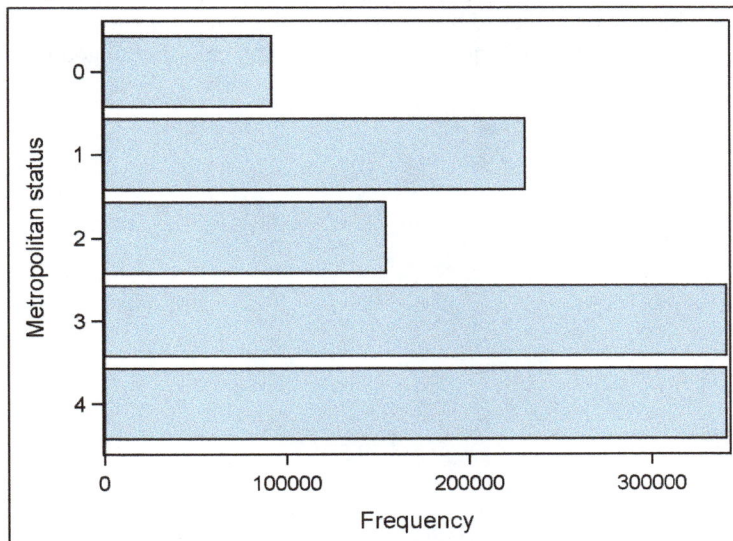

As with other procedure statements that treat variables as categorical, HBAR and VBAR use a format to create the categories when a format is supplied. In its default format, the MortgagePayment variable is a poor choice for a charting variable, just as it is a poor choice for a variable in a CLASS or TABLE statement in Chapter 2. However, the same format used in Program 2.5.6 to make it a useful CLASS variable also makes it a suitable charting variable; Program 3.3.3 illustrates this concept.

Program 3.3.3: Using a Format to Set Charting Categories

```
proc format;
  value Mort
    0="None"
    1-350="$350 and Below"
    351-1000="$351 to $1000"
    1001-1600="$1001 to $1600"
    1601-high="Over $1600"
  ;
run;

proc sgplot data=BookData.Ipums2005Basic;
  hbar MortgagePayment;
  format MortgagePayment Mort.;
run;
```

Output 3.3.3: Formatting Mortgage Payment for Use as a Charting Variable

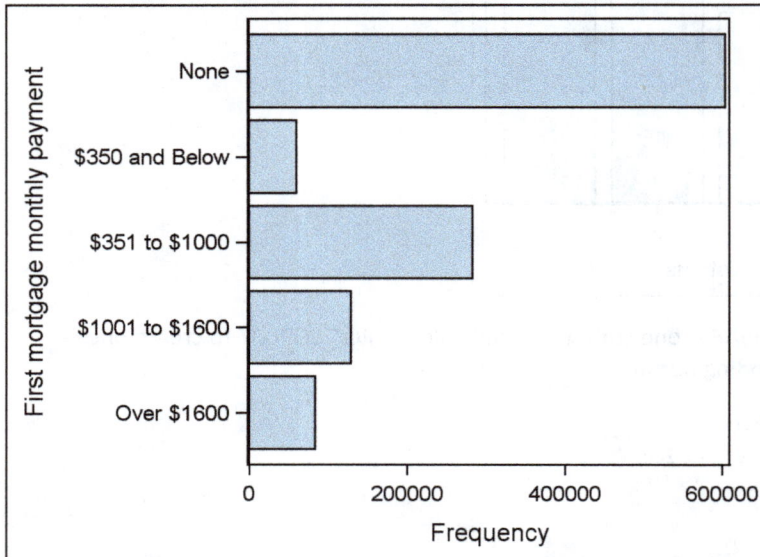

Thus, changing categories for the charting variable can be achieved by simply modifying a format and using it in the corresponding SGPLOT procedure call. For bar charts on frequencies, the summary can be changed to relative frequency (percentage) with the STAT= option as shown in Program 3.3.4.

Program 3.3.4: Using the STAT= Option to Display Percentages

```
proc sgplot data=BookData.Ipums2005Basic;
  hbar MortgagePayment/stat=percent;
  format MortgagePayment Mort.;
run;
```

Output 3.3.4: Percent as the Summary Statistic

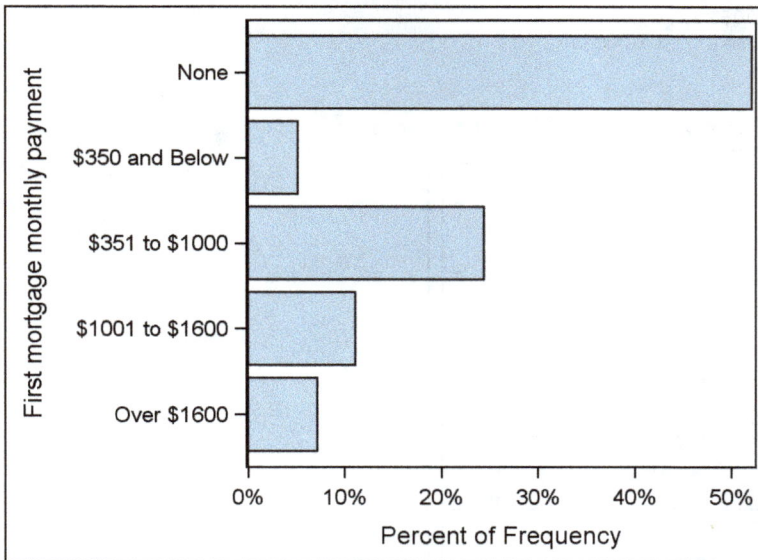

3.3.2 Setting a Response Variable for HBAR or VBAR

The RESPONSE= option, in either the HBAR or VBAR statements, chooses a numeric (quantitative) variable that is summarized in each category. As seen in Output 3.3.5, the default summary statistic is the sum of all values in the category.

Program 3.3.5: Choosing a Response Variable for a Bar Chart

```
proc sgplot data=BookData.Ipums2005Basic;
  hbar metro/response=MortgagePayment;
run;
```

Output 3.3.5: Mortgage Payment as the Response Across Metro Categories

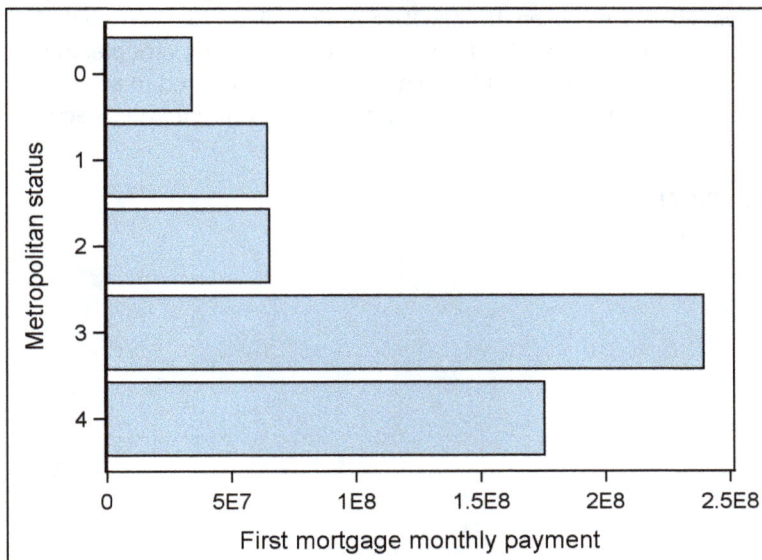

Since the sums in certain Metro areas are large, the numbers displayed on the axis are in scientific notation. Here the STAT= option can be used to set the summary statistic to be the mean or median, and Program 3.3.6 modifies the chart to show means.

Program 3.3.6: Choosing a Response Variable for a Bar Chart

```
proc sgplot data=BookData.Ipums2005Basic;
  hbar metro/response=MortgagePayment stat=mean;
run;
```

Output 3.3.6: Mortgage Payment Means Summarized Across Metro Categories

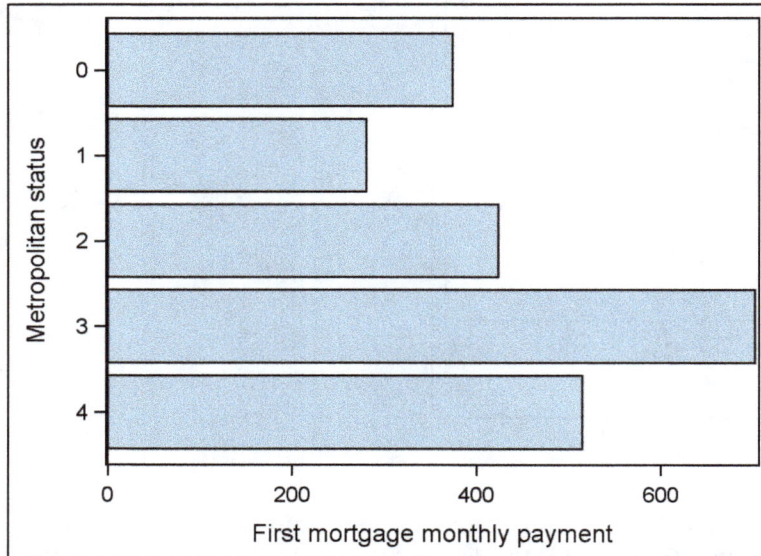

The limitation of the STAT= option to merely the sum, mean, and median may at first appear to be problematic; however, many more statistics can be charted by using other procedures in a preceding step. Procedures like MEANS can deliver their results to data sets, and those data sets can be used as the input to other procedures, such as SGPLOT. This concept is discussed in Section 3.5.

3.3.3 Grouped Bar Charts in HBAR or VBAR

No matter what charting and/or response variables are chosen, the GROUP= option allows for each level of the charting variable to be broken down across levels of a specified variable. Like the charting variable, the GROUP= variable is treated as categorical, so it should be naturally of that form or be formatted to an appropriate set of categories. Program 3.3.7 groups the frequencies for the MortgagePayment groups across levels of the Metro variable.

Program 3.3.7: Using a GROUP= Variable in a Bar Chart

```
proc sgplot data=BookData.Ipums2005Basic;
  hbar MortgagePayment/group=metro;
  format MortgagePayment Mort.;
run;
```

Output 3.3.7: Mortgage Payment Frequencies Split Across Metro Status

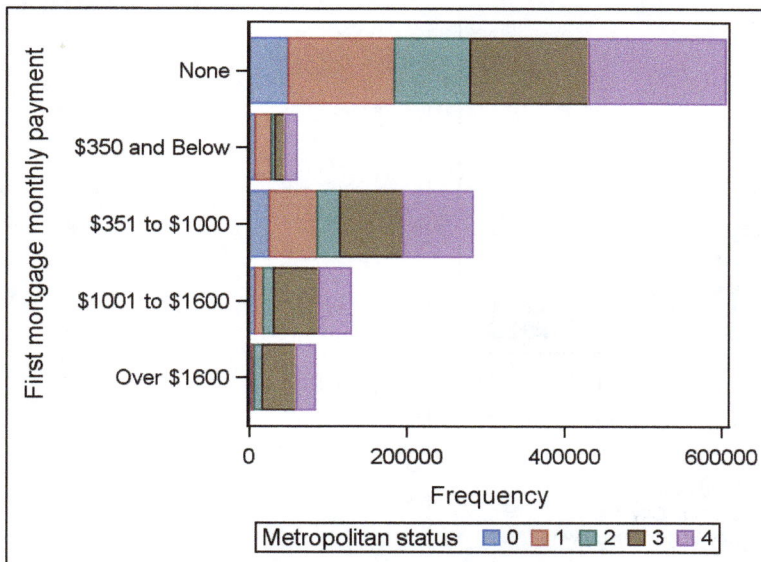

By default, bars are split into sections representing each level of the group variable. This is intuitively reasonable for statistics like frequency and sum, but much less so for percent, mean, or median. The structure of the groups within each charting variable level can be altered with the GROUPDISPLAY= option. STACK is the default option, but the CLUSTER option places bars side-by-side, as shown in Program 3.3.8.

Program 3.3.8: Changing the Group Structure with GROUPDISPLAY=

```
proc format;
  value METRO
    0 = "Not Identifiable"
    1 = "Not in Metro Area"
    2 = "Metro, Inside City"
    3 = "Metro, Outside City"
    4 = "Metro, City Status Unknown"
  ;
run;

proc sgplot data=BookData.Ipums2005Basic;
  hbar MortgagePayment/group=metro groupdisplay=cluster response=HHIncome stat=mean;
  format MortgagePayment Mort. metro metro.;
  where metro between 2 and 4;
run;
```

Output 3.3.8: Group Bars Displayed Side-By-Side

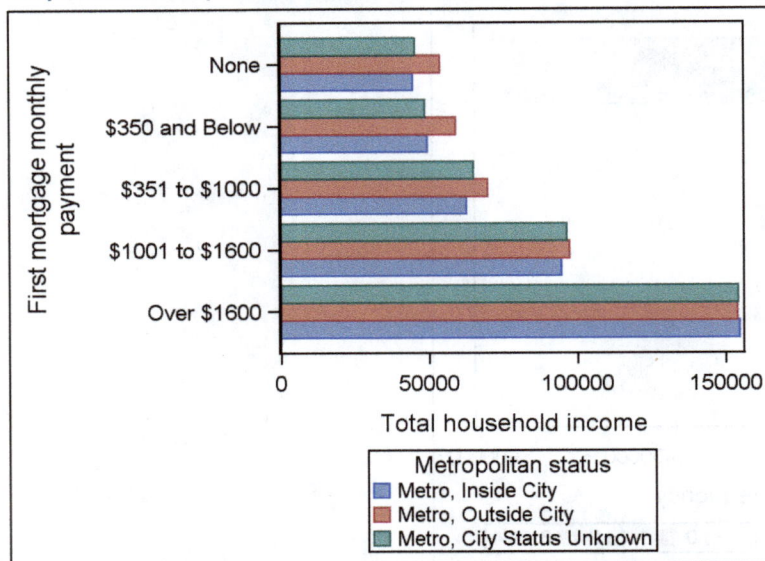

Be sure to note all of the programming concepts in use in this example beyond the concepts for PROC SGPLOT introduced in this section. Formats create a categorical set from the quantitative variable MortgagePayment and improve the display of Metro, while a WHERE statement limits the chart to only metropolitan cases. To become a well-rounded programmer, it is important to continually think about how ideas covered previously are applicable to new concepts as they are introduced.

3.4 Options and Statements to Style Bar Charts

While default fonts, colors, and other style elements may provide suitable styling for many charts, PROC SGPLOT provides several statements and options to give detailed control over these elements. Many of these apply to a wide variety of chart and graph types, and they are structured to provide a measure of logical consistency in the syntax. Some of the statements covered in this section are directly applicable in the same or a similar form to other graph types covered later. It is as important to focus on the logic behind the syntax as on the actual syntax itself.

3.4.1 HBAR and VBAR Options

Previously, options in the HBAR and VBAR statements have been used to change the displayed statistic, set a response variable or a group variable, and control the display of the group levels. Other options are available to modify the appearance of the graph; including bar fill and outline styles, error bars, data labels, and other elements. In many SGPLOT options, the suffix ATTRS is used to set options for style attributes, with the prefix providing a reference to the graphical element being modified. Program 3.4.1 creates a bar graph similar to Output 3.3.4 (the records with MortgagePayment values of 0 are removed) with some styling changes noted.

Program 3.4.1: Setting Styles for Fills and Outlines in HBAR or VBAR

```
proc sgplot data=BookData.Ipums2005Basic;
  hbar MortgagePayment / stat=percent fillattrs❶=(color=cx36CF36❷)
                         outlineattrs❸=(color=gray3F❹ thickness=3pt❺);
  format MortgagePayment Mort.;
  where MortgagePayment gt 0;
run;
```

❶ FILLATTRS= modifies the bar fill—for fills, COLOR= and TRANSPARENCY= are available style elements. Transparency is discussed later in this section for graphs involving overlays.

❷ SAS supports several color models, including RGB (red-green-blue) mixing, with colors named in the form cxRRGGBB. The cx characters are a prefix indicating an RGB hexadecimal code follows, the first two digits are the red saturation or intensity, the next two are green, and the final two are blue. Several web resources exist to demonstrate which colors are achieved by various mixes.

❸   OUTLINEATTRS= modifies the bar outline—for lines COLOR=, THICKNESS=, and STYLE= are the commonly available style elements. STYLE= allows for various dashed lines to be used, but this attribute is not available for bar outlines. STYLE= is demonstrated in plots later in this text (starting in Chapter 5) that use lines as part of the data plot.

❹   SAS also supports a grayscale color model. Specifications are of the form grayXY, where X and Y are hexadecimal digits that determine the lightness of the gray level. Therefore, gray00 is black, grayFF is white, and levels between are an intermediate gray.

❺   Here, point size (pt) is used to specify thickness. Common units for thickness (and other sizes) are pt, cm, mm, in, pct (or %), and px.

Output 3.4.1: Modifying Bar Fills and Outlines

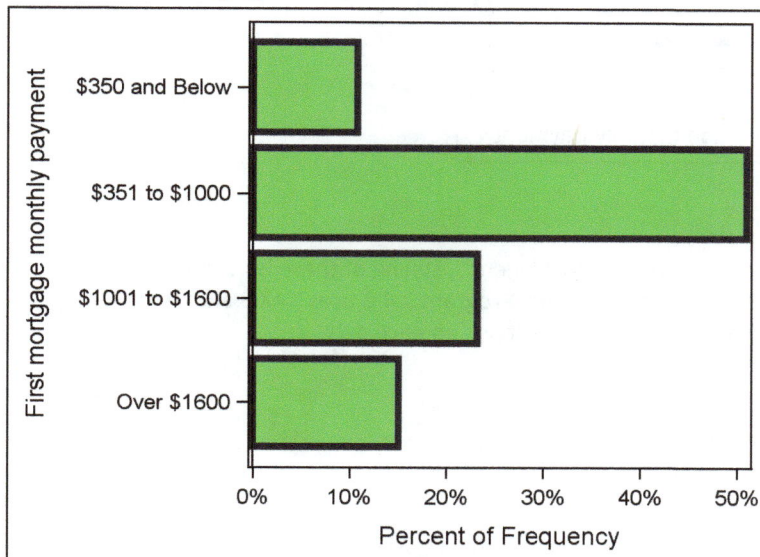

Many options are available in the HBAR and VBAR statements and, while this section is not intended to be a comprehensive review of those options, several are shown in Program 3.4.2.

Program 3.4.2: Some Other Bar Chart Appearance Modifications

```
proc sgplot data=BookData.Ipums2005Basic;
  hbar Metro / response=MortgagePayment stat=mean categoryorder=respasc❶
               dataskin=gloss❷ limits=upper❸ limitstat=stderr❹;
  format Metro Metro.;
  where metro between 2 and 4 and HHIncome ge 500000❺;
run;
```

❶   The default order of the chart variable categories is alpha-numeric on the names. CATEGORYORDER= permits options of RESPASC and RESPDESC for ascending and descending order of the response statistic, respectively.

❷   DATASKIN= applies an effect to the bar fills. The default value of NONE can be modified to CRISP, GLOSS, MATTE, PRESSED, or SHEEN.

❸   When STAT=MEAN, error bars can be included with the LIMITS= option. Values include BOTH, LOWER, and UPPER, and the default statistic is the 95% confidence limit. ALPHA= is an available option to allow for different confidence levels.

❹   The statistic represented by the error bar can be modified to either standard error (STDERR) or standard deviation (STDDEV). By default, bars corresponding to a single standard deviation or error are displayed, but NUMSTD= can be used to set a multiplier (which must be a positive number).

❺   Output 3.4.2 is limited to Household Income at $500,000 or more. If the full data set is used, the standard error bars or confidence limit bars are quite small given the full number of observations.

Output 3.4.2: Various Bar Chart Modifications

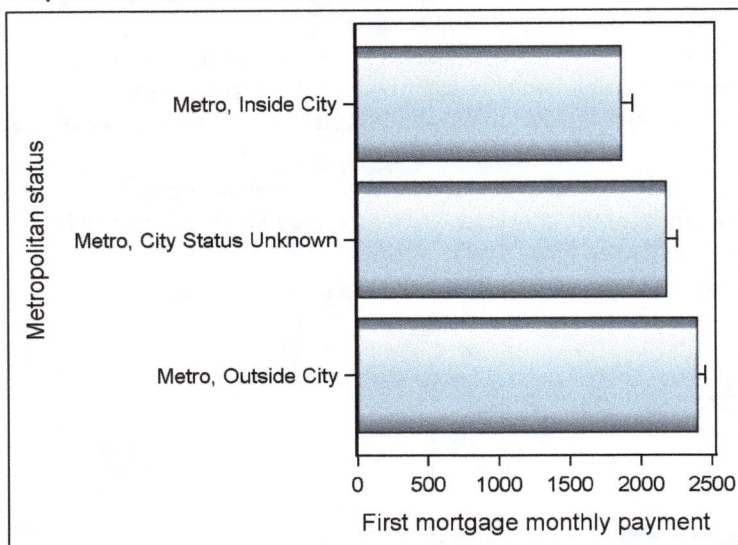

3.4.2 Statements to Alter Axis and Legend Styles

Several different graph and chart types produce axes and/or legends, styling of these is controlled by statements that apply to several different SGPLOT statements. Program 3.4.3 uses XAXIS and YAXIS statements to modify Output 3.4.2 into Output 3.4.3, altering styles on each axis.

Program 3.4.3: Axis Modifications

```
proc sgplot data=BookData.Ipums2005Basic;
  hbar Metro / response=MortgagePayment stat=mean categoryorder=respasc
            dataskin=gloss limits=upper limitstat=stderr;
  format Metro Metro.;
  where metro between 2 and 4 and HHIncome ge 500000;
  yaxis display=(nolabel) ❶ valueattrs=(family='Papyrus' size=8pt) ❷;
  xaxis values=(0 to 2750 by 250)❸ offsetmax=0❹ valuesformat=comma5.❺
      label='Avg. Mortgage Payment'❻;
run;
```

❶ The default value for the DISPLAY= option is ALL, meaning all axis elements are displayed. Therefore, DISPLAY= is used primarily to remove the display of certain axis elements with values including NONE, NOLABEL, NOLINE, NOTICKS, and NOVALUES. ALL or NONE can only be specified individually (no other values can be used with them) and must not be enclosed in parentheses. Any other set of values specified must be enclosed in parentheses, even if only one is used.

❷ Values occur at the major tick marks on any axis; in this case, values correspond to the categories on the chart variable axis (Y) and the numeric values listed on the response axis (X). VALUEATTRS= modifies the value text—text attributes available for modification are COLOR=, FAMILY=, SIZE=, STYLE=, and WEIGHT=. STYLE and WEIGHT are used to set italic and bold, respectively, and FAMILY is used to choose the font.

❸ The VALUES= option is used to set the values at the major ticks. The form *start-value* TO *stop-value* BY *increment-value* is only legal for numeric values, with the increment set to 1 if the BY and increment are omitted. Other methods, such as lists of individual values, are legal for both character and numeric values. The requested values must be enclosed in parentheses no matter what form is used.

❹ Offsets are available at both ends of an axis to set the distance from the beginning of the axis to the first major tick mark (OFFSETMIN=) or from last tick mark to the end of the axis (OFFSETMAX=). For the response axis of a bar chart, zero is the default value OFFSETMIN=, so the first tick is in line with the other vertical axis. Here OFFSETMAX=0 aligns the final major tick with the right side of the graph border. The offset value is given as a proportion of the axis that comprises the offset and is always a value between 0 and 1.

❺ VALUESFORMAT= assigns a format to the values at the major ticks.

❻ The LABEL= option sets the axis label to the provided literal value.

Output 3.4.3: Changing Various Axis Attributes

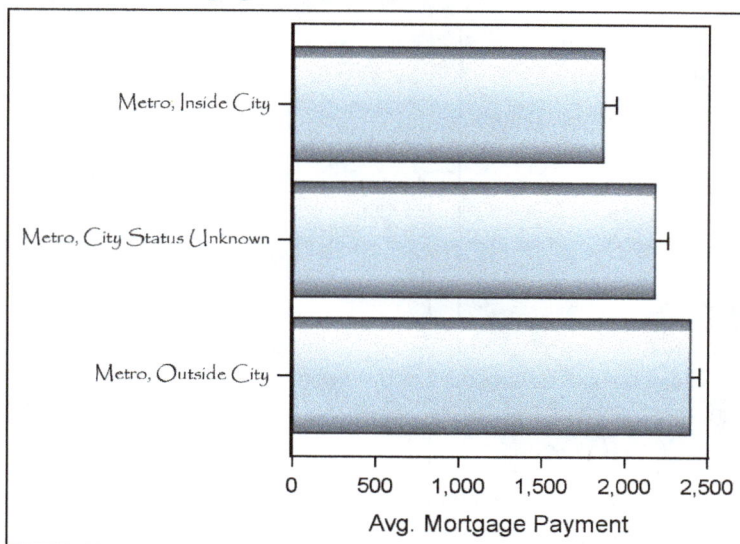

The options listed here are not intended to be a comprehensive set of all available axis options, others are discussed in subsequent sections of the book. Consult the SAS Documentation to view the set of available options.

For any graph that generates a legend (or multiple legends), KEYLEGEND statements are available to modify the legend(s). Program 3.4.4 revisits the chart generated in Output 3.3.8, making some modifications to the style of the legend, which is automatically produced as a result of using the GROUP= option.

Program 3.4.4: Legend Modifications

```
proc sgplot data=BookData.Ipums2005Basic;
  hbar MortgagePayment/group=metro groupdisplay=cluster response=HHIncome stat=mean;
  format MortgagePayment Mort. metro metro.;
  where metro between 2 and 4;
  keylegend /❶ location=inside❷ position=topright❸ across=1❹
              title='Metro Status'❺ noborder❻
              valueattrs=(family='Georgia')❼;
run;
```

❶ The statements controlling axis styles did not use the forward slash (/) character to separate options; however, it is possible for a graph to generate multiple legends, which can be named distinctly. Thus, the first item specified in a KEYLEGEND statement is the legend name(s) that the style specifications apply to, with the slash separating the name list from the options. For this graph, there is only one legend, so no name is required but the slash must be present.

❷ LOCATION= specifies whether the legend appears INSIDE or OUTSIDE the graph border; OUTSIDE is the default.

❸ POSITION= specifies a place around the perimeter of the graph to locate the legend. See the SAS Documentation for the set of possible values.

❹ ACROSS= controls the number of columns available in the legend. DOWN= can be used to set the number of rows. Generally, only one of these two options is specified in any KEYLEGEND statement.

❺ The string naming the legend is modified by TITLE=, which is set to a literal value. Avoid confusing it with the LABEL= attribute for axes.

❻ By default, the legend is drawn with a bounding box, NOBORDER removes this box.

❼ VALUEATTRS= is used to modify the text that corresponds to each value described by the legend.

Output 3.4.4: Changing Various Legend Attributes

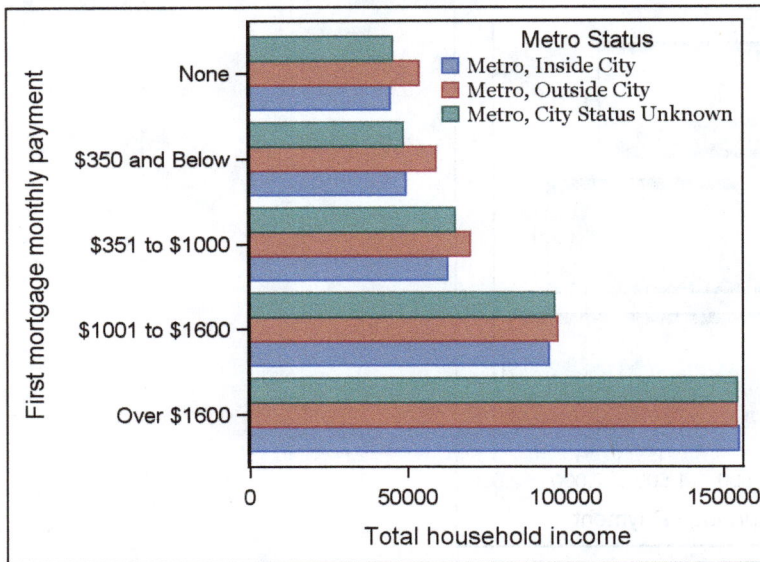

Many more style options are available in the KEYLEGEND statement with some covered in subsequent sections; however, a review of the SAS Documentation for KEYLEGEND is recommended.

3.5 Creating and Using Output Data Sets from MEANS and FREQ

In many cases, it is desirable to chart a statistic that is not available in the STAT= option for the HBAR and VBAR statements. In those cases, using procedures like MEANS and FREQ to deliver their output to SAS data sets allows for a variety of statistics to be charted. Procedures in SAS have methods for generating output, some of which include use of: an OUTPUT statement, data set output options, or the ODS OUTPUT statement applied to the desired output object(s).

3.5.1 The OUTPUT Statement

Program 3.5.1 computes the mean and median for Household Income across the levels of the Metro variable, using the OUTPUT statement to send those statistics to a data set. Both of these statistics can be requested in the HBAR statement; however, the SGPLOT procedure need not be used for any statistical calculations.

Program 3.5.1: Using the OUTPUT Statement in PROC MEANS

```
ods _all_ close; ❶
proc means data=BookData.Ipums2005Basic;
  class metro; ❷
  var HHincome; ❷
  output out=stats❸ mean=avg median=median❹;
run;
```

❶ Creating an output table is not necessary, here the ODS _ALL_ CLOSE statement is used to turn off all output destinations.

❷ The variable that takes on the role of the charting variable in the HBAR or VBAR statement is set as the class variable and the analysis variable is put in the VAR statement.

❸ OUT= specifies the data set that contains the results. Here, the data set Stats is created in the Work library.

❹ Irrespective of the statistic keywords listed in the PROC MEANS statement, any statistic keyword can be listed here followed by a legal variable name that is assigned to it. By default, the median is not computed for the procedure output, but it is available for the output data set. Keywords form legal variable names and can be used in both roles, take care not to confuse the two.

Now either, or both, of these statistics can be summarized in a vertical or horizontal bar chart. Program 3.5.2 summarizes both statistics with an overlay of two horizontal bar charts.

Program 3.5.2: Charting the Results from the OUTPUT Statement in PROC MEANS

```
ods listing; ❶
proc sgplot data=work.stats;
   hbar❷ metro / response=median❸ fillattrs=(color=green)
                 legendlabel='Median'❹ barwidth=.9❹;
   hbar❷ metro / response=avg fillattrs=(transparency=0.3❹ color=orange)
                 outlineattrs=(color=black) legendlabel='Mean' barwidth=.7;
   where metro between 2 and 4;
   format metro metro.;
   xaxis label='Household Income' valuesformat=dollar8.;
   yaxis display=(nolabel);
   keylegend / position=topright across=1 location=inside
               valueattrs=(size=8pt) noborder; ❺
run;
```

❶ The listing destination is the delivery destination for the graph file. Since ODS _ALL_ CLOSE was used previously, an appropriate output destination must be restored in order for the graph to be generated.

❷ Multiple graphing statements in SGPLOT cause the graphs to be overlaid (when compatible). The graphing calls can be thought of as layers, with the first call on the bottom and the last as the top layer.

❸ The response variable is now the named statistic. While the default statistic for this variable is the sum, it is irrelevant since the pre-summarized data contains a single value.

❹ These options, which have not been previously discussed, are helpful in making this chart more readable, consult the SAS Documentation to find details about what they are and how they work.

❺ Options previously discussed for modifying axes and the legend are also useful in improving the readability of this chart.

Output 3.5.2: Using Statistics in SGPLOT Generated by PROC MEANS

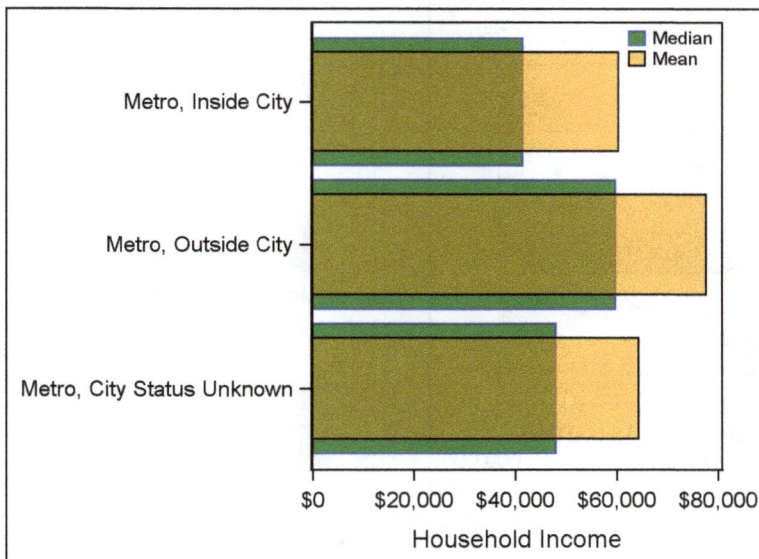

Statistics on more than one analysis variable can be computed and output simultaneously; in fact, the VAR statement is unnecessary. Program 3.5.3 computes the third quartile for both the MortgagePayment and HomeValue variables and delivers them to a data set, subsequently placing the two into a single chart.

Program 3.5.3: Computing and Charting the Results for Two Variables in PROC MEANS

```
ods listing close;
proc means data=BookData.Ipums2005Basic;
   class metro;
   output out=work.twoStats q3(MortgagePayment HomeValue)=MortQ3 ValueQ3❶;
run;
```

```
ods listing;
proc sgplot data= work.twoStats;
  hbar metro / response=MortQ3 legendlabel='Mortgage' barwidth=.4
              discreteoffset=.2❷ x2axis❸;
  hbar metro / response=ValueQ3 legendlabel='Value' barwidth=.4
              discreteoffset=-.2;
  where metro between 2 and 4;
  format metro metro.;
  xaxis label='Value Q3' valuesformat=dollar12. fitpolicy=stagger❹;
  x2axis❸ label='Mortgage Q3' valuesformat=dollar8.;
  yaxis display=(nolabel);
  keylegend / position=bottomright across=1 location=inside title=''
     valueattrs=(size=8pt) noborder;
run;
```

❶ A list of analysis variables can be attached to any statistic keyword in a set of parentheses. A list of legal variables is given on the right side of the equals sign to name these with no parentheses used for this list. The order of the analysis variables determines the summary values stored in each of the output variables.

❷ While these two HBAR calls produce an overlay, DISCRETEOFFSET= (in conjunction with BARWIDTH=) is used to make a chart that mimics the style of a grouped bar chart with GROUPDISPLAY=STAGGER. Consult the SAS Documentation for details about permitted values for these options.

❸ Given the vastly different scales for these two statistics, the Mortgage statistic would appear incredibly small on the HomeValue scale. It is possible to use the top of the frame as a second X axis with the X2AXIS option, and this axis is styled using the X2AXIS statement. A Y2AXIS option and statement are also available to place and style an axis on the right side of the graph frame.

❹ If values on an axis have difficulty fitting in the allowed space, FITPOLICY= can remedy such problems. Several values are available, consult the SAS Documentation for details.

Output 3.5.3: Chart of Two Custom Statistics, Staggered with Separate Axes

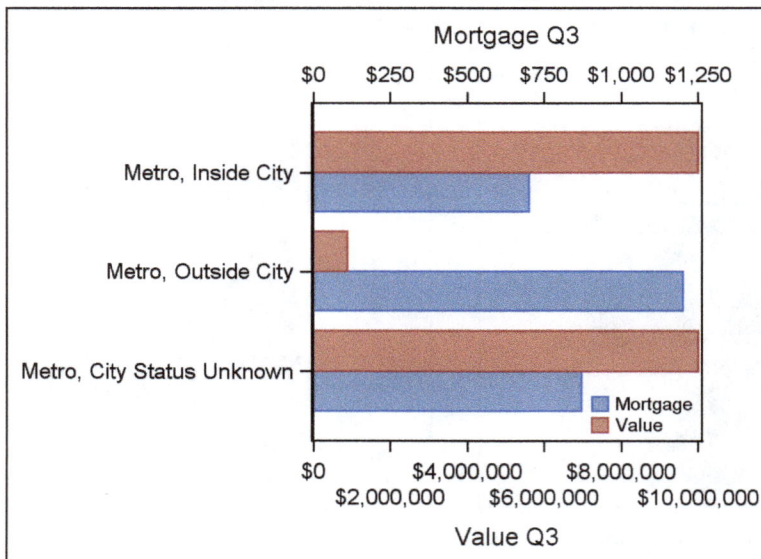

3.5.2 The ODS OUTPUT Statement

The OUTPUT statement in PROC FREQ is designed primarily to output inferential statistics, not summaries contained in the table; however, an OUT= option is available in the TABLE statement to create such a data set. This section focuses on using the ODS OUTPUT statement, which is able to create data sets from output tables in both the MEANS and FREQ procedures and also applies to output tables generated by other SAS procedures. In order to use ODS OUTPUT for any procedure, the name of the output table must be known. It can be determined by checking the table properties in the Results window or, for many procedures, by checking the details section for the procedure in the SAS Documentation—with PROC MEANS being a notable exception as it does not include a details section in its documentation. More commonly, submitting the procedure with ODS TRACE set to ON produces the list of tables generated by the procedure, as seen in

Section 1.5. Program 3.5.4 reproduces Output 3.5.2, resulting from Programs 3.5.1 and 3.5.2, using ODS OUTPUT in place of the OUTPUT statement.

Program 3.5.4: Using ODS OUTPUT to Replicate Output 3.5.2

```
ods _all_ close;
proc means data=BookData.Ipums2005Basic mean median❶;
 class metro;
 var HHincome; ❷
 ods output summary= work.odsStats; ❸
run;

ods listing;
proc sgplot data= work.odsStats;
   hbar metro / response=HHIncome_Median❹ fillattrs=(color=green)
                legendlabel='Median' barwidth=.9;
   hbar metro / response=HHIncome_Mean❹ fillattrs=(transparency=0.3
                color=orange) outlineattrs=(color=black)
                legendlabel='Mean' barwidth=.7;
   where metro between 2 and 4;
   format metro metro.;
   xaxis label='Household Income' valuesformat=dollar8.;
   yaxis display=(nolabel);
   keylegend / position=topright across=1 location=inside;
run;
```

❶　The output data set is built from the information in the procedure output. Therefore, any statistics to be used must be specified in the statistic keyword list in the PROC MEANS statement.

❷　The variables to be summarized must be listed in the VAR statement.

❸　The general form of the ODS OUTPUT statement name has *table-name* = *data-set-name*, which can include a list for several output table assignments to data sets.

❹　The ODS OUTPUT data set from the Summary table in MEANS names its summary statistic variables in the form: *analysis-variable-name_stat-keyword*. (Truncation of the name occurs if the length exceeds 32 characters.)

Depending on the request in the TABLE statement, PROC FREQ can produce several different output tables. Program 3.5.5 requests a one-way frequency table for the categorized values of MortgagePayment, generating an output table with the name OneWayFreqs. This table includes all of the statistics generated in the usual PROC FREQ output, and the remainder of Program 3.5.5 charts the cumulative percentage across the categories.

Program 3.5.5: Using ODS OUTPUT on a One-Way Frequency Table

```
ods _all_ close;
proc freq data=BookData.Ipums2005Basic;
   table MortgagePayment;
   format MortgagePayment Mort.;
   ods output OneWayFreqs❶= work.Freqs;
run;

ods listing;
proc sgplot data= work.Freqs noborder❷;
   vbar MortgagePayment / response=CumPercent❸ barwidth=1❹;
   xaxis label='Mortgage Payment';
   yaxis label='Cumulative Percentage' offsetmax=0;
run;
```

❶　When using ODS OUTPUT in any procedure, it is always important to know the names of any tables to be converted to data sets.

❷　By default, the final bar gets very close to the bounding box of the graph, and touches it when OFFSETMAX=0 on the Y axis. The SGPLOT statement contains several options for broader styling of the graph area, including the ability to remove this border. Other options that are global to the plot, such as NOAUTOLEGEND are available here, see the SAS Documentation for details.

❸ For any tables used, the naming of variables is automatic; investigating the data set to determine these names is often necessary.

❹ The chart created is effectively a histogram, and the gaps between the bars can be removed by setting BARWIDTH= to 1 (100%). Chapter 4 discusses the HISTOGRAM statement available in PROC SGPLOT.

Output 3.5.5: Chart of Cumulative Percentage Statistics from PROC FREQ

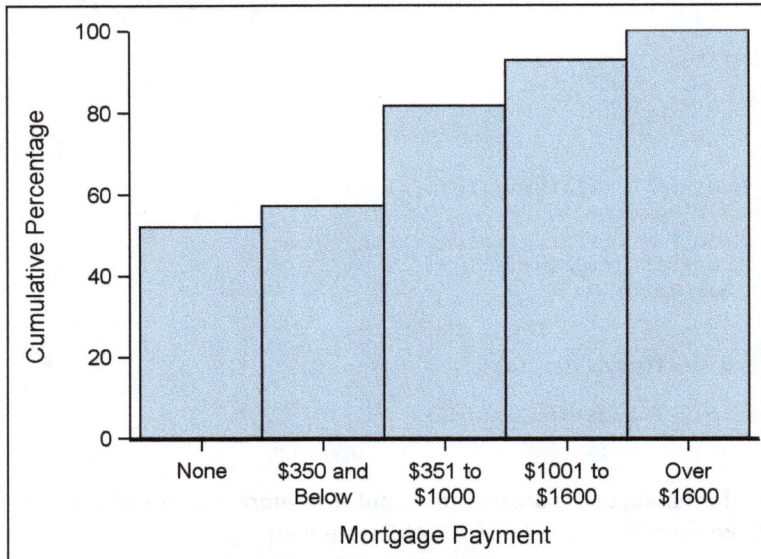

SAS gives a different name, CrossTabFreqs, to any multi-way tables (two-way, three-way, and higher dimensional tables). To revisit an example similar to a table constructed for the IPUMS case study in Chapter 2, Program 3.5.6 builds a two-way table on a subset of HHIncome and MortgagePayment categories. This is delivered to a data set in order to chart the conditional distribution of the MortgagePayment categories within each HHIncome category, as shown in Output 3.5.6.

Program 3.5.6: Using ODS OUTPUT on a Two-Way Frequency Table

```
ods _all_ close;
proc format;
  value income
  low-<0='Negative'
  0-45000='$0 to $45K'
  45001-90000='$45K to $90K'
  90001-high='Above $90K'
  ;
run;

proc freq data=BookData.Ipums2005Basic;
  table HHIncome*MortgagePayment;
  format HHIncome income. MortgagePayment mort.;
  where MortgagePayment gt 0 and HHIncome ge 0;
  ods output CrossTabFreqs❶= work.TwoWay;
run;

ods listing;
proc sgplot data= work.TwoWay;
  hbar HHIncome❷ / response=RowPercent❷ group=MortgagePayment❸
                   groupdisplay=cluster;
  xaxis label='Percent within Income Class' grid gridattrs=(color=gray66)❹
        values=(0 to 65 by 5) offsetmax=0;
  yaxis label='Household Income';
  keylegend / position=top title='Mortgage Payment';
  where HHIncome is not missing and MortgagePayment is not missing❺;
run;
```

❶ For any tables used, ODS OUTPUT automatically names the variables based on the defined conventions for the table in use; investigating the resulting data set to determine these names is often necessary.

❷ Here, the goal is to chart the conditional distribution of MortgagePayment across bins for HHIncome. HHIncome is the charting variable, and the table statistic desired is the conditional percentage in each row: RowPercent. Note, HHIncome and MortgagePayment are formatted into categories in PROC FREQ, with those categories preserved in the output table. No additional formatting is required in PROC SGPLOT.

❸ MortgagePayment must then be the group variable (and the bars are displayed as clusters).

❹ The graph is a bit easier to read with reference lines, GRID places reference lines at every major tick on the requested axis. GRIDATTRS= sets modifications for these line elements.

❺ The output table includes row and column totals. RowPercent is missing for these; however, it can be ensured that they are explicitly skipped with an appropriate WHERE statement.

Output 3.5.6: Chart of Conditional Percentage Statistics from PROC FREQ

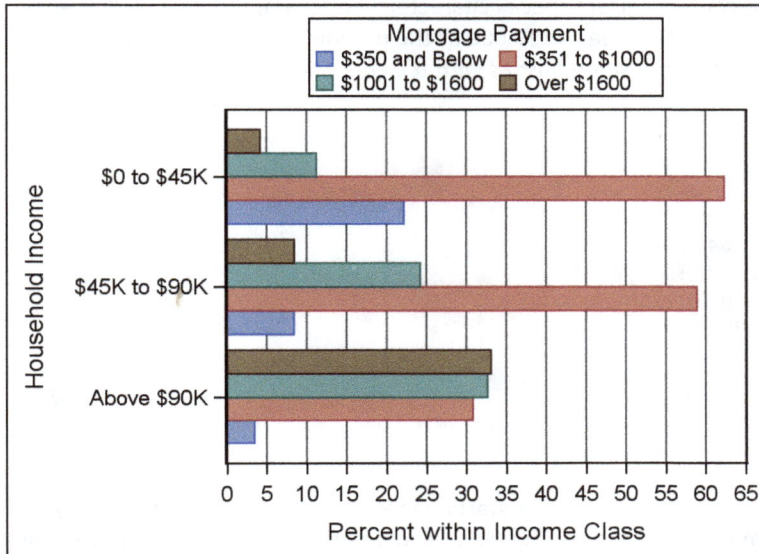

In general, results generated in any procedure can be used in a subsequent procedure or DATA step by use of an OUTPUT or ODS OUTPUT statement. This utility is exploited in various ways in subsequent sections of this book. This section also demonstrates some ODS statements for controlling output objects generated by SAS procedures. For more information about the differences between these ODS statements, see the SAS Documentation and Chapter Note 1 in Section 3.11.

3.6 Reading Raw Data with Informats

It is often necessary to create a SAS data set from a raw text file, and Section 2.8 introduced two techniques for reading raw data: simple list input and column input. Both styles have one characteristic in common – they are designed to read standard numeric and character values only. However, programmers frequently encounter nonstandard data and need to provide special instructions for reading it—these special instructions typically include SAS informats.

3.6.1 Formatted Input

Input Data 3.6.1 is similar to the file given in Input Data 2.8.8 in that it is fixed-position data; however, the columns corresponding to MortgagePayment and HHIncome contain dollar signs and commas. SAS does not allow the dollar sign and comma to appear in standard numeric data, so the column input style shown in Section 2.8 is not effective for this file. As stated in Chapter 2, it is always a good programming practice to begin by investigating the raw file to determine what type of input is appropriate.

Input Data 3.6.1: Fixed Position Raw File (Partial Listing: Columns 79 through 152)

```
        1         1         1         1         1         1
 -8----+----9----+----0----+----1----+----2----+----3----+----4----+----5--
 4 73RentedN/A                                            $0      $12,000
 1  0RentedN/A                                            $0      $17,800
 4 73Owned Yes, mortgaged/ deed of trust or similar debt  $900   $185,000
 1  0RentedN/A                                            $0       $2,000
 3 97Owned No, owned free and clear                       $0      $72,600
```

Program 3.6.1 shows one way to read the complete file, using an alternate input style, formatted input, which uses SAS informats to allow for correct interpretation of nonstandard values. Each variable listed in the INPUT statement is followed by what appears to be a format name; however, it is an informat name in this context. Some informats used in Program 3.6.1 are for standard data, such as 7. and $25. The COMMA informat is used to extract numeric values from those that contain certain special characters. For example, Comma12. is given as an instruction to MortgagePayment to read the value of $12,000 as the number 12000. When using formatted input SAS reads data based on column position, just as with the column input style shown in Section 2.8.3.

Program 3.6.1: Reading the Formatted 2005 Basic IPUMS CPS Data

```
data work.ipums2005formatted;
  infile RawData("IPUMS2005formatted.txt"); ❶
  input Serial 7. State $25. City $40. CityPop comma6. ❷
        Metro 1. CountyFips 3. Ownership $6. ❸
        MortgageStatus $40. MortgagePayment comma12.
        HhIncome comma12. Homevalue comma12.;
run;

proc print data= work.ipums2005formatted(obs=5);
  var State CountyFips Ownership MortgageStatus HHIncome;
run;
```

❶ For this code to execute successfully, as with the examples starting in Section 2.8 that used the *fileref* RawData, include a FILENAME statement which assigns the *fileref* and points to the folder where the raw data is stored.

❷ The INPUT statement now uses informats to change how SAS parses the input buffer. The default starting position is column 1, and the informat widths control how the column pointer moves to successive starting positions in the input buffer. In this example, the informat widths also set the stored widths for character variables, much like column ranges do for column input. For additional information about the interaction between informats and length, see Chapter Note 2 in Section 3.11.

❸ Like formats, informats follow the convention that the $ indicates a character informat, but the character versus numeric roles are a bit different for informats. Like formats, all informats include a period.

Output 3.6.1 Reading the Formatted 2005 Basic IPUMS CPS Data

Obs	State	CountyFips	Ownership	MortgageStatus	HHIncome
1	Alabama	73	Rented	N/A	12000
2	Alabama	0	Rented	N/A	17800
3	Alabama	73	Owned	Yes, mortgaged/ deed of trust or similar	185000
4	Alabama	0	Rented	N/A	2000
5	Alabama	97	Owned	No, owned free and clear	72600

Informats

It may seem strange that the Comma informat is used in Program 3.6.1 for values that have dollar signs instead of Dollar, but formats and informats have different roles and no direct correspondence between them is guaranteed. SAS formats change the appearance of data values, while informats provide instructions for how to interpret data values. For example, the inclusion of a dollar sign in a format name indicates it applies to character values, while in an informat it indicates an instruction to create character values. Also, note how HHIncome in Output 3.6.1 does not include any dollar signs or commas, as the informat is a reading/interpretation instruction only. If a default display of these values as dollar amounts is desired, a FORMAT statement that assigns a Dollar format to HHIncome can be included in the DATA step.

Since the purpose of an informat is to give SAS instructions on how to interpret a nonstandard value, it is important to be familiar with the different types of SAS informats, and how they affect the data being read. SAS provides many informats, and users can create custom informats with PROC FORMAT. The three most common categories of informats are character, numeric, and date and time. The syntax for an informat requires only specification of the name of a valid informat and the period. However, for all informats a width can be specified, and for numeric informats a decimal scaling factor can be included.

Character Informats

All character informats begin with a dollar sign such as $w, $CHAR, and $QUOTE. Each different informat provides SAS with different instructions for how to parse values into SAS data. For these common character informats, a summary of their instructions is below.

- $w removes leading/trailing blanks and converts a field with a single period to a missing value.
- $CHAR removes trailing blanks only and does not convert a single period to a missing value.
- $QUOTE removes matched quotation marks (single or double) from around a field.

When using formatted input, the inclusion of a width in a character informat sets the length of the variable (if it is not already set) and determines the exact number of columns read from the input buffer. If no width is specified, then SAS reads the default number of columns. For $CHAR and $QUOTE, the default length is eight columns; for $w the width must be specified. To see the effect of the width when using formatted input, consider the partial listing shown in Input Data 3.6.2 that displays the Destination (columns 1-3), Date (columns 4-13), FirstClass (columns 14-15), and EconClass (columns 16-18).

Input Data 3.6.2: Example Flight Data

```
----+----1----+----2
LAX12/11/200020137
DFW12/11/200020131
DFW12/12/200015170
```

Compare the following two INPUT statements and their results, shown in Table 3.6.1, to see the effect of the width for character informats in formatted input.

```
1.  input Destination $3. Date $3. FirstClass $3. EconClass $3.;
2.  input Destination $3. Date $10. FirstClass $2. EconClass $3.;
```

Setting the widths to three in each of the informats has the predictable effect shown in the first row of Table 3.6.1—each variable only contains three characters. However, note which three characters – in each case the INPUT statement parses the input buffer by reading the next three characters. As a result, only Destination is read in correctly. In contrast, the second INPUT statement uses the correct widths, correctly reading the values for each variable.

Table 3.6.1: Comparing Effects of Informat Widths

INPUT Statement	Destination	Date	FirstClass	EconClass
1	LAX	12/	11/	200
2	LAX	12/11/2000	20	137

Numeric Informats

As with character informats, numeric informats provide SAS instructions on how to parse the values in the input buffer. Numeric informats have more options compared to character formats because of the variety of ways numeric data is encoded in raw files. Like numeric formats, numeric informats can include a whole number after the period to indicate decimal digits; however, the behavior is very different for informats. Other than the standard numeric informat, two of the most common numeric informats are COMMA and PERCENT, with their rules shown below, assuming no value is specified after the period.

- COMMA removes embedded characters (commas, blanks, dollar signs, percent signs, closing parentheses), and converts an opening parenthesis that precedes the first digit to a minus sign.

- PERCENT takes the same actions as COMMA and also divides the value by 100 if a percent sign is present anywhere in the value.

Including a value after the period in a numeric informat results in decimal scaling, and Table 3.6.2 shows its effects for both formats on several values.

Table 3.6.2: Comparison of COMMA and PERCENT Informats for Various Values

Raw Value	COMMA5.	COMMA5.1	PERCENT5.	PERCENT5.2
1,234	1234	123.4	1234	12.34
$250	250	25	250	2.5
%350	350	35	3.5	0.035
(100)	-100	-10	-100	-1
12.0%	12	12	0.12	0.12
%9.7$	9.7	9.7	0.097	0.097

COMMA5. operates as described above, while COMMA5.1 takes the additional step of dividing by 10 for those values that did not already contain a decimal point. In general, for raw values without a decimal point, SAS first strips out the extraneous characters and then divides by 10^d where d is the scaling value provided in an informat such as COMMA$w.d$.

The PERCENT informat strips out the same characters as the COMMA informat. If one of those characters is the percent sign, as it is for the value of %350, then SAS also converts the number from a percent to a proportion by dividing the value by 100. If, in addition, a decimal scaling factor is provided, then all values without a decimal point are further scaled. As seen by looking at the %350 value, when using PERCENT5.2 as the informat, the value is scaled by $10^2 = 100$ in addition to converting from percentages to proportions. As a result, be careful when specifying a decimal scaling factor as part of an informat.

Date and Time

Another important category of informats applies to SAS dates, times, and datetimes. While dates, times, and datetimes can always be read as character data, this severely limits the utility of those fields. For example, sorting dates stored as character values places 17APR2018 after 01AUG2018, and July 4, 1776 comes before June 4, 1776 (and after August 1, 2019) due to the alphanumeric sorting of character data. Also, it is impossible to compute any span of time from date or time values stored as character. SAS has the ability to interpret numeric values as dates, and informats provide instructions for reading various types of dates into a numeric variable. SAS interprets dates as the number of days since January 1, 1960; times as the number of seconds from midnight; and datetimes as the number of seconds since midnight on January 1, 1960. Though there are many common date, time, and datetime informats, Table 3.6.3 displays some of the most common, along with the date form expected and the interpreted value.

Table 3.6.3: Demonstration of Common Date, Time, and Datetime Informats

Informat	Raw	Stored	Raw	Stored	Raw	Stored
DATE9.	01JAN1960	0	31DEC1959	-1	04MAR2018	21247
DDMMYY10.	01/01/1960	0	31-12-1959	-1	04:03:2018	21247
MMDDYY10.	01/01/1960	0	12/31/1959	-1	04/03/2018	21277
TIME8.	12.56	46560	125.6	450360	11:13 pm	83580
DATETIME18.	03MAR2018T16:49:00	1835714940				

For a discussion of additional types of informats, see the SAS Documentation and Chapter Note 3 in Section 3.11.

3.6.2 Column Pointer Controls

As discussed in Section 3.6.1, formatted input uses column positions to determine what columns to read from the input buffer for each variable. Unlike column input where starting and ending columns to read are explicitly stated, thus also defining how many columns to read, the informat in formatted input only specifies the number of columns to read. As such, the starting column position is implicitly defined relative to the ending column of the previous variable. Table 3.6.4 shows the position of the column pointer for the data in Input Data 3.6.2 when using the following INPUT statement.

```
input Destination $3. Date $10. FirstClass $2. EconClass $3.;
```

Table 3.6.4: Column Positions Determined By Formatted Input

Variable	Starting Column	Ending Column
Destination	1	3
Date	4	13
FirstClass	14	15
EconClass	16	18

Recall that for all input styles, SAS automatically advances the column pointer one column after reading each variable. However, to produce a more human-readable file, it is common for fixed-position files to have data where the fields are separated by one or more spaces. Input Data 3.6.3 below shows such a structure.

Input Data 3.6.3: Partial Listing of Flight Data with Spacing

```
----+----1----+----2---+
LAX   12/11/2000   20   137
DFW   12/11/2000   20   131
DFW   12/12/2000   15   170
```

Below is one possible version of the INPUT statement that uses only informats to read Input Data 3.6.3 correctly.

```
input Destination $3. Date $12. FirstClass $4. EconClass $5.;
```

Note the informat widths are defined in a manner that covers the additional spacing separating the fields. However, to provide more flexibility when reading raw data, SAS provides column pointer controls to position the column pointer in the input buffer. Note that these controls can be used with any input style, not just formatted input, and they are always executed after the one-column advancement of the pointer that SAS automatically carries out after reading a variable. There are two types of column pointer controls: absolute and relative. The syntax for both controls, along with examples, are shown in Table 3.6.5.

Table 3.6.5: Syntax and Examples of Column Pointer Controls

Control Type	Generic Syntax	Example	Initial column pointer location	Subsequent column pointer location	Description
Absolute	@n	@6	10	6	Moves column pointer to column 6
Relative	+n	+7	10	17	Moves column pointer right 7 columns
	+(-n)	+(-5)	10	5	Moves column pointer left 5 columns
	+(expression)	+(x-y)	10	10+(x-y)	Moves column pointer x-y columns to the left/right based on sign of x-y

In addition to providing a method to read fixed-position files containing additional spacing, column pointer controls also give the ability to read variables from the input buffer in any order. For the data shown in Input Data 3.6.3, the INPUT statement below reads the data set so that Date comes first, Destination comes last, and the passenger counts remain in the same order.

```
input ❶ @6 ❷ Date mmddyy10. ❸
           +2 ❹ FirstClass 2.
           +2 EconClass 3.
       @1 ❺ Destination $3.;
```

❶ Vertical alignment of pointer controls improves code readability, but is not required. However, pointer controls must precede the target variable in the INPUT statement.

❷ @6 moves the column pointer to column 6 before Date is read.

❸ Use MMDDYY to read in dates of this form as numeric values, the width of 10 specifies that columns 6 through 15 are read.

④ After SAS automatically increments by 1 column to column 16 when the parsing of date is complete, +2 moves the column pointer to the right 2 columns to column 18, which is the correct position to begin reading FirstClass.

⑤ @1 moves the column pointer back to column 1 before Destination is read.

3.6.3 Modified List Input

Just as formatted input reads fixed-position fields that require informats, modified list input reads delimited fields that require informats. Thus, modified list input is a more flexible input style for handling delimited data than simple list input. However, as shown in Section 3.6.2, the default behavior of an informat in the INPUT statement is to read the field based on column position and informat width. To use modified list input, an additional piece of syntax is required: a format modifier. Table 3.6.6 shows the three format modifiers available in SAS.

Table 3.6.6: Format Modifiers and Their Descriptions

Modifier	Symbol	Description
Colon	:	Read until next delimiter or end-of-line marker, whichever comes first
Ampersand	&	Read until next delimiter followed by a space or end-of-line marker, whichever comes first
Tilde	~	Read until next delimiter not enclosed in quotation marks or end-of-line marker, whichever comes first When the DSD option (first introduced in Section 2.8) is not present, the tilde modifier is equivalent to the colon modifier

Recall that simple list input reads columns from the input buffer until the INPUT statement finds a delimiter or end of line, whichever comes first. Thus, the format modifiers all exhibit this behavior because modified list input also reads all values from the input buffer. Note that the format modifiers do not impact the length of the stored value—that is controlled by the informat itself, by LENGTH statements, or by other statements that control variable length.

Input Data 3.6.4 shows a partial view of the comma-delimited IPUMS 2005 basic data. It shows modified values of MortgageStatus, Ownership, and HHIncome for one record. Reading this record using the three informat modifiers helps to clarify the differences between them.

Input Data 3.6.4: Partial Listing of IPUMS 2005 Basic Data

```
----+----1----+----2----+----3----+----4----+----5----+----6----+
"Yes,mortgaged/ deed of trust or similar debt", Owned,$185,000
```

```
A.  input MortgageStatus : $50. Ownership : $10. HHIncome : comma8.;
B.  input MortgageStatus & $50. Ownership : $10. HHIncome & comma8.;
C.  input MortgageStatus ~ $50. Ownership : $10. HHIncome & comma8.;
D.  input MortgageStatus $50. Ownership $10. HHIncome comma8.;
```

Table 3.6.7 shows the results of applying the format modifiers using the above INPUT statements, assuming the delimiter is specified as a comma in the INFILE statement, and assuming this single record is read in from an external file (rather than using instream data). The table also indicates how the results differ based on whether the DSD option in the INFILE statement is in effect except for INPUT statement D where DSD has no effect since list input is not in use.

Table 3.6.7: Results of Various INPUT Statements on the Data from Input Data 3.6.4

	INPUT	DSD?	MortgageStatus	Ownership	HHIncome
1	A	No	"Yes	mortgaged/	.
2	B	No	"Yes,mortgaged/ deed of trust or similar debt"	Owned	185000
3	C	No	"Yes	mortgaged/	.
4	A	Yes	Yes,mortgaged/ deed of trust or similar debt	Owned	185
5	B	Yes	Yes,mortgaged/ deed of trust or similar debt	Owned	185000
6	C	Yes	"Yes,mortgaged/ deed of trust or similar debt"	Owned	185000
7	D	N/A	"Yes,mortgaged/ deed of trust or similar debt", Ow	ned,$185,0	.

Focusing on the first two rows of Table 3.6.7, the difference between the colon and ampersand modifiers is evident. Because the commas contained in the MortgageStatus and HHIncome fields are not followed by a space, the ampersand modifier forces SAS to keep reading past the fields. Of course, this is only effective because those two commas are not followed by a space *and* MortgageStatus is separated from Ownership by a comma followed by a space. The third row shows that without the DSD option in use, the tilde format modifier works exactly as the colon modifier does.

Rows 4 through 6 show the effect of including the DSD option with the three format modifiers. As a result of DSD ignoring delimiters in a quoted string, the value of MortgageStatus is read in its entirety in all three rows and is stored without the quotation marks in rows 4 and 5 since the tilde modifier is not used. Row 6 shows that the tilde modifier prevents DSD from stripping the quotation marks. Since the HHIncome field is not quoted, the DSD option does not prevent the comma in that field from being interpreted as a delimiter, so the ampersand modifier is necessary to read that field correctly.

Finally, row 7 highlights the difference between modified list input and formatted input. In modified list input, as in simple list input, SAS reads until it encounters a delimiter (or end-of-line marker). Thus, the width specified in the informat does not determine how many characters are read from the input buffer—that is determined by the position of the delimiters. However, in formatted input the informat widths do specify how many columns SAS reads from the input buffer. Leaving out the format modifier does not cause a syntax error; however, it changes the input style from delimiter-based to position-based. SAS does provide an INFORMAT statement when using modified list input; see Chapter Note 4 in Section 3.11 for additional notes on its usage.

Row 7 also provides an example of another issue when reading raw data. Comparing the results to Input Data 3.6.4 may be confusing at first since the value of HHIncome appears as missing in Table 3.6.7 despite there being two characters – both zeros – left for the INPUT statement to read from the input buffer. This occurs because this is an incomplete record; there are only two characters left in the raw record after parsing MortgageStatus and Ownership, but the informat requests eight. The next section discusses multiple techniques for handling incomplete records.

3.7 Handling Incomplete Records

Regardless of the input style used—simple list, modified list, column, or formatted—often the source data has missing values. If the missing values are explicitly denoted, such as with a period or other character, then it is often possible to read the data with only minor adjustments to the input styles presented earlier in this chapter and in Chapter 2. In general, mishandling of missing values results in serious data integrity issues for the resulting data set.

3.7.1 Using DSD to Read Delimited Data with Missing Values

Consider the data shown in Input Data 3.7.1 below, which revisits the flight data seen earlier with several values omitted. Despite some missing values, delimiters allow for identification of where missing data occurs. For the second record, no data follows the final comma; indicating a missing value at the end of the line. For the third record, the sequential commas indicate a value is missing that should appear between them.

Input Data 3.7.1: Example Flight Data with Some Values Set to Missing

```
----+----1
20,LAX,137
20,DFW,
15,,170
5,DF,
14,LAX,196
```

Program 3.7.1 below uses the DSD option in the INFILE statement and simple list input to read the data set in Input Data 3.7.1 correctly. By default, SAS ignores sequential delimiters—both simple and modified list input scan to the next non-delimiter character—but the DSD option interprets sequential delimiters (or a delimiter followed by the end of the line) as an indication that a value is missing.

Program 3.7.1: Using DSD to Read Input Data 3.7.1

```
data work.miss01;
   infile RawData('FlightsMiss01.txt') dsd;
   input FirstClass Destination $ EconClass;
run;

proc print data = work.miss01;
run;
```

Output 3.7.1: Using DSD to Read Input Data 3.7.1

Obs	FirstClass	Destination	EconClass
1	20	LAX	137
2	20	DFW	.
3	15		170
4	5	DF	.
5	14	LAX	196

Alternatively, consider the data shown in Input Data 3.7.2, which is nearly identical to the data in Input Data 3.7.1; the only change is the removal of the comma at the end of the second line. Reading this data set with Program 3.7.2 yields the data set shown in Output 3.7.2.

Input Data 3.7.2: Example Flight Data with Some Values Missing at the End of a Line Not Denoted

```
----+----1
20,LAX,137
20,DFW
15,,170
5,DF,
14,LAX,196
```

Program 3.7.2: Using DSD to Read Input Data 3.7.2

```
data work.miss02;
  infile RawData('FlightsMiss02.txt') dsd;
  input FirstClass Destination $ EconClass;
run;

proc print data = work.miss02;
run;
```

Output 3.7.2: Using DSD to Read Input Data 3.7.2

Obs	FirstClass	Destination	EconClass
1	20	LAX	137
2	20	DFW	15
3	5	DF	.
4	14	LAX	196

As noted earlier, failing to denote the missing value leads to consequences in the resulting data unless the DATA step is modified accordingly. In this case, the consequence is quite clear; not all records appear to have never been read. However, the first note included in Log 3.7.2 indicates SAS read five records from the raw file.

Log 3.7.2: Partial Log from Program 3.7.2

```
NOTE: 5 records were read from the infile RAWDATA('FlightsMiss02.txt').
      The minimum record length was 5.
      The maximum record length was 10.
NOTE: SAS went to a new line when INPUT statement reached past the end of
      a line.
```

Recall that both simple and modified list input read a value until it encounters the next delimiter or end-of-line marker. In the second record, SAS reads all the way to the end-of-line marker in order to completely read the value of Destination. Thus, when SAS tries to parse the input buffer to determine the value of EconClass, there are no columns remaining for it to use. When reading raw data from an external file, if no columns remain in the input buffer to populate remaining variables in the INPUT statement, the default behavior is to move to the next record and continue parsing from there. The second note included in Log 3.7.2 is a result of this behavior; SAS prints this note to the log any time it reads past the end of a line when populating the PDV.

3.7.2 Reading Past the End of a Line

When reading Input Data 3.7.2 with Program 3.7.2, the unexpected results arise as SAS is instructed to read three variables from each record, while the second record only has two values indicated. Any time a raw record has insufficient data to populate all variables, SAS looks for an instruction about what to do when it encounters an end-of-line marker before one is expected. The behavior noted for Program 3.7.3 is a result of the default option of FLOWOVER for the INFILE statement. Other common options to control how SAS interprets the end-of-line marker in raw data are MISSOVER and TRUNCOVER.

FLOWOVER

When the FLOWOVER option is in effect, SAS continues reading records until all variables across all INPUT statements are populated. Returning to the example shown in Program 3.7.2, Output 3.7.2, and Log 3.7.2, the following steps begin at the second record and describe why the final data set contains four records, even though five are read from the raw file.

1. SAS reads the second record into the input buffer.
2. Simple list input assigns the value prior to the first delimiter, 20, to FirstClass.
3. Simple list input now assigns the value prior to the end-of-line marker, DFW, to Destination.

4. Given that EconClass is not populated, and the end-of-line marker is reached, FLOWOVER causes SAS to read in the 3rd record and places the column pointer in column 1.
5. Simple list input assigns the value prior to the first delimiter, 15, to EconClass.
6. The fully populated PDV is output to the data set, and SAS returns to the top of the DATA step and resets the input buffer and PDV.
7. The next record from the raw file is loaded into the input buffer and parsing begins on it.

Step 6 causes the remaining information about the third record to be lost, with the third iteration of the DATA step using only information from the fourth record of the external file. By default, once SAS loads a raw record into the input buffer, SAS only uses it to create a single record in the resulting SAS data set – even if that record does not use the input buffer in its entirety.

MISSOVER and TRUNCOVER

MISSOVER and TRUNCOVER are alternatives to the default INFILE statement option, FLOWOVER. As their names imply, they differ in how they handle raw records containing insufficient information – either assigning missing or truncated values. Some additional INFILE statement options relating to these issues are discussed briefly in Chapter Note 5 in Section 3.11. To demonstrate the usage of MISSOVER and TRUNCOVER, and to compare them to FLOWOVER, consider the data set shown in Input Data 3.7.3. It shows a modification of Input Data 3.7.2, with a fixed-position structure instead of a delimited structure.

Note the missing value in the second record is not denoted; multiple spaces were used to maintain the column alignment. Using DSD is of no help since it causes the INPUT statement to interpret those five spaces (columns 3-7) as denoting four missing values, one between each pair of spaces. Instead, use column or formatted input to specify which columns SAS should read. When no data values are present in the specified columns, SAS sets the value to missing using either of column or formatted input

Input Data 3.7.3: Fixed-Position Flight Data with Some Missing Values

```
----+----1
20 LAX 137
15     170
20 DFW
14 LAX 196
5  DF
```

Program 3.7.3 attempts to read in this file using column input with no other adjustments; for example, without changing the default option from FLOWOVER. Output 3.7.3 shows the results from this attempt.

Program 3.7.3: Reading Input Data 3.7.3 With Column Input and FLOWOVER

```
data work.miss03a;
   infile RawData('FlightsMiss03.txt');
   input FirstClass 1-2 Destination $ 4-6 EconClass 8-10;
run;

proc print data = work.miss03a;
run;
```

Output 3.7.3: Incorrect Results Due to Incomplete Records

Obs	FirstClass	Destination	EconClass
1	20	LAX	137
2	15		170
3	20	DFW	14

First, note the INPUT statement in Program 3.7.3 correctly determines Destination is missing for the second record since columns four through six are all blanks. However, similar to the results from Program 3.7.2, SAS moves the column pointer to the beginning of the third raw record in an attempt to complete the second observation in the SAS data set due to the incompleteness of the second record in the raw file. Here, the end-of-line marker occurs immediately after column six, so there no blanks for the INPUT statement to recognize as a missing value. Similarly, in the fifth record, the end-of-line marker occurs immediately after column 5 and results in a note in the log as well as a data error since there are no more records from which the INPUT statement can read data. Log 3.7.3 shows the relevant portions of the log due to the effects of FLOWOVER on this final record. As always, be sure to check the log regularly—notes can be more serious than warnings.

Log 3.7.3: Results of FLOWOVER When the Final Record is Incomplete

```
NOTE: LOST CARD.
FirstClass=5 Destination=  EconClass=. _ERROR_=1 _N_=4
```

Recall column input (and formatted input) are position-based. As such, the column specification (8-10) for EconClass instructed SAS to read exactly three columns from the input buffer. However, since the input buffer for the third record stopped at column 6, the FLOWOVER option causes SAS to move to the next record and place the column pointer in column one. Rather than advancing to column eight, SAS reads the next three columns it encounters, selecting the value 14 from the fourth raw record as the EconClass value for the third record in the output data set. When returning to the top of the DATA step, SAS resets the input buffer and loads the next record from the raw file into it when the execution phase reaches the INPUT statement again, which then reads in the next record. Program 3.7.4 provides a second, and more successful, approach to reading Input Data 3.7.3 by employing the MISSOVER option.

Program 3.7.4: Reading Input Data 3.7.3 Using MISSOVER

```
data work.miss03b;
  infile RawData('FlightsMiss03.txt') missover;
  input FirstClass 1-2 Destination $ 4-6 EconClass 8-10;
run;

proc print data = work.miss03b;
run;
```

Output 3.7.4: Reading Input Data 3.7.3 Using MISSOVER

Obs	FirstClass	Destination	EconClass
1	20	LAX	137
2	15		170
3	20	DFW	.
4	14	LAX	196
5	5		.

Several differences between Output 3.7.3 and Output 3.7.4 are apparent—notably in the third and fifth records. Using MISSOVER results in a correct reading of the third record by preventing the INPUT statement from reading information from the next raw record; instead it sets the value of EconClass to missing. Similarly, in the fifth record EconClass is again set to missing since its associated columns do not exist in the input buffer. However, note that Destination is also missing—the value DF from the input buffer is interpreted as missing as well. To further explore the difference between FLOWOVER, MISSOVER, and TRUNCOVER, compare and contrast the results of Program 3.7.4, which uses MISSOVER, to the results of Program 3.7.5, which uses TRUNCOVER.

Program 3.7.5: Reading Input Data 3.7.3 Using TRUNCOVER

```
data work.miss03c;
  infile RawData('FlightsMiss03.txt') truncover;
  input FirstClass 1-2 Destination $ 4-6 EconClass 8-10;
run;

proc print data = work.miss03c;
run;
```

Output 3.7.5: Reading Input Data 3.7.3 Using TRUNCOVER

Obs	FirstClass	Destination	EconClass
1	20	LAX	137
2	15		170
3	20	DFW	.
4	14	LAX	196
5	5	DF	.

Unlike FLOWOVER, both MISSOVER and TRUNCOVER prevent SAS from advancing past the end-of-line marker when it encounters an incomplete record. Notice that the only difference between Outputs 3.7.4 and 3.7.5 occurs in the fifth record where the Destination is incomplete—DF is present instead of DFW.

When MISSOVER encounters an incomplete record, it sets the current variable to missing as well as all remaining variables in the INPUT statement. In the fifth record, that corresponds to Destination and EconClass, respectively. In contrast, when TRUNCOVER encounters an incomplete record, it reads any columns that are available, even if there are fewer than the number requested by the INPUT statement. SAS assigns a value to the current variable after parsing the partial set of columns as best as possible for the given instructions in the INPUT statement, and all further variables are set to missing. In the fifth record, that corresponds to setting Destination equal to the two columns read (even though the column input requests three columns) and EconClass is still set to missing. Neither MISSOVER nor TRUNCOVER produce the data error and lost card note that results from FLOWOVER.

When using list input (simple or modified) it is not possible to read a partial value. List input determines values only by the position of delimiters and end-of-line markers in these cases so there is no pre-defined number of columns for SAS to read. As a result, there is no difference between MISSOVER and TRUNCOVER when using list input. Only when using column or formatted input is it possible for SAS to read a partial value. In those cases, a determination of whether to include a partial value (TRUNCOVER) or set it to missing (MISSOVER) is necessary.

3.8 Reading and Writing Raw Data with the IMPORT and EXPORT Procedures

In some cases, the IMPORT procedure provides a method to read raw data in addition to the DATA step-based techniques in Sections 2.8 and 3.7. Specifically, PROC IMPORT provides the ability to read delimited data files, subject to a few constraints. Program 3.8.1 revisits Program 2.8.4 and demonstrates the basic usage of PROC IMPORT.

Program 3.8.1: Simple Application of PROC IMPORT

```
filename rawCSV "—insert path here--\ipums2005basic.csv"; ❶

proc import file = rawCSV ❷ dbms = csv ❸ out = work.Import01 ❹ replace ❺;
run;

proc contents data = work.Import01 varnum; ❻
run;
```

❶ Unlike the INFILE statement in the DATA step, the IMPORT procedure does not support using a *fileref* that points to a folder such as the RawData *fileref* used in this chapter and in Chapter 2.

❷ The FILE= option identifies the raw file via either a *fileref* or a quoted path and filename. Recall, if a relative path is present, SAS builds the full path off from the working directory. While the quotation marks on a file specification are not required in some circumstances, their presence reduces the need for program maintenance when file names change to a value that includes prohibited characters. For a full discussion, see the SAS Documentation.

❸ DBMS= specifies the style of data referenced by the FILE= option. In this case, the CSV keyword denotes a comma-separated file. As a result, the delimiter is automatically changed from the default – a blank – to a comma. In Base SAS, PROC IMPORT supports only delimited data or JMP program files. Reading additional file types, such as Excel spreadsheets, requires a SAS/ACCESS license.

❹ The OUT= option names the SAS data set in which PROC IMPORT saves the imported records.

❺ By default, PROC IMPORT does not overwrite an existing data set. If the Import01 data set already exists in the Work library, then this IMPORT procedure produces a note in the log indicating SAS canceled the procedure.

❻ The CONTENTS procedure displays the descriptor portion of the Import01 data set. The VARNUM option ensures the variables appear in the same order they appear in the data set.

Unlike the DATA step, PROC IMPORT does not provide a statement for naming the newly created variables. Instead, it determines names, types, lengths, and all other attributes automatically. Output 3.8.1 shows the partial results of this IMPORT procedure—specifically, it shows the Variables table containing the variable-level metadata.

Output 3.8.1: Variables Table from Imported CSV Data Set

Variables in Creation Order					
#	Variable	Type	Len	Format	Informat
1	_2	Num	8	BEST12.	BEST32.
2	Alabama	Char	7	$7.	$7.
3	Not_in_identifiable_city__or_siz	Char	40	$40.	$40.
4	_0	Num	8	BEST12.	BEST32.
5	_4	Num	8	BEST12.	BEST32.
6	_73	Num	8	BEST12.	BEST32.
7	Rented	Char	6	$6.	$6.
8	N_A	Char	47	$47.	$47.
9	_12000	Num	8	BEST12.	BEST32.
10	_9999999	Num	8	BEST12.	BEST32.

As Output 3.8.1 shows, the data set has several issues. The variable names are derived from the intended first record, with modifications made to the values to ensure they follow SAS variable naming conventions. These include: adding the underscore as the first character for those that start with a digit, replacing special characters with underscores, and truncating those longer than 32 characters. The second variable—legally but inappropriately named Alabama—represents the state names, but is assigned only a length of seven, which is insufficient to handle longer names such as North Carolina. If two or more fields have the same value, SAS uses names in the form VAR*n* where *n* is the position of the variable during the import – for example, VAR9 for the ninth variable.

These issues occur because when the IMPORT procedure reads a delimited file it simply generates a DATA step, which SAS writes to the log for easy review. This DATA step defines the attributes, such as lengths and formats, based on what SAS finds from a limited look into the raw file. The INFILE statement that appears in this DATA step contains familiar options such as MISSOVER, DSD, and FIRSTOBS= in addition to other options such as DELIMITER=, which is an alias for the DLM= option from Section 2.8. By default, SAS scans the first 20 rows to determine variable attributes. If a numeric or date/time informat is applicable, SAS uses it and sets the variable as numeric. Otherwise PROC IMPORT stores the values in a character variable. Program 3.8.2 modifies Program 3.8.1 to read a tab-delimited version of the file and adds options to produce more desirable results.

Program 3.8.2: Customizing PROC IMPORT

```
filename rawTab "—insert path here--\ipums2005basic.txt"; ❶

proc import file = rawTab dbms = tab ❷ out = work.Import02 replace;
   getnames = no; ❸
   guessingrows = 250000; ❹
run;

proc contents data = work.Import02;
run;

proc print data = work.Import02(obs = 5); ❺
   var var1-var3 var9-var11; ❻
run;
```

❶ Instead of ipums2005basic.csv, Program 3.8.2 uses the tab-delimited version, so a new *fileref* is assigned.

❷ Using the TAB keyword in the DBMS= option reads tab-delimited files without the need to specify the delimiter using its hexadecimal representation as in Section 2.8.

❸ PROC IMPORT expects the first row of the raw file to include variable names. Use the GETNAMES statement to specify that no variable names appear in the raw file (the default is GETNAMES = YES).

❹ The GUESSINGROWS= option specifies how many rows the IMPORT procedure should scan to determine how to best set the variable attributes. In Program 3.8.1, the variable containing state names has a length of seven because the first 20 records are all from Alabama. In this case, 250,000 records are necessary before identifying long values such as District of Columbia.

❺ After printing the descriptor portion with PROC CONTENTS, the PRINT procedure writes out the first five records.

❻ As discussed in Chapter Note 3 in Section 1.7, the single-dash selects all variables with the specified prefix and for which the numeric suffix is in the specified range, inclusive. Here, the single-dash selects Var1, Var2, and Var3 then selects Var9, Var10, and Var11.

The GETNAMES statement interacts with another statement: DATAROW. In PROC IMPORT, the DATAROW statement performs the same action as the FIRSTOBS= option from the INFILE statement in the DATA step. When the default GETNAMES = YES is in effect, the default value of DATAROW is two. When GETNAMES = NO is in use, the default value of DATAROW is one. Using DATAROW = *n* to provide a user-specified value is valid regardless of the inclusion of the GETNAMES statement; however, the value cannot be set to one when GETNAMES=YES. Output 3.8.2A displays the results of the CONTENTS procedure (Variables table only) and Output 3.8.2B contains the results of the PRINT procedure.

Output 3.8.2A: Variables Table from Imported Tab-Delimited Data Set

Alphabetic List of Variables and Attributes					
#	Variable	Type	Len	Format	Informat
1	VAR1	Num	8	BEST12.	BEST32.
2	VAR2	Char	20	$20.	$20.
3	VAR3	Char	40	$40.	$40.
4	VAR4	Num	8	BEST12.	BEST32.
5	VAR5	Num	8	BEST12.	BEST32.
6	VAR6	Num	8	BEST12.	BEST32.
7	VAR7	Char	6	$6.	$6.
8	VAR8	Char	45	$45.	$45.
9	VAR9	Num	8	BEST12.	BEST32.
10	VAR10	Num	8	BEST12.	BEST32.
11	VAR11	Num	8	BEST12.	BEST32.

Output 3.8.2B: Content Portion of Imported Tab-Delimited Data Set (First Five Records)

Obs	VAR1	VAR2	VAR3	VAR9	VAR10	VAR11
1	2	Alabama	Not in identifiable city (or size group)	0	12000	9999999
2	3	Alabama	Not in identifiable city (or size group)	0	17800	9999999
3	4	Alabama	Not in identifiable city (or size group)	900	185000	137500
4	5	Alabama	Not in identifiable city (or size group)	0	2000	9999999
5	6	Alabama	Not in identifiable city (or size group)	0	72600	95000

Since no variable names appear in the raw file, and no statement is available to name them in PROC IMPORT, SAS indexes the variables as VAR1 through VAR11. (Chapter 4 introduces a method for manually renaming each of the variables.) In addition, the IMPORT procedure offers no FORMAT, LABEL, or similar statements to customize the resulting data set. However, since the DATA step code is placed directly into the log, it is possible to copy that code to the Editor window and modify it to produce more desirable results. One final example appears in Program 3.8.3 to demonstrate how to read other delimited files.

Program 3.8.3: Reading General Delimited Files

```
proc import file = rawTab dbms = dlm ❶ out = work.Import02 replace;
  getnames = no;
  guessingrows = 250000;
  delimiter = '09'x; ❷
run;
```

❶ Instead of the CSV or TAB keywords, Program 3.8.3 uses the DLM keyword to indicated delimited data without specifying the delimiter.

❷ The DELIMITER statement allows specification of the delimiter(s). In this case, the only delimiter is the tab, and the data set created in Program 3.8.3 is identical to the data set from Program 3.8.2. However, this approach is more general in that any delimiter(s) can be used as with the DLM= option in Section 2.8. See the Chapter Note 6 in Section 3.11 for further discussion.

PROC IMPORT offers a simple method for reading standard delimited files, but at the expense of flexibility. The IMPORT procedure cannot read fixed-position data as was done in Section 2.8 with column input or Section 3.6 with formatted input. Furthermore, there is no control over setting variables as character when numeric values are used – as with the CountyFIPS information in the IPUMS data – or setting informats and formats, including custom versions defined with PROC FORMAT. In those cases, as with the more advanced data reading in Chapters 6 and 7, the DATA step is invaluable.

If SAS/ACCESS is licensed, then PROC IMPORT is also capable of reading a single sheet from a Microsoft Excel workbook or a single table from a Microsoft Access database. However, if SAS/ACCESS is licensed, the LIBNAME statement supports engines that can directly connect to all sheets in an Excel workbook, all tables from an Access database, or similar sets for other spreadsheet and database products. The DATA step can then be used to read any of this data, including simultaneously using multiple tables from potentially different sources, making it more versatile than PROC IMPORT in this scenario as well.

The data-reading process provided by PROC IMPORT is easy to reverse by using PROC EXPORT. It is subject to the same limitations as PROC IMPORT in that Base SAS can only create delimited files or JMP files. With SAS/ACCESS, other options, such as Excel Workbooks, are available. The syntax of PROC EXPORT is nearly identical to that of PROC IMPORT, as shown in Program 3.8.4.

Program 3.8.4: Writing a CSV with PROC EXPORT

```
proc export outfile = "IpumsOut.csv" ❶ dbms = csv ❷
            data = work.Import01 ❸ replace ❹;
run;
```

❶   Instead of the FILE= option for reading a file in PROC IMPORT, PROC EXPORT uses the OUTFILE= option to name the raw file to which the procedure writes the data. If the OUTFILE= option only includes a file name, then PROC EXPORT places this file in the working directory.

❷   The DBMS= option works exactly as in PROC IMPORT. If DBMS=DLM, then the DELIMITER= option must also appear in the PROC EXPORT statement.

❸   PROC EXPORT writes the contents of the data set in the DATA= option to the file named in OUTFILE=.

❹   The REPLACE statement instructs SAS to overwrite the file named in OUTFILE= if it already exists. If the file exists and the REPLACE option does not appear, then SAS does not overwrite the file and thus it does not export any data.

In addition to PROC IMPORT, SAS provides an Import Wizard accessible via **File → Import Data** in the SAS windowing environment or via the **Import Data Utility** in SAS Studio. It provides an interactive user interface that imports the same data types as PROC IMPORT, but with the additional limitation that multiple delimiters are not allowed. However, the wizard does provide an option to save the generated PROC IMPORT code to a user-specified location. The EXPORT procedure has a similar wizard.

3.9 Simple Data Inspection and Cleaning

Inspecting data sets for inconsistencies and anomalies is often required to produce appropriate results. The MEANS and FREQ procedures are used in this chapter and in Chapter 2 to produce various summaries; however, those summary methods are also effective diagnostic tools. Program 3.9.1 reads a tab-delimited text file and builds a SAS data set from it (presuming a FILENAME statement correctly defines the *fileref* RawData).

Program 3.9.1: Reading IPUMS2015Dirtied.dat

```
data work.Dirty2005basic;
  infile RawData('IPUMS2005Dirtied.dat') dlm='09'x dsd;
  input Serial $ CityPop : comma. Metro CountyFips Ownership : $50.
        MortgageStatus : $50. HHIncome : comma. HomeValue : comma.
        City : $50. MortgagePayment : comma.;
  format hhIncome homeValue mortgagePayment dollar16.;
run;
```

A check of the log, which good programming practice dictates should always be done, does reveal several invalid data notes; however, there are problems with this data that go beyond those notes. Even with the invalid data noted, the information in the log may not be sufficient to understand exactly what problems have arisen.

3.9.1 MEANS and FREQ as Diagnostic Tools

Character data can be difficult to work with at times because what humans see as unimportant differences are taken by SAS as clear and absolute distinctions among character data values. Items like casing and spacing can cause two character values that are intended to be the same to register as entirely different data values. Since the FREQ procedure is designed to summarize the membership in all categories for a given variable, it is an excellent tool for determining what distinct values of a variable are present in the data set. Program 3.9.2 looks at the City, Ownership, and MortgageStatus variables, and the associated output reveals some unwanted distinctions.

Program 3.9.2: A Check of Three Character Variables in Dirty2015basic

```
ods exclude all; ❶
proc freq data= work.Dirty2005basic;
  table city;
  ods output onewayfreqs= work.freqs;
run;

ods exclude none; ❶
proc print data= work.freqs(obs=10) label noobs; ❷
  var city--cumPercent;
run;
```

```
proc freq data= work.Dirty2005basic;
  table ownership MortgageStatus;
run;
```

❶ The use of ODS EXCLUDE is an output-limiting strategy, the one-way frequency table is still generated and sent to an output data set by the ODS OUTPUT statement, but to no other output destinations.

❷ A limited portion of the one-way table is shown in Output 3.9.2A by using OBS= with PROC PRINT, showing the first ten rows of the one-way frequency table. Using OBS=10 in PROC FREQ limits the frequency analysis to the first ten records in the original data, which is not the desired result.

Output 3.9.2A: Partial Listing of Categories for the City Variable

City	Frequency	Percent	CumFrequency	CumPercent
Akron, OH	74	0.01	74	0.01
Akron, OH	691	0.06	765	0.07
Albany, NY	38	0.00	803	0.07
Albany, NY	330	0.03	1133	0.10
Alexandria, VA	69	0.01	1202	0.10
Alexandria, VA	554	0.05	1756	0.15
Allentown, PA	38	0.00	1794	0.15
Allentown, PA	253	0.02	2047	0.18
Anaheim, CA	95	0.01	2142	0.18
Anaheim, CA	873	0.08	3015	0.26

Each city appears twice and, looking carefully, it appears when spaces are required that some instances use more than one space. While a distinction between these values is likely unintentional, character strings are compared on a character-by-character basis as shown when using PROC COMPARE in Output 2.10.2B.

Output 3.9.2B: Listing of Categories for the Ownership Variable

Ownership	Frequency	Percent	Cumulative Frequency	Cumulative Percent
OWNED	25901	2.23	25901	2.23
Owned	795913	68.67	821814	70.90
RENTED	21098	1.82	842912	72.72
Rented	270210	23.31	1113122	96.04
owned	33906	2.93	1147028	98.96
rented	12034	1.04	1159062	100.00

In this data set, the casing of the values for the Ownership variable causes distinctions among values that are likely not intended either. Some values are in proper case, while others are all lower or upper case. What should be three categories is seen by SAS as seven, as shown in Output 3.9.2B.

Output 3.9.2C: Listing of Categories for the Mortgage Status Variable

MortgageStatus	Frequency	Percent	Cumulative Frequency	Cumulative Percent
N/A	288225	24.87	288225	24.87
N\A	15117	1.30	303342	26.17
No, owned free and clear	300349	25.91	603691	52.08
Yes, contract to purchase	9756	0.84	613447	52.93
Yes, mortgaged-deed of trust or similar	13630	1.18	627077	54.10
Yes, mortgaged/ deed of trust or similar	518505	44.73	1145582	98.84
Yes, mortgaged\ deed of trust or similar	13480	1.16	1159062	100.00

For the MortgageStatus variable, differing uses of forward and backward slashes, or dashes, once again make distinctions that are not likely to have been intended. Output 3.9.2C shows that while four categories are desired, seven actually exist. Section 3.9.2 looks at functions that can be used during the DATA step to help fix the inconsistencies present in each of these variables.

Note that the MortgagePayment variable generates several invalid data notes in the log, and it is read in as a numeric variable. The MEANS procedure, with particular choices for summary statistics, can be employed as a diagnostic tool here, as shown in Program 3.9.3.

Program 3.9.3: A Check of the Mortgage Payment Variable in Dirty2005basic

```
proc means data= work.Dirty2005basic n nmiss min q1 median q3 max maxdec=1;
  var mortgagePayment;
run;
```

Output 3.9.3: Statistics on Mortgage Payment from Dirty2005basic

Analysis Variable : MortgagePayment						
N	N Miss	Minimum	Lower Quartile	Median	Upper Quartile	Maximum
1112983	46079	-1500.0	0.0	0.0	780.0	7900.0

The minimum value certainly shows something that is not to be expected, and the NMISS keyword is helpful for getting the count of missing values. While it was not certain at the outset of this section, it is true that the MortgagePayment variable should have a minimum of zero and should never be missing. Given that many homes in this data are not mortgaged, Program 3.9.4 gives a more detailed look at the distribution of nonzero mortgage payments. Output 3.9.4 shows that at least 5% of the nonzero Mortgage Payment values are negative.

Program 3.9.4: A Check of the Nonzero Mortgage Payments in Dirty2005basic

```
proc means data= work.Dirty2005basic n nmiss min p5 p10 q1 median max maxdec=1;
  var mortgagePayment;
  where mortgageStatus contains 'Yes';
run;
```

Output 3.9.4: Statistics on the Nonzero Mortgage Payments in Dirty2005basic

					Lower		
N	N Miss	Minimum	5th Pctl	10th Pctl	Quartile	Median	Maximum
533380	21991	-1500.0	-660.0	150.0	470.0	800.0	7900.0

Analysis Variable : MortgagePayment

3.9.2 Application of Functions for Data Cleaning

SAS provides many built-in functions, and several are useful for manipulating character values. For the City variable in Dirty2005basic, there is inconsistent spacing, and a look through the available SAS functions shows that COMPRESS can be used to remove blanks (among other uses). Program 3.9.5 uses COMPRESS in an effort to repair the values of the City variable.

Program 3.9.5: Modifying the City Variable when Reading IPUMS2005Dirtied.dat

```
data work.IPUMS05CleanA;
   infile RawData('IPUMS2005Dirtied.dat') dlm='09'x dsd;
   input Serial $ CityPop : comma. Metro CountyFips Ownership : $50.
         MortgageStatus : $50. HHIncome : comma. HomeValue : comma.
         City : $50.  MortgagePayment : comma.;

   City=compress(City);
   format hhIncome homeValue mortgagePayment dollar16.;
run;

ods exclude all;
proc freq data= work.IPUMS05CleanA;
   table city;
   ods output onewayfreqs= work.freqs;
run;

ods exclude none;
proc print data= work.freqs(obs=10) label noobs;
   var city--cumPercent;
run;
```

Output 3.9.5: Summary of the Updated City Variable (Partial Output)

City	Frequency	Percent	CumFrequency	CumPercent
Akron,OH	765	0.07	765	0.07
Albany,NY	368	0.03	1133	0.10
Alexandria,VA	623	0.05	1756	0.15
Allentown,PA	291	0.03	2047	0.18
Anaheim,CA	968	0.08	3015	0.26
Anchorage,AK	696	0.06	3711	0.32
AnnArbor,MI	351	0.03	4062	0.35
Arlington,TX	1233	0.11	5295	0.46
Arlington,VA	848	0.07	6143	0.53
Augusta-RichmondCounty,GA	760	0.07	6903	0.60

Output 3.9.5 shows that compressing out all spaces is suboptimal. The lack of a space between the comma and the state postal code is perhaps acceptable, but the lack of spaces between multiple words in the city name, such as Ann Arbor, likely is not. SAS provides another function that better serves this role; COMPBL compresses any series of multiple blanks into a single blank, leaving single blanks as they are. Program 3.9.6

uses COMPBL to prevent unwanted duplication of city names and ensure the remaining names are spelled in a reasonable fashion.

Program 3.9.6: A Better Modification for the City Variable when Reading IPUMS2005Dirtied.dat

```
data work.IPUMS05CleanB;
   infile RawData('IPUMS2005Dirtied.dat) dlm='09'x dsd;
   input Serial $ CityPop : comma. Metro CountyFips Ownership : $50.
         MortgageStatus : $50. HHIncome : comma. HomeValue : comma.
         City : $50. MortgagePayment : comma.;

   City=compbl(City);
   format hhIncome homeValue mortgagePayment dollar16.;
run;

ods exclude all;
proc freq data= work.IPUMS05CleanB;
   table city;
   ods output onewayfreqs= work.freqs;
run;

ods exclude none;
proc print data= work.freqs(obs=10) label noobs;
   var city--cumPercent;
run;
```

Output 3.9.6: Summary of a Better Update to the City Variable (Partial Output)

City	Frequency	Percent	CumFrequency	CumPercent
Akron, OH	765	0.07	765	0.07
Albany, NY	368	0.03	1133	0.10
Alexandria, VA	623	0.05	1756	0.15
Allentown, PA	291	0.03	2047	0.18
Anaheim, CA	968	0.08	3015	0.26
Anchorage, AK	696	0.06	3711	0.32
Ann Arbor, MI	351	0.03	4062	0.35
Arlington, TX	1233	0.11	5295	0.46
Arlington, VA	848	0.07	6143	0.53
Augusta-Richmond County, GA	760	0.07	6903	0.60

The Ownership variable displays inconsistent casing, and SAS provides multiple functions that are useful for repairing this, including UPCASE, LOWCASE, and PROPCASE. The first two functions are self-explanatory, UPCASE and LOWCASE convert any letter to uppercase or lowercase, respectively. PROPCASE is for proper casing, which has the first letter of any word in uppercase, all other letters in lowercase, where words are determined by a set of delimiters including characters like spaces, dashes, commas and a host of others. The SAS Documentation supplies details about this set of delimiters. Any of the three fix the issue of distinct categories that are not different; however, to be consistent with the other IPUMS data provided with the book, PROPCASE is the best choice, as shown in Program and Output 3.9.7.

Program 3.9.7: Using PROPCASE to Fix the Ownership Variable when Reading IPUMS2005Dirtied.dat

```
data work.IPUMS05CleanC;
   infile RawData('IPUMS2005Dirtied.dat) dlm='09'x dsd;
   input Serial $ CityPop : comma. Metro CountyFips Ownership : $50.
         MortgageStatus : $50. HHIncome : comma. HomeValue : comma.
         City : $50. MortgagePayment : comma.;
```

```
  City=compbl(City);
  Ownership=propcase(Ownership);
  format hhIncome homeValue mortgagePayment dollar16.;
run;

proc freq data= work.IPUMS05CleanC;
  table ownership;
run;
```

Output 3.9.7: Summary of the Modified Ownership Variable

Ownership	Frequency	Percent	Cumulative Frequency	Cumulative Percent
Owned	855720	73.83	855720	73.83
Rented	303342	26.17	1159062	100.00

For MortgageStatus, the inconsistency between forward and backward slashes, along with dashes, must be resolved. The TRANWRD function allows for a partial string to be replaced by an alternate value, and Program 3.9.8 uses it to modify MortgageStatus to be consistent with the IPUMS data used previously.

Program 3.9.8: Using TRANWRD to Repair Mortgage Status when Reading IPUMS2005Dirtied.dat
```
data work.IPUMS05CleanD;
  infile RawData('IPUMS2005Dirtied.dat) dlm='09'x dsd;
  input Serial $ CityPop : comma. Metro CountyFips Ownership : $50.
      MortgageStatus : $50. HHIncome : comma. HomeValue : comma.
      City : $50. MortgagePayment : comma.;

  City=compbl(City);
  Ownership=propcase(Ownership);
  MortgageStatus=tranwrd(tranwrd❶(MortgageStatus,'\','/'❷),'-','/ '❸);
  format hhIncome homeValue mortgagePayment dollar16.;
run;

proc freq data= work.IPUMS05CleanD;
  table MortgageStatus;
run;
```

❶   This operation could have been achieved by applying TRANWRD in one statement and reapplying it in a subsequent statement, but nesting of functions is permitted and is useful here.

❷   The inner TRANWRD swaps the backward slash for the forward slash, with no spacing.

❸   The replacement of the dash in the outer TRANWRD function must include the space after the slash as no space was originally present after the dash character.

Output 3.9.8: Summary of the Modified Mortgage Status Variable

MortgageStatus	Frequency	Percent	Cumulative Frequency	Cumulative Percent
N/A	519751	37.79	519751	37.79
No, owned free and clear	343716	24.99	863467	62.78
Yes, contract to purchase	7703	0.56	871170	63.34
Yes, mortgaged/ deed of trust or similar	504311	36.66	1375481	100.00

Repairing the MortgagePayment values requires one assumption along with some more investigation in the DATA step. Good programming practices dictate the existence of the negative values should be investigated to determine how they appeared. For this example, it is assumed that the value is correct and the negative sign was simply added in error. SAS provides several mathematical functions, including ABS for absolute value, which can fix the negative values, but does not ensure all values have been read as valid numeric values. Finding a way to repair the invalid values is aided by the DATA step shown in Program 3.9.9.

Program 3.9.9: Displaying MortgagePayment as Both Character and Numeric

```
data work.IPUMS05CleanE;
   infile RawData('IPUMS2005Dirtied.dat') dlm='09'x dsd;
   input Serial $ CityPop : comma. Metro CountyFips Ownership : $50.
         MortgageStatus : $50. HHIncome : comma. HomeValue : comma.
         City : $50. MortgagePaymentC : $20.❶;

   City=compbl(City);
   Ownership=propcase(Ownership);
   MortgageStatus=tranwrd(tranwrd(MortgageStatus,'\','/'),'-','/ ');
   MortgagePayment=input(MortgagePaymentC,dollar20.)❷;
   format hhIncome homeValue mortgagePayment dollar16.;
run;

proc print data= work.IPUMS05CleanE;
   var MortgagePaymentC;
   where MortgagePayment eq .❸;
run;
```

❶ Given that some values of MortgagePayment do not convert to a valid numeric form even when using an informat, it is initially read as a character value.

❷ The INPUT function allows for the conversion of values from character to numeric with an informat, operating in much the same fashion as using the informat in modified list input.

❸ The strategy used here is inspection of the character values on the records where the numeric value conversion fails and gives a missing result.

Output 3.9.9: Mortgage Payment Character Values that Fail to Convert to Numeric (Partial Listing)

Obs	MortgagePaymentC
30	$O
87	$71O
92	$68O
97	$9O
103	$29O
175	$1,6OO
179	$52O
197	$45O
223	$O
246	$O
254	$6OO
296	$O

These values may look reasonable at first glance, but the zero character is actually an uppercase letter o. Upon finding this, one solution to the problem is to modify the code used for investigation to actually change the invalid character before converting to a number, and then fixing the sign errors after the numeric conversion is successful. Program 3.9.10 extends Program 3.9.9 to complete these tasks.

Program 3.9.10: Fixing Invalid and Negative Values for Mortgage Payment

```
data work.IPUMS05CleanF;
    infile RawData('IPUMS2005Dirtied.dat) dlm='09'x dsd;
    input Serial $ CityPop : comma. Metro CountyFips Ownership : $50.
          MortgageStatus : $50. HHIncome : comma. HomeValue : comma.
          City : $50. MortgagePaymentC : $20.;

    City=compbl(City);
    Ownership=propcase(Ownership);
    MortgageStatus=tranwrd(tranwrd(MortgageStatus,'\','/'),'-','/ ');
    MortgagePaymentC=tranwrd(MortgagePaymentC,'O','0') ❶;
    MortgagePayment=abs(input(MortgagePaymentC,dollar20.)) ❷;
    format hhIncome homeValue mortgagePayment dollar16.;
run;

proc means data= work.IPUMS05CleanF n nmiss min median max maxdec=1;
    var mortgagePayment;
run; ❸
```

❶ Since the initial read of MortgagePayment is character, TRANWRD can be used to swap the zero in for the incorrect characters.

❷ The INPUT function can now convert all values correctly, and is nested inside the ABS function to also correct the negative values.

❸ MEANS should now show no missing values and a minimum of zero.

Output 3.9.10: Mortgage Payment Summary After Repairs

Analysis Variable : MortgagePayment				
N	N Miss	Minimum	Median	Maximum
1159062	0	0.0	0.0	7900.0

For the examples in this section, a uniform application of functions to all records repaired the problems noted. In certain instances, different interventions are required on different records based on the nature of the problem encountered. The conditional logic techniques introduced in Chapter 4 allow for application of these data cleaning methods to more precise targets.

To close this section, note that PROC SORT with the NODUPKEY option can be used to replace the one-way frequency tables from PROC FREQ as diagnostic tool. Program 3.9.11 replicates a portion of the initial diagnostic given in Program 3.9.2.

Program 3.9.11: Using NODUPKEY in PROC SORT for Diagnostics

```
proc sort data= work.Dirty2005basic out= work.OwnerVals❶ nodupkey❷;
    by ownership;
run;

proc sort data= work.Dirty2005basic out= work.MortStatVals nodupkey;
    by MortgageStatus;
run;

proc print data= work.OwnerVals;
    var ownership; ❸
run;

proc print data= work.MortStatVals;
    var MortgageStatus;
run;
```

❶ The use of NODUPKEY or NODUP potentially causes removal of records from the data set being sorted. As the default behavior of PROC SORT is to replace the input data set with the sorted data, use of OUT= is extremely important to prevent data loss.

❷ NODUPKEY removes records with replicate values on the key, which is defined as the set of variables listed in the BY statement. NODUP is also available, which removes records that are duplicates across all variables in the data set.

❸ In this instance, only the value of the key variable is relevant.

Output 3.9.11: Using NODUPKEY in PROC SORT for Diagnostics

Obs	Ownership
1	OWNED
2	Owned
3	RENTED
4	Rented
5	owned
6	rented

Obs	MortgageStatus
1	N/A
2	N\A
3	No, owned free and clear
4	Yes, contract to purchase
5	Yes, mortgaged-deed of trust or similar
6	Yes, mortgaged/ deed of trust or similar
7	Yes, mortgaged\ deed of trust or similar

3.10 Wrap-Up Activity

Use the lessons and examples contained in this and previous chapters to complete the activity shown in Section 3.2.

Data

Various data sets are available for use to complete the activity. This data is similar in structure and contents to the 2005 data used for many of the in-chapter examples. Completing this activity also requires at least one of the following files:

1. Ipums2010Basic.sas7bdat
2. Ipums2010Formatted.csv
3. Ipums2010Formatted.txt
4. Ipums2010Dirtied.dat

Scenario

Read in the raw files and validate the data sets created from the raw files match the provided SAS data set. To maximize the benefit of this case study, it is suggested that each of Files 2 and 3 be read and validated against 1. File 4 contains data integrity issues, read that file and repair it to match any of the previous three. Summaries on any of the four data sets should be identical to the summaries in Section 3.2.

3.11 Chapter Notes

1. *Managing ODS Destinations.* As shown previously in Chapter 1 and here in Chapter 3, the Output Delivery System provides several techniques for controlling which objects SAS creates when executing a procedure. In practice, there are three basic techniques for suppressing output: NOPRINT options, ODS SELECT or ODS EXCLUDE statements, and ODS CLOSE statements. Since the NOPRINT option is incompatible with saving any output objects via ODS, only the other two approaches are discussed here.

   Output objects are sent to all compatible, open ODS destinations. The difference between the use of ODS SELECT/EXCLUDE or ODS CLOSE is in how they interact with the open destinations. Using ODS RTF EXCLUDE ALL temporarily suspends writing output objects to the RTF destination. However, using ODS RTF CLOSE finishes writing objects to the destination and closes it – making it accessible to other programs such as Microsoft Word – and thus no new objects can be sent to this destination. Attempting to write to the same file would cause SAS to overwrite it rather than append new objects to it.

   Similarly, using ODS _ALL_ CLOSE instructs SAS to close every open destination, while ODS EXCLUDE ALL temporarily stops writing to the destinations without closing them. In the case of the LISTING destination for table output, the distinction is typically less important as new objects can be appended there without issue unless the LISTING results are also directed to an external file, for example, using the .lst extension. However, the LISTING destination is no longer the default destination for non-graphics results in the SAS windowing environment and is not the default destination for any results in SAS Studio or SAS University Edition.

   As a result, using ODS LISTING CLOSE may not be sufficient to create code that functions in the same way across multiple SAS sessions, and ODS _ALL_ CLOSE may have unintended consequences, such as closing an RTF or PDF file prematurely. In modern SAS coding, it is a better practice to use ODS EXCLUDE instead of ODS CLOSE. However, be aware that the effect of ODS EXCLUDE ALL is global and remains in effect until either the SAS session ends or SAS encounters an ODS EXCLUDE NONE statement (or an ODS SELECT statement). This is not the case when ODS EXCLUDE is applied to a single table—for example, ODS EXCLUDE SUMMARY—which is only in effect until the next step boundary; however, that behavior can be modified using the PERSIST option. For more information about managing ODS destinations in different versions of SAS, see the SAS Documentation.

2. *Informats and LENGTH Statement.* Because informats can set the length of a character variable if the length is not previously defined, it is a good programming practice to ensure the LENGTH statement precedes the use of informats. If the length has already been defined, then the informat is not used to define the length and no errors, warnings, or notes are written to the log. However, similar to Program 2.8.6, if the LENGTH statement appears after the informat, then SAS issues the following warning.

   ```
   WARNING: Length of character variable x has already been set.
   Use the LENGTH statement as the very first statement in the DATA STEP to declare
   the length of a character variable.
   ```

3. *Other Informat Categories.* In addition to numeric, date/time, and character informats, SAS provides two additional categories: ISO8601 and binary. ISO8601 informats are designed to read date-, time-, or datetime-related values that are stored using the International Organization for Standardization's 8601 style. This format is designed to allow unambiguous transfer of such data between organizations that may represent such information using various local standards. Binary informats are for reading data stored in column-binary files. To learn more about both categories of informats, see the SAS Documentation.

4. *INFORMAT Statement.* The INFORMAT statement assigns informats in much the same way as the FORMAT statement assigns formats. If used, place the INFORMAT statement prior to the INPUT statement(s) in which the variables appear. However, use of the INFORMAT statement is simply an alternative way to employ modified list input. It is not possible to use formatted input if the informat appears in the INFORMAT statement instead of the INPUT statement.

5. *PAD.* In addition to FLOWOVER, MISSOVER, and TRUNCOVER, SAS provides other INFILE options to control the behavior of the INPUT statement when dealing with too-short records. While FLOWOVER is the default when reading data from an external file, another option is the default when reading instream data: PAD. The PAD option adds spaces to the end of each record causing all records to be the same length. The amount of padding is determined by the record length SAS is expecting; the expected record

length is controlled by the LRECL= option in the INFILE statement. SAS assigns a default value to LRECL= based on the operating system and SAS version, see the SAS Documentation for additional details on PAD, LRECL=, and other alternatives to FLOWOVER such as STOPOVER and SCANOVER.

6. *DELIMITER Statement.* When DBMS = TAB or DBMS = CSV, the DELIMITER statement is allowed but not necessary. However, when DBMS = DLM, the syntax of PROC IMPORT requires the DELIMITER statement. In any of these cases, PROC IMPORT determines the number of variables present in the input file based on the indicated delimiters. PROC IMPORT then reads the fields between each of the delimiters and stores those values. However, when representing multiple delimiters using hexadecimal notation it is important to consider the order of the delimiters.

Unlike the DATA step where the order of delimiters is not important, unexpected behavior occurs in the IMPORT procedure if the last hexadecimal delimiter is a space. In that case, the space is not used when determining the number of variables, but is used when reading in the values. As such, it is necessary to ensure the space is not the final delimiter in the DELIMITER statement. This issue is not present if delimiters are not represented using hexadecimal notation such as ', |' (comma, space, and pipe), or if space is the only delimiter in hexadecimal notation ('20'x).

3.12 Exercises

Concepts: Multiple Choice

1. Consider the following code and raw record (the ruler is not part of the raw file). What columns are read when creating the variable Weight?

   ```
   Input Age 2. Weight 3. Height 3.;

           ----+----1
           16 110 601
   ```

   a. 1-3
   b. 3-5
   c. 4-6
   d. 3-6

2. Consider the raw file shown below and referenced by the *fileref* Pats (the ruler is not part of the raw file). If the following program is submitted, how many records appear in the Patient data set?

   ```
   data Patients;
      infile pats truncover;
      input name : $10. age;
   run;

   ----+----1----+
   Cho 45
   Daniels
   Rodruiguez 15
   Smith 92
   ```

   a. 1
   b. 2
   c. 3
   d. 4

3. Which of these informats should be used to read a raw value of 27-12-2018 as a SAS date?
   a. MMDDYY10.
   b. DDMMYY10.
   c. DDMMYYYY10.
   d. DATE10.

4. Consider the raw file shown below (the ruler is not part of the raw file). Which of the following INPUT statements should be used to successfully read in this file?

```
----+----1----+----2----+
Kabul      +034.53+069.17
Tirana     +041.33+019.82
Algiers    +036.75+003.04
Pago Pago -014.28-170.70
```

   a. `input Name $9. +1 Latitude 7. +1 Longitude 7.;`
   b. `input Name $9. +2 Latitude 7. +1 Longitude 7.;`
   c. `input Name $9. +1 Latitude 7. Longitude 7.;`
   d. `input Name $9. Latitude 7. Longitude 7.;`

5. If a DATA step uses the INPUT statement shown below to read in the provided raw data set (the ruler is not part of the raw file), then to achieve the result shown, which of the following options must be used in the INFILE statement?

```
input Hours 1. +1 Department $2. +1 Course $7.;
```

```
----+----1----+
3 ST 101
4 ST 305-001
4 ST 305-002
3 ST
```

Hours	Department	Course
3	ST	101
4	ST	305-001
4	ST	305-002
3	ST	

   a. DSD
   b. TRUNCOVER
   c. MISSOVER
   d. Either MISSOVER or TRUNCOVER

6. Which of the following assignment statements correctly produces the string JAN\FEB\MAR\APR as the value for Months?

   a. `Months=tranwrd(tranwrd('JAN\FEB/MAR-APR','/','\'),'-','\');`
   b. `Months=tranwrd(tranwrd('JAN\FEB/MAR-APR','\','/'),'\','-');`
   c. `Months=tranwrd(tranwrd('JAN\FEB/MAR-APR','\','/'),'-','\');`
   d. `Months=tranwrd(tranwrd('JAN\FEB/MAR-APR','-','/'),'-','\');`

7. Consider a value 'LOS   Alam9s,   NM' that was read in for the variable Location. Of course, the value likely should have been 'Los Alamos, NM' instead. Which of the following functions is NOT useful for correcting this data entry error?

   a. COMPBL
   b. TRANWRD
   c. COMPRESS
   d. All of the above must be used

8. Which of the following programs generates the following graph?

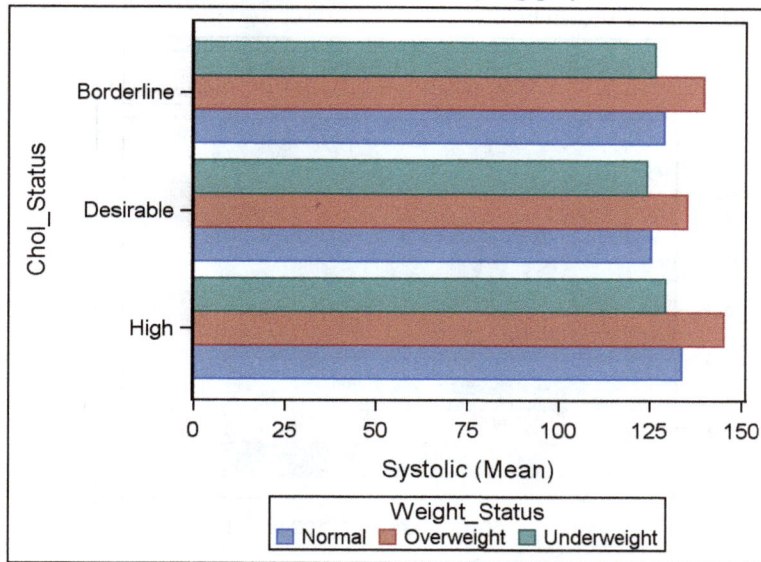

a.

```
proc sgplot data = sashelp.heart;
  hbar Weight_Status / group= Chol_Status groupdisplay = cluster
  response = systolic stat = mean;
run;
```

b.

```
proc sgplot data = sashelp.heart;
  hbar Chol_Status / group=Weight_Status groupdisplay = cluster
  response = systolic stat = mean;
run;
```

c.

```
proc sgplot data = sashelp.heart;
  hbar Chol_Status / group=Weight_Status groupdisplay = overlay
  response = systolic stat = mean;
run;
```

d.

```
proc sgplot data = sashelp.heart;
  hbar Chol_Status / group=Weight_Status groupdisplay = cluster
  response = systolic;
run;
```

9. Which of the following KEYLEGEND statements can be used to modify the graph given in the previous question to match the one shown below?

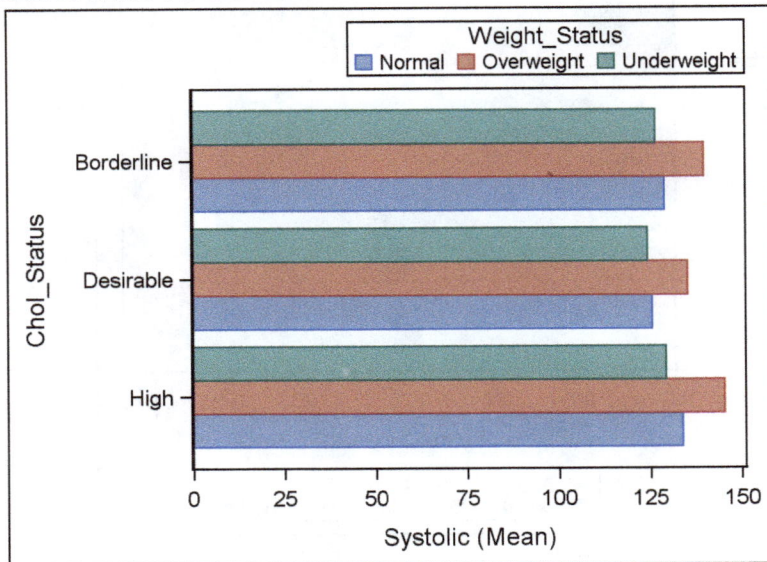

a. `keylegend / position = outside location = topright;`

b. `keylegend position = outside location = topright;`

c. `keylegend / location = outside position = topright;`

d. `keylegend location = outside position = topright;`

10. Which of the following axis-related statements can be used to modify the graph given in the previous question to match the one shown below?

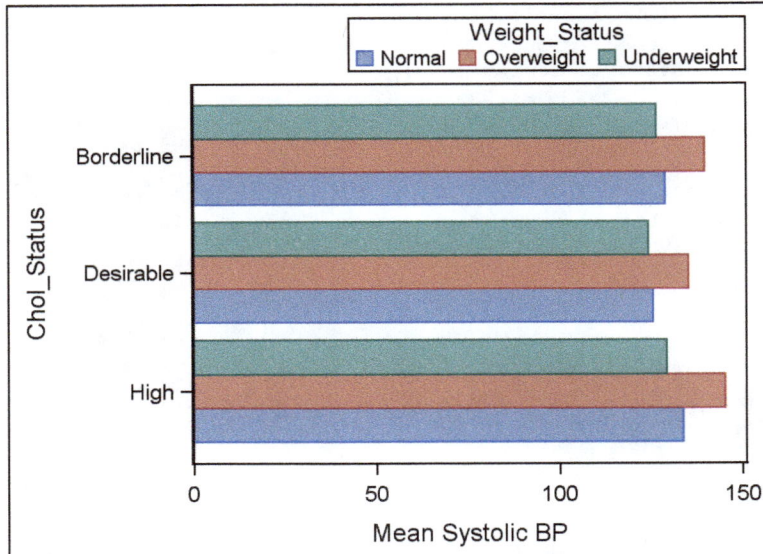

a. `xaxis values = (0 to 150 by 50) label = 'Mean Systolic BP';`

b. `xaxis / values = (0 to 150 by 50) label = 'Mean Systolic BP';`

c. `yaxis values = (0 to 150 by 50) label = 'Mean Systolic BP';`

d. `yaxis / values = (0 to 150 by 50) label = 'Mean Systolic BP';`

Concepts: Short-Answer

1. Explain the difference between the MISSOVER and TRUNCOVER options as well as how they differ from the default FLOWOVER option. In what cases, if any, do they produce the same results? In what cases, if any, do they produce different results? Be specific.

2. What are the attributes available for modification for each of the following graph elements?

    a. Text

    b. Fill

    c. Line

3. Why is PROC FREQ a useful tool when investigating the values of a categorical variable?

4. Why is it sometimes necessary to pre-summarize data prior to constructing a plot?

5. When using the following in the INPUT statement to read in the value of a variable, explain how SAS determines what characters to read from the input buffer.

    a. Formatted input

    b. Modified list input with the colon format modifier

    c. Modified list input with the ampersand format modifier

6. Consider the comma-delimited data set shown below.

```
----+----1----+----2----+----3
439,12/11/2000,LAX,20,137
921,12/11/2000,DFW,20,131
114,12/12/2000,LAX,15,170
982,12/12/2000,dfw,5,85
439,12/13/2000,LAX,14,196
982,12/13/2000,DFW,15,116
431,12/14/2000,LaX,17,166
982,12/14/2000,DFW,7,88
114,12/15/2000,LAX,,187
982,12/15/2000,DFW,14,31
```

Provide the contents of the input buffer and program data vector for the ninth observation in this data set when each of the following INFILE statements is used to complete the DATA step shown here.

```
data PartD;
   <--infile statement-->
   input FlightNum $ Date $ Destination $ FirstClass EconClass;
run;
```

   a. `infile myData dsd;`

   b. `infile myData dsd missover;`

   c. `infile myData dsd dlm=' ';`

   d. `infile myData dsd dlm=' ' missover;`

   e. `infile myData dlm=',';`

   f. `infile myData dlm=',' missover;`

7. Suppose the data set from the previous question is modified so that the ninth observation is replaced by the observation shown below. Repeat parts (a) through (f) of the previous question with this new record.

```
----+----1----+----2----+----3
114,12/15/2000,LAX
```

8. Do parts (b), (d) and (f) from the previous two questions have different answers if TRUNCOVER is used instead? Why or why not?

Programming Basics

1. Program 3.9.5 introduced the COMPRESS function to remove all blanks from a string and indicated that it can be used to modify strings in other ways as well. Use the following statements to modify the given strings and answer the associated questions.

    a. What value is generated when using COMPRESS('Patient X: is your ID# 17?','#?')?

    b. What does that value suggest is the purpose of the second argument in the COMPRESS function?

    c. What value is generated when using COMPRESS('Patient X: is your ID# 17?', ,'D')?

    d. What does that value suggest is the purpose of the D (or d) in the third argument in the COMPRESS function?

    e. What value is generated when using COMPRESS('Patient X: is your ID# 17?', ,'kd')?

    f. What does that value suggest is the purpose of the k (or K) in the third argument in the COMPRESS function?

    g. What value is generated when using COMPRESS('Patient X: is your ID# 17?', '#' ,'kd')?

    h. What does that value suggest is the relationship between the arguments of the COMPRESS function?

    i. Using the SAS Help Documentation, determine the correct second and third arguments to produce the string 'PX:ID#17' using the same first argument from part (a).

2. Use the Sashelp.Cars data set to produce the following graphs.

    a.

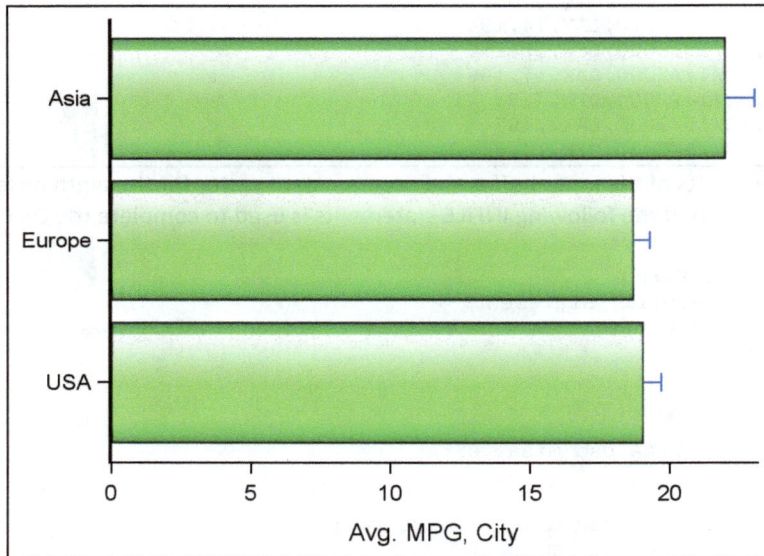

b.

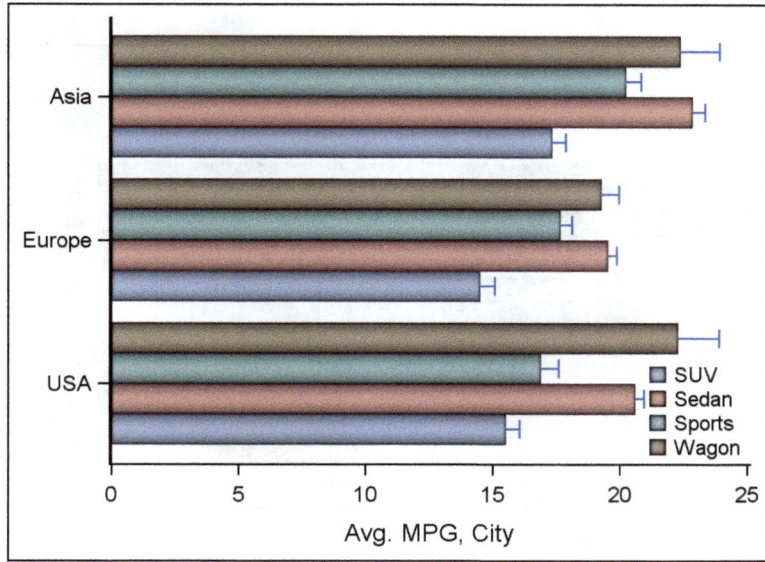

3. Use the Sashelp.Cars data set to produce the following graphs.

a.

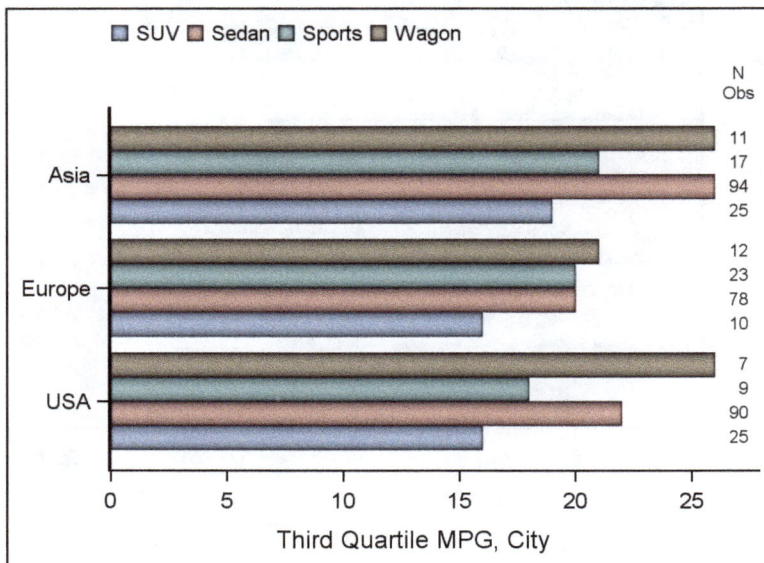

b.

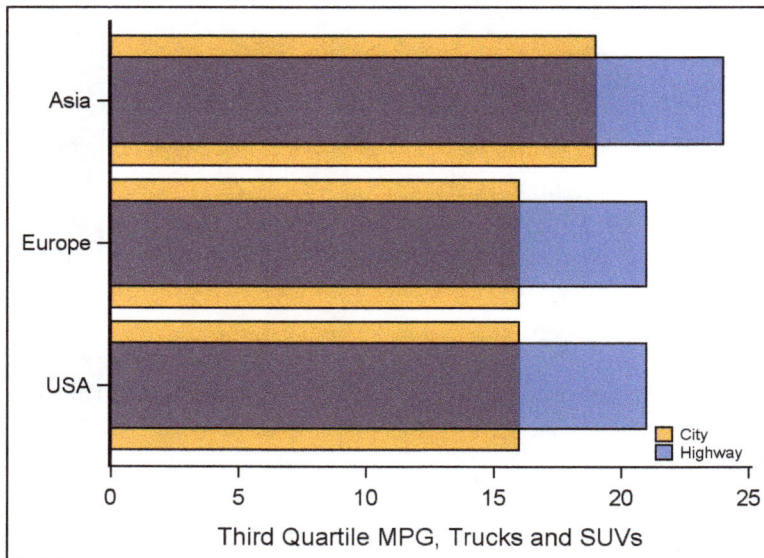

4. The raw file Density Estimation v1.txt contains hourly observations from a network data monitoring center as well as the company's predictions.

   a. Write a program that reads the data in, calculates the difference between predicted and observed values, and validates the results against the provided SAS data set named DensityFull.

   b. Using the validated data set from part (a), produce the following graph.

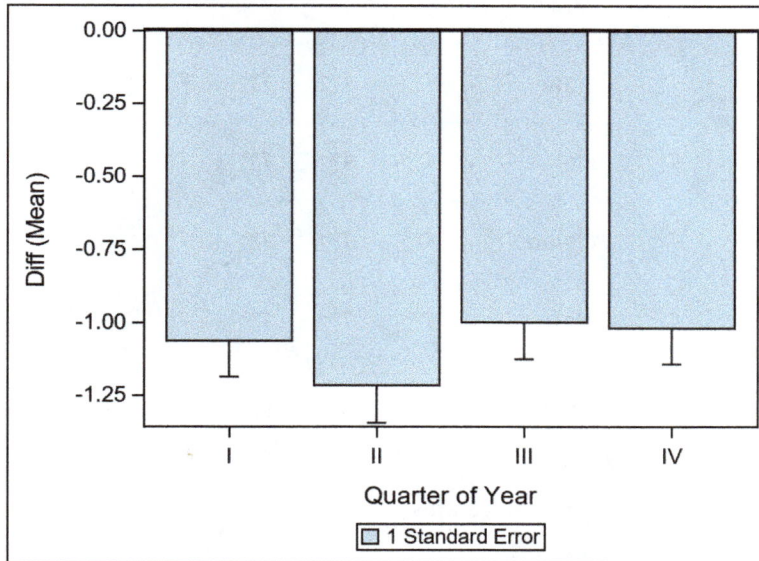

5. The raw file Density Estimation v2.txt is a modified version of the raw file from the previous exercise. Write a program that reads in the data from this modified file without reading in the variable Time, calculates the difference between observed and predicted values, and validates the results against the provided SAS data set named DensityPartial.

6. Input Data 3.6.1 shows a partial listing of the data in Ipums2005Formatted.txt and a full template for the data layout is below. Start refers to the column in which the variable begins and Stop refers to the column in which the variable ends.

Variable	Start	Stop
Serial	1	7
State	8	27
City	33	72
CityPop	73	78
Metro	79	79
CountyFIPS	80	82
OwnershipStatus	83	88
MortgageStatus	89	133
MortgagePayment	135	140
HHIncome	143	152
Homevalue	155	164

Use the template to help develop the following programs.

a. Modify Program 3.6.1 to read the following subset of variables, in the given order, using formatted input: Serial, State, MortgagePayment, HomeValue, City, CityPop, Metro, and CountyFips. After reading, format MortgagePayment, Homevalue, and CityPop to match the raw file.

b. Formatted input and column input can be used in the same INPUT statement. Repeat (a) using formatted input only on variables with nonstandard values, and use column input on those with standard numeric or character values. Repeat part (a) without using formatted input more than once.

c. It is also possible to combine list input, column input, and formatted input into a single INPUT statement. Read the file using list input, simple or modified, on any variable where it is applicable to do so and read all others with column or formatted input, as appropriate.

Case Studies

For additional practice, multiple case studies are available in addition to the IPUMS CPS case study used in the chapters. See Section 8.3 to apply the skills from this chapter to the Clinical Trials Case Study. For additional case studies including extensions to the IPUMS CPS case study, see the author pages.

Chapter 4: Combining Data Vertically in the DATA Step

4.1 Learning Objectives

At the conclusion of this chapter, mastery of the concepts covered in the narrative includes the ability to:

- Differentiate between concatenation and interleaving and apply the correct technique in a given scenario

- Formulate a strategy for selecting only the necessary rows and columns when processing a SAS data set

- Apply conditional logic to create a new variable

- Develop sound strategies for the use of IF-THEN, IF-THEN/ELSE, and SELECT when applying conditional logic in a given scenario

- Describe how SAS stores date and time values; apply functions and arithmetic operations to perform calculations on date and time variables; apply formats to control how date and time values appear

- Apply the UNIVARIATE and SGPLOT procedures to explore a SAS data set

Use the concepts of this chapter to solve the problems in the wrap-up activity. Additional exercises and case-studies are also available to test these concepts.

4.2 Case Study Activity

For a further continuation of the case study covered in the previous chapters, the output shown below provides summaries on mortgage payments across the years of 2005, 2010, and 2015 from the IPUMS CPS Basic data sets. The objective, as explained in the Wrap-Up Activity in Section 4.9, is to assemble the data from the individual files for these three years and produce the results shown.

Output 4.2.1: Basic Statistical Summaries on Nonzero Mortgage Payments

Variable: MortgagePayment (First mortgage monthly payment)
Year = 2005

Basic Statistical Measures			
Location		Variability	
Mean	1043.929	Std Deviation	754.34069
Median	860.000	Variance	569030
Mode	1200.000	Range	7896
		Interquartile Range	750.00000

Variable: MortgagePayment (First mortgage monthly payment)
Year = 2010

Basic Statistical Measures			
Location		Variability	
Mean	1228.807	Std Deviation	890.39318
Median	1000.000	Variance	792800
Mode	1200.000	Range	7396
		Interquartile Range	970.00000

Variable: MortgagePayment (First mortgage monthly payment)
Year = 2015

Basic Statistical Measures			
Location		Variability	
Mean	1227.864	Std Deviation	871.51891
Median	1000.000	Variance	759545
Mode	1200.000	Range	6896
		Interquartile Range	960.00000

Output 4.2.2: Histograms Across Years for Nonzero Mortgage Payments

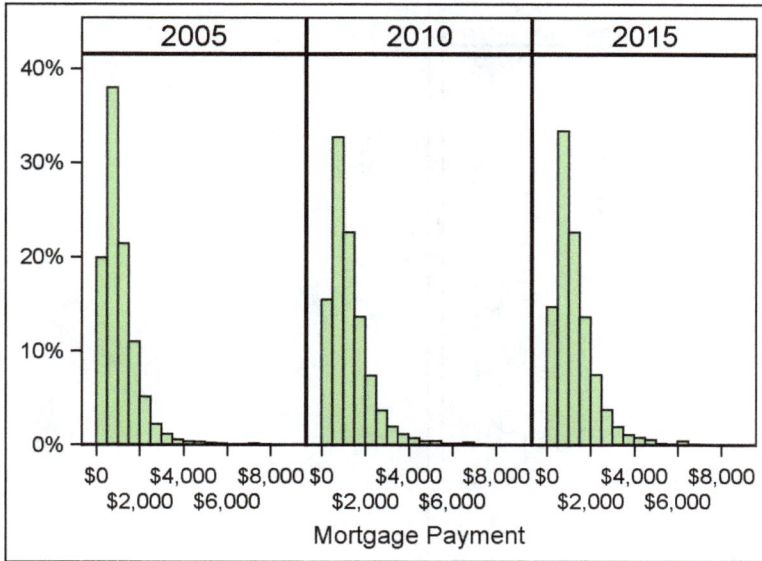

Output 4.2.3: Boxplots Across Years for Nonzero Mortgage Payments

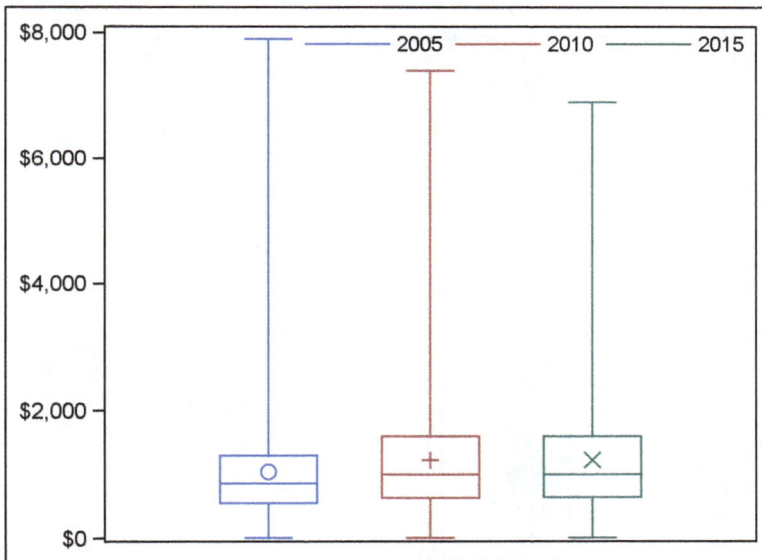

Output 4.2.4: Boxplots Across Years, Separated by Metro Status

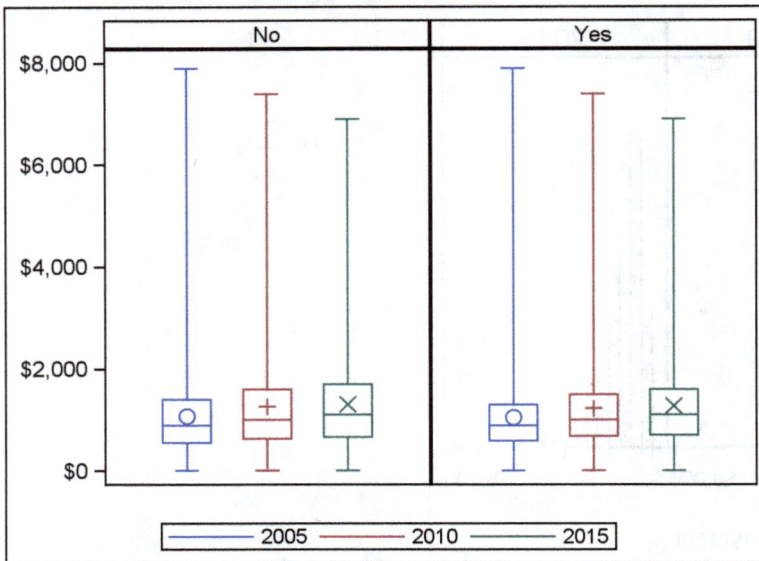

Output 4.2.5: Customized Distribution Plots Across Years for Nonzero Mortgage Payments

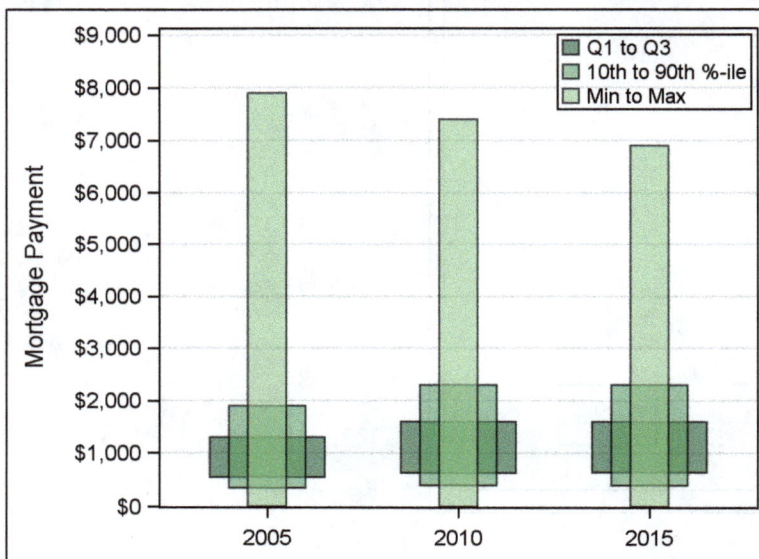

4.3 Vertically Combining SAS Data Sets in the DATA Step

Chapters 2 and 3 show how to read raw files with various structures using the INFILE and INPUT statements and Chapter 1 has examples on how to read a single SAS data set into a DATA step by using the SET statement. Sometimes the information needed for an analysis comes from multiple files. This scenario can occur naturally when data is collected over time or when related information for a single record is stored in separate data tables.

There are several ways to describe methods available in SAS for combining data sets. The two basic criteria used to classify techniques are:

- *Orientation*. Orientation is either vertical, a stacking of rows from multiple data sets based on matching columns, or horizontal, joining columns together on a criterion that matches rows.

- *BY-Grouping*. BY-grouping can either be used or not used.

These result in four classifications of data combination, with the SAS terminology for each shown in Table 4.3.1.

Table 4.3.1: Overview of the Four Methods for Combining Data via the DATA Step

		Orientation of Combination	
		Vertical	**Horizontal**
Grouping Used?	**No**	Concatenation	One-to-One Read One-to-One Merge
	Yes	Interleave	Match-Merge

The following subsections provide details about each of the vertical methods, with horizontal methods discussed in Chapter 5.

4.3.1 Concatenation

In the SAS DATA step, concatenation is a vertical combination method that creates a data set by stacking records from the data sets in the SET statement, with Table 4.3.2 illustrating the concept of concatenating records for a single matching variable. From this illustration, the vertical concept implies a "taller" data set—the number of records is the sum of those from the contributing data sets. No grouping of records is used, the records from one data set are simply appended to the other.

Table 4.3.2: Illustration of Concatenation

Amount		Amount		Amount
1		2		1
2		4		2
3	+	4	=	3
4		9		4
5		9		5
				2
				4
				4
				9
				9

Regardless of the number of data sets to concatenate, use a single SET statement and place all the data sets in the same statement, as shown in Program 4.3.1, which concatenates two of the IPUMS CPS data sets into a single data set.

Program 4.3.1: A Simple Concatenation

```
data work.Ipums0105Basic;
  set BookData.Ipums2001Basic(obs=3)
      BookData.Ipums2005Basic(obs=3); ❶
run;

proc report data= work.Ipums0105Basic; ❷
  column Serial HHIncome HomeValue State MortgageStatus Metro;
run;
```

❶ The OBS=3 data set option is used to limit the number of records input for this example only to make the display of results easier. For other examples and wrap-up activities, full data sets are used.

❷ Instead of using PROC PRINT to display the results, this program uses the REPORT procedure. This portion of the code is replicated in Program 4.3.2 with some introductory information about using PROC REPORT.

Output 4.3.1: A Simple Concatenation

Household serial number	Total household income	House value	state	MortgageStatus	Metropolitan status
1	6400	9999999	Alabama	N/A	.
2	30000	45000	Alabama	No, owned free and clear	.
3	37500	12500	Alabama	No, owned free and clear	.
2	12000	9999999	Alabama	N/A	4
3	17800	9999999	Alabama	N/A	1
4	185000	137500	Alabama	Yes, mortgaged/ deed of trust or similar debt	4

Recall the compilation phase is responsible for creating the PDV and, in this case, during the compilation phase the data from 2001 is encountered first and the data from 2005 second. As a result, SAS adds variables from the 2001 data set to the PDV first, followed by any variables in the 2005 data set that are not present in the 2001. Metro is not present in the 2001 data, but is present in the concatenated result since it is in the 2005 data. Because all values are initialized to missing in the PDV for each iteration of this DATA step, Metro is missing for those first three records read from the 2001 data.

During execution, records are added to the output data set in the order in which they are encountered. SAS reads left-to-right in the SET statement and top-to-bottom within each data set – records from the second data set are not accessed until the first data set is exhausted. The result is a data set with all records from 2001 stacked on top of all records from 2005.

Program 4.3.2: Using PROC REPORT to Display Data

```
proc report data= work.Ipums0105Basic nowd ❶;
  columns ❷ Serial HHIncome HomeValue State MortgageStatus Metro;
run;
```

❶ Prior to SAS 9.4, the REPORT procedure default is to deliver results in an interactive report window instead of producing a static report file. Starting with SAS 9.4, the default is to produce a static file instead. To ensure a PROC REPORT step produces the same output regardless of software version, explicitly include an option to produce the desired results—NOWINDOWS (or NOWD) to produce the static file or WINDOWS (or WD) to deliver the report to the interactive report window.

❷   By default, PROC REPORT includes every variable from the data set in its output. The COLUMN statement names specific variables and sets the order in which they appear in the results—in that role, it is very similar to the VAR statement in the PRINT procedure. Both the COLUMN and COLUMNS keywords are valid syntax, and there is no difference in their effects in the REPORT procedure.

PROC REPORT provides a substantial increase in functionality for creating reports compared to the PRINT procedure. However, due to that additional functionality, one important characteristic of PROC REPORT needs to be demonstrated, as shown in Program 4.3.3.

Program 4.3.3: Behavior of PROC REPORT with All Numeric Columns
```
proc report data= work.Ipums0105Basic;
  columns Serial HHIncome HomeValue;
run;
```

Because Program 4.3.3 contains a subset of the columns from Program 4.3.2, it is reasonable to assume that the report generated by Program 4.3.3 simply contains fewer columns than Output 4.3.2 but otherwise has the same structure. However, Output 4.3.3 shows this is not the case. Since the COLUMNS statement in Program 4.3.3 now only contains numeric variables, PROC REPORT produces a summary report—specifically with the sums of the columns—instead of reporting the individual values. This is an example of how the REPORT procedure assigns a usage to each variable, which can be controlled with a DEFINE statement. The examples in Chapters 4 and 5 introduce this and other small additions to the PROC REPORT syntax in order to present results. Chapter 6 provides a more thorough introduction to PROC REPORT and Chapter 7 provides a more in-depth look. For now, as long as a character variable is present in the COLUMNS statement, no summarization occurs.

Output 4.3.3: Behavior of PROC REPORT with All Numeric Columns

Household serial number	First mortgage monthly payment	Total household income	House value
15	900	288700	30194997

4.3.2 Interleaving

Interleaving in the DATA step is similar to concatenation; both combine two or more data sets vertically and use the SET statement. However, unlike concatenation which stacks observations sequentially across data sets and records within each, an interleave stacks records vertically based on the values of key variables. In order to interleave data sets, they must be sorted or indexed by the key variables. Key variables are often referred to as BY variables in SAS because they are typically specified in a BY statement. Table 4.3.3 illustrates the concept of interleaving using Amount as the key variable.

Table 4.3.3: Illustration of Interleaving

Amount		Amount		Amount
1		2		1
2		4		2
3	+	4	=	2
4		9		3
5		9		4
				4
				4
				5
				9
				9

In Table 4.3.1, the interleave is classified as a vertical, grouped technique. Comparing Table 4.3.3 to Table 4.3.2, the vertical component is the same—the number of records is the sum of the contributing records from the two data sets. Unlike Table 4.3.2, these records are in groups based on the value of Amount, rather than the records from the second table simply following the first. The grouping is not unique as there are ties on the key variable—these ties are then resolved by the concatenation rules. Effectively, interleaving can be viewed as a method to concatenate sorted data sets in a manner that preserves the sorting, up to any ambiguity introduced by ties.

Program 4.3.4 demonstrates a simple interleave on two of the IPUMS CPS data sets into a single file using a single key variable.

Program 4.3.4: A Simple Interleave

```
proc sort data = BookData.Ipums2001Basic out = work.Ipums2001Basic;
  by serial; ❶
run;

proc sort data = BookData.Ipums2005Basic out = work.Ipums2005Basic;
  by serial; ❶
run;

data work.Ipums0105Basic;
  set work.Ipums2001Basic work.Ipums2005Basic; ❷
  by serial; ❸
run;

proc report data= work.Ipums0105Basic(obs = 5);
  column Serial MortgagePayment HHIncome HomeValue State MortgageStatus;
run;
```

❶ Sort all data sets to be interleaved using the same key variables.

❷ Place all data sets to be interleaved in the same SET statement.

❸ Use key variables compatible with previous SORT procedures. In this case, each of the BY statements uses the same key variable.

Output 4.3.4: A Simple Interleave

Household serial number	First mortgage monthly payment	Total household income	House value	state	MortgageStatus
1	0	6400	9999999	Alabama	N/A
2	0	30000	45000	Alabama	No, owned free and clear
2	0	12000	9999999	Alabama	N/A
3	0	37500	12500	Alabama	No, owned free and clear
3	0	17800	9999999	Alabama	N/A

As with concatenation, the PDV is built by adding the variables from 2001 first and the variables from 2005 second, and variable attributes are based on the first encounter of the variable during the compilation phase. The interleave differs from concatenation in that the resulting data set records are ordered by the key variables. In the case of Program 4.3.4, the Ipums0105Basic data set is ordered by Serial, while the results of Program 4.3.1 contain the same records, but not sorted by Serial.

To further explore interleaving, consider Program 4.3.5 which combines the 2001 and 2005 Basic IPUMS CPS data by both HHIncome and MortgagePayment using sorting principles discussed in Section 2.3.

Program 4.3.5: Interleaving Using Two Variables

```
proc sort data = BookData.Ipums2001Basic out = work.Sort2001Basic;
  by HHIncome descending ❶ mortgagePayment;
  where HHIncome gt 84000 and state eq 'Vermont';
run;

proc sort data = BookData.Ipums2005Basic out = work.Sort2005Basic;
  by HHIncome descending mortgagePayment;
  where HHIncome gt 84000 and state eq 'Vermont';
run;

data work.Ipums0105Basic;
 set work.Sort2001Basic work.Sort2005Basic;
 by HHIncome descending❷ mortgagePayment; ❸
run;

proc report data= work.Ipums0105Basic(obs=8);
 column Serial HHIncome MortgagePayment HomeValue MortgageStatus;
run;
```

❶ DESCENDING changes the sort order from the default value of ASCENDING. It must precede the variable name; here it is being used to sort MortgagePayment from largest to smallest within each value of HHIncome.

❷ Sort order for all variables included in the BY statement in the DATA step must match across all data sets named in the SET statement. If any variable included in the BY statement is sorted in descending order, then the DESCENDING option must accompany that variable.

❸ The BY statement used in the DATA step does not have to be identical to those used in the SORT procedures. However, they must be compatible. The BY statement can successfully use HHIncome alone, since MortgagePayment is sorted within each level of HHIncome, but cannot use MortgagePayment alone as the primary sort of the data set is not on its values.

Output 4.3.5: Interleaving Using Two Variables

Household serial number	Total household income	First mortgage monthly payment	House value	MortgageStatus
469684	84020	1300	137500	Yes, mortgaged/ deed of trust or similar debt
467495	84100	1400	225000	Yes, mortgaged/ deed of trust or similar debt
467086	84100	800	275000	Yes, mortgaged/ deed of trust or similar debt
469463	84110	990	112500	Yes, mortgaged/ deed of trust or similar debt
1148721	84110	290	85000	Yes, mortgaged/ deed of trust or similar debt
468355	84141	1200	95000	Yes, mortgaged/ deed of trust or similar debt
469870	84180	890	112500	Yes, mortgaged/ deed of trust or similar debt
1148731	84200	650	112500	Yes, mortgaged/ deed of trust or similar debt

4.4 Managing Data Sets During Combination

When combining data sets it is often necessary to manage variables, records, or both. Variables with different names may contain comparable information, for example, MortPay and MortgagePayment may both contain the amounts homeowners pay on their mortgages. However, it is not possible to use these as a key variable during an interleave because they have different names and, as shown in Program 4.4.5, they do not align in the same column. Furthermore, after either a concatenation or an interleave, there is no way to track which record came from which data set. Outputs 4.3.1 and 4.3.4 are results from combining data from the years 2001 and 2005, but that information is lost in the final data set (it was only present in the data set names). This section introduces several tools for managing variables and records that have a wide variety of applications and which are available during the concatenation and interleaving processes.

4.4.1 Selecting Variables with KEEP and DROP

When concatenating or interleaving data sets, it is not uncommon to encounter a situation in which the data sets have only some columns in common. This is one example of where it is beneficial to select variables using a KEEP or DROP list. Program 4.4.1 demonstrates the usage of the DROP= data set option.

Program 4.4.1: Selecting Variables with DROP=

```
data work.Ipums0105Basic;
  set BookData.Ipums2001Basic
     BookData.Ipums2005Basic ❶(drop=❷  CountyFips Metro CityPop City);
run;

proc contents data= work.Ipums0105Basic;
  ods select variables;
run;
```

❶ Data set options are valid any time a program references a SAS data set. Use parentheses immediately after the data set name to indicate what options to apply. Use of multiple options simultaneously is demonstrated later in this section.

❷ The DROP= option provides a space-delimited list of variables to omit when accessing the data set. The list is taken as exhaustive; that is, any variable not in this list is read into the step. The KEEP= option specifies an exhaustive list of variables to include in the current step. This DROP= option is equivalent to the following.

```
KEEP= Serial MortgagePayment HHIncome HomeValue State MortgageStatus Ownership
```

Output 4.4.1: Viewing Selected Variables with PROC CONTENTS

Alphabetic List of Variables and Attributes			
# Variable	Type	Len	Label
3 HHINCOME	Num	8	Total household income
4 HomeValue	Num	8	House value
2 MortgagePayment	Num	8	First mortgage monthly payment
6 MortgageStatus	Char	45	
7 Ownership	Char	6	
1 SERIAL	Num	8	Household serial number
5 state	Char	57	

When DROP= and KEEP= are used, their effects are local to the current step and its results. For example, in Program 4.4.1, the four listed variables are not included when the Ipums2005Basic data set is read and, therefore, are not in the output data set. However, the Ipums2005Basic data set itself remains unchanged; that is, the variables are not dropped from the data set—they are just not processed by the current DATA step.

In the DATA step, there are two ways to apply KEEP and DROP lists: via the KEEP= and DROP= options as shown above, or via the KEEP and DROP statements. Program 4.4.2 demonstrates the use of a KEEP statement. The syntax is nearly identical to that of the KEEP= option with two notable differences: no parentheses or equals sign are needed since this is a stand-alone statement. The results of Program 4.4.2 are identical to Output 4.4.1.

Program 4.4.2: Using a KEEP Statement

```
data work.Ipums2005Basic;
  set BookData.Ipums2005Basic;
  keep Serial MortgagePayment HHIncome HomeValue State MortgageStatus Ownership;
run;

proc contents data= work.Ipums2005Basic;
  ods select variables;
run;
```

In Program 4.4.2 there is no difference between using the KEEP statement as demonstrated or using a KEEP= option in the SET statement. This is because in both programs only one data set is read in and only one data set is created. However, KEEP and DROP lists in the data set options are local to the data set they modify, while the DROP and KEEP statements apply to all data sets created by the DATA step. For illustration, Program 4.4.3 compares the effects of the KEEP statement, the KEEP= option in the SET statement, and the KEEP= option in the DATA statement.

Program 4.4.3: Contrasting the KEEP Statement and KEEP= Options

```
data work.Ipums2005Basic work.Serial2005Basic(keep = Serial) ❶;
  set BookData.Ipums2005Basic(keep = Serial MortgagePayment State Ownership) ❷;
  keep Serial MortgagePayment State; ❸
run;

ods select variables; ❹
proc contents data= work.Ipums2005Basic;
run;

ods select variables; ❹
proc contents data= work.Serial2005Basic;
run;
```

❶ Listing multiple data set names in the DATA statement instructs SAS to create multiple data sets. This KEEP= option directs the DATA step to omit every variable in the PDV except Serial when writing to the Serial2005Basic data set. As such, it is considered a write-condition since it only has an effect when SAS writes records to the data set.

❷ This KEEP= option directs the SET statement to omit any variable not in this list when it reads variables into the PDV. As such, it is considered a read-condition since it only has an effect when SAS reads records into the current data set.

❸ The KEEP statement is like the KEEP= option from ❶, except it controls the variables when writing to all data sets created in the current DATA step. Thus, it affects both Ipums2005Basic and Serial2005Basic while ❶ only affects Serial2005Basic. As such, it is also considered a write-condition since it only has an effect when SAS writes records to those data sets.

❹ Because ODS statements are global, they are valid outside of the PROC steps as well. By default, ODS SELECT and EXCLUDE are only in effect until they encounter a step boundary.

Output 4.4.3A: Contrasting the KEEP Statement and KEEP= SET Statement Option

Alphabetic List of Variables and Attributes			
# Variable	Type	Len	Label
2 MortgagePayment	Num	8	First mortgage monthly payment
1 SERIAL	Num	8	Household serial number
3 state	Char	57	

Output 4.4.3B: Using the KEEP Statement and KEEP= Option in the SET and DATA Statements

Alphabetic List of Variables and Attributes			
# Variable	Type	Len	Label
1 SERIAL	Num	8	Household serial number

The DATA step in Program 4.4.3 simultaneously creates two data sets in the Work library: Ipums2005Basic and Serial2005Basic. Each data set contains the same records; however, they do not contain the same variables. The first data set created, Ipums2005Basic, contains the three variables shown in Output 4.4.3A since the KEEP= option in the SET statement adds four variables to the PDV, and the KEEP statement only flags three of them for use in the output data sets. However, the Serial2005Basic data set is further altered by a KEEP= option that names only the Serial variable; thus, this data set then contains only one variable, Serial. Just as in previous programs in this section, Program 4.4.3 could be modified by using drop lists in place of any of the keep lists.

4.4.2 Changing Variable names with RENAME

Just as variables that do not exist in one data set can cause issues when combining data sets, so can variables with mismatched names. Common examples are cases where variables have different names but contain comparable information (MortPay versus MortgagePayment) or where variables have the same name but contain different information. The latter case is of particular concern if the variables have a different type, preventing the data sets from being combined. To handle the former case, SAS provides the ability to rename data set variables.

Program 4.4.4: Demonstrating the RENAME= Option

```
data work.Ipums2005Basic work.Serial2005Basic ❶(rename = ❷(Id = IdNum❸));
  set BookData.Ipums2005Basic;
  rename Serial = Id MortgagePayment = MortPay;❹
run;

proc contents data= work.Serial2005Basic varnum ❺;
  ods select position ❻;
run;
```

❶ As noted with the DROP= and KEEP= options, data set options must be enclosed in parentheses and follow immediately after the target data set.

❷ The RENAME= option requires a second set of parentheses to enclose its arguments.

❸ When specifying the new name, note the old variable name appears on the left of the equals sign, and the new variable name appears on the right.

❹ Note the RENAME statement goes into effect prior to the RENAME= option, which is why the option specifies Id=IdNum and not Serial=IdNum—the variable Serial had already been renamed.

❺ The VARNUM option instructs PROC CONTENTS to produce a table of variable names listed in position order, rather than the default alphanumeric order.

❻ When using the VARNUM option, PROC CONTENTS no longer creates an ODS table named Variables and instead creates an ODS table named Position. This is a good example of when ODS TRACE is useful for determining ODS table names.

Output 4.4.4 shows the position-ordered table created by PROC CONTENTS. Note that a much simpler alternative for renaming Serial to IdNum is to either use only the RENAME statement or RENAME= option in Program 4.4.4 and complete the rename in a single location. However, there are certain situations where a variable may need to change names multiple times and so it is important to understand how the RENAME statement and RENAME= option affect the results of a DATA step.

Output 4.4.4: Demonstrating the RENAME= Option

#	Variable	Type	Len	Format	Label
			Variables in Creation Order		
1	IdNum	Num	8		Household serial number
2	COUNTYFIPS	Num	8		County (FIPS code)
3	METRO	Num	8	BEST12.	Metropolitan status
4	CITYPOP	Num	8		City population
5	MortPay	Num	8		First mortgage monthly payment
6	HHINCOME	Num	8		Total household income
7	HomeValue	Num	8		House value
8	state	Char	57		
9	MortgageStatus	Char	45		
10	City	Char	43		
11	Ownership	Char	6		

The following examples show how renaming is potentially useful for fixing alignment of columns during the concatenation process, but also show some pitfalls to avoid. Program 4.4.5 reads raw CSV files for each of 2005 and 2010 and then concatenates them, with subsequent output from PROC CONTENTS and PROC PRINT revealing problems with the concatenation.

Program 4.4.5: Concatenation with Mismatched Variable Names

```
data work.Basic2005;
   infile RawData("IPUMS2005formatted.csv") dsd obs=10; ❶
   input serial : 7. state : $25. city : $50. citypop : comma6.
        metro : 1. countyfips : 3. ownership : $6. MortgageStatus : $50.
        MortgagePayment : dollar12. HHIncome : dollar12. HomeValue : dollar12.; ❷
run;

data work.Basic2010;
   infile RawData("IPUMS2010formatted.csv") dsd obs=10; ❶
   input serial : 7. state : $25. city : $50. metro : 1. countyfips : 3.
        citypop : comma6. ownership : $6. MortStat : $50.
        MortPay : dollar12. HomeValue : dollar12. Inc : dollar12.; ❷
run;

data work.Basic05And10;
   set work.basic2005 work.basic2010;
run;

proc contents data= work.basic05And10 varnum; ❸
   ods select position;
run;

proc print data= work.Basic05And10(obs=1); ❹
   var MortgageStatus MortStat MortgagePayment MortPay HHIncome Inc;
run;

proc print data= work.Basic05And10(firstobs=11 obs=11); ❹
   var MortgageStatus MortStat MortgagePayment MortPay HHIncome Inc;
run;
```

❶ Reading of the raw files is limited to a few observations in order to make diagnostics on the result more manageable. When working with large data sets, it is a good programming practice to check the code logic on smaller subsets before committing to the execution time required for the complete data.

❷ Each of these CSV files is read using modified list input. The columns in the raw files are ordered differently so the variables in the INPUT statements are ordered differently as well. Also, different variable names are used for three of the fields.

❸ Here, the CONTENTS procedure is used to show all the variables in their creation order, as shown in Output 4.4.5A. Recall that creation order is the order in which the variables are encountered during compilation of the DATA step.

❹ The invocations of PROC PRINT display the variables with mismatched names containing matching information, showing the first record read from each of the data sets in Output 4.4.5B.

Output 4.4.5A: Variable Set from Concatenation with Mismatched Columns

Variables in Creation Order			
#	Variable	Type	Len
1	serial	Num	8
2	state	Char	25
3	city	Char	50
4	citypop	Num	8
5	metro	Num	8
6	countyfips	Num	8
7	ownership	Char	6
8	MortgageStatus	Char	50
9	MortgagePayment	Num	8
10	HHIncome	Num	8
11	HomeValue	Num	8
12	MortStat	Char	50
13	MortPay	Num	8
14	Inc	Num	8

Output 4.4.5B: Data in Columns Mismatched During Concatenation

MortgageStatus	MortStat	MortgagePayment	MortPay	HHIncome	Inc
N/A		0	.	12000	.

MortgageStatus	MortStat	MortgagePayment	MortPay	HHIncome	Inc
	No, owned free and clear		.	0	.	7500

Output 4.4.5A shows that during the compilation of the DATA step in Program 4.4.5 that concatenates Basic05 and Basic10, the 11 variables from Basic05 are added to the PDV first, followed by the three variables from Basic10 that are not in Basic05. The fact that variables such as CityPop are not in the same column position in Basic05 and Basic10 does not affect the matching of those columns because it is based on the variable name, with the additional requirement that the type also matches. Of course, one fix for the name mismatches is to use the same variable names in the two INPUT statements. However, the RENAME methods discussed

previously offer a method to fix the alignment without having to alter existing data sets or any previous code generating them. Program 4.4.6 uses the RENAME statement in an attempt to align the columns—several renaming choices are available to align names, this renames the variables in Basic10 to match those in Basic05.

Program 4.4.6: Attempting to Align Columns During Concatenation with the RENAME Statement

```
data work.Basic05And10Try2;
  set work.basic2005 work.basic2010;
  rename MortStat=MortgageStatus MortPay=MortgagePayment Inc=HHIncome;
run;
```

Running the same PROC CONTENTS and PRINT procedures from Program 4.4.5 on this data set reveals the same result, and the SAS log contains the following warnings:

```
WARNING: Variable MortStatus cannot be renamed to MortgageStatus because
         MortgageStatus already exists.
WARNING: Variable MortPayment cannot be renamed to MortgagePayment because
         MortgagePayment already exists.
WARNING: Variable Income cannot be renamed to HHIncome because HHIncome already
         exists.
```

The RENAME statement sets up renaming flags in the PDV to rename the variables as they are output to the final data set, which is a conflict between two different variables having the same name, so the RENAME statement is ignored. Using the RENAME option on the output data set is effectively the same—the solution is to apply the RENAME option to the input data set(s) so that the alignment occurs during construction of the PDV, as in Program 4.4.7.

Program 4.4.7: Renaming Mismatched Variables During Concatenation in the SET Statement

```
data work.Basic05And10Try3;
  set work.basic2005
      work.basic2010(rename=(MortStat=MortgageStatus MortPay=MortgagePayment
                             Inc=HHIncome));
run;

proc contents data= work.basic05And10Try3 varnum;
  ods select position;
run;
```

Using the same strategy of renaming variables in Basic10 to match those in Basic05 requires the RENAME option to be attached to the Basic2010 data set in the SET statement. As Output 4.4.7 shows, the 11 variables in the final data set match the names (and order) from Basic05.

Output 4.4.7: Successfully Renaming Mismatched Variables During Concatenation

#	Variable	Type	Len
	Variables in Creation Order		
1	serial	Num	8
2	state	Char	25
3	city	Char	50
4	citypop	Num	8
5	metro	Num	8
6	countyfips	Num	8
7	ownership	Char	6
8	MortgageStatus	Char	50
9	MortgagePayment	Num	8
10	HHIncome	Num	8
11	HomeValue	Num	8

Program 4.4.8 shows a case of mismatching columns on type—fixing this problem requires intervention in the data sets prior to the concatenation.

Program 4.4.8: Type Mismatches During Concatenation

```
data work.Basic2010B;
  infile RawData("IPUMS2010formatted.csv") dsd obs=100;
  input serial : $7. state : $25. city : $50. metro : 1. countyfips : 3.
       citypop : comma6. ownership : $6. MortStat : $50.
       MortPay : dollar12. HomeValue : dollar12. Inc : dollar12.;
run;

data work.Basic05And10B;
  set work.basic2005
     work.basic2010B(rename=(MortStatus=MortgageStatus MortPayment=MortgagePayment
                            Income=HHIncome));
run;
```

Submission of this code results in the following error in the SAS log:

```
ERROR: Variable serial has been defined as both character and numeric.
```

The concatenation DATA step fails to compile (and execute) because the difference in types for Serial, numeric in Basic05 and character in Basic2010B, cannot be resolved. When aligning columns for concatenation or interleaving, it is important to determine whether columns that are expected to align match on name and type. Matching on name can be accomplished with the RENAME option on the input data—matching type requires an intervention in the data itself prior to concatenation. Other variable attributes, such as length, format, and label, should be inspected as well to ensure the desired result occurs; however, mismatches on these do not affect the column alignment. For more information about how these attribute mismatches are resolved, see Chapter Note 1 in Section 4.10.

4.4.3 Selecting Records with WHERE and Subsetting IF

Previous chapters demonstrate various applications of the WHERE statement for filtering records when using a SAS data set. However, WHERE conditioning is available in a statement or an option just like DROP, KEEP, and RENAME, but WHERE is different in that the statement is valid in procedures as well as in the DATA step. The following examples demonstrate the WHERE= option and introduce an alternative approach to filtering records—the subsetting IF, which is only available in the DATA step. Program 4.4.9 demonstrates several instances of conditional logic via WHERE= options.

Program 4.4.9: WHERE= Option Applications to Input and Output Data Sets in a DATA Step

```
data work.Metro4(where=(metro eq 4)) ❶ work.AllMetro ❷;
  set BookData.Ipums2001Basic(obs = 3) ❸
    BookData.Ipums2005Basic(where=(countyfips eq 73) obs = 3) ❹;
run;

proc report data = work.metro4;
  column Serial MortgagePayment HHIncome HomeValue State MortgageStatus;
run;

proc report data = work.allmetro;
  column Serial MortgagePayment HHIncome HomeValue State MortgageStatus;
run;
```

❶ As with the RENAME= option, the WHERE= option requires a set of inner parentheses to separate its arguments from any other data set options in use. In this case, the Metro4 data set is created by only writing out records from the PDV if the current value of Metro is 4.

❷ The work.AllMetro data set contains all records loaded into the PDV.

❸ The 2001 data set does not have a WHERE= option, so all three requested observations are sent to the PDV.

❹ The 2005 data set has a WHERE= option that only sends the first 3 records with CountyFips = 73 to the PDV.

Output 4.4.9A: WHERE= Option Applications to Input and Output Data Sets in a DATA Step

Household serial number	First mortgage monthly payment	Total household income	House value	state	MortgageStatus
2	0	12000	9999999	Alabama	N/A
4	900	185000	137500	Alabama	Yes, mortgaged/ deed of trust or similar debt
10	0	10200	9999999	Alabama	N/A

Output 4.4.9B: WHERE= Option Applications to Input and Output Data Sets in a DATA Step

Household serial number	First mortgage monthly payment	Total household income	House value	state	MortgageStatus
1	0	6400	9999999	Alabama	N/A
2	0	30000	45000	Alabama	No, owned free and clear
3	0	37500	12500	Alabama	No, owned free and clear
2	0	12000	9999999	Alabama	N/A

Household serial number	First mortgage monthly payment	Total household income	House value	state	MortgageStatus
4	900	185000	137500	Alabama	Yes, mortgaged/ deed of trust or similar debt
10	0	10200	9999999	Alabama	N/A

As shown in Program 4.4.9, the WHERE= options can be used to filter out records before they enter the PDV, as was done for the 2005 data, or after they enter the PDV but before they are sent to the data set, as was done for the Metro4 data set. If the records are not needed in the current DATA step, then using a WHERE= option on the incoming data set allows for a more efficient DATA step because it prevents unnecessary records from being processed. However, if the records are needed for processing in the current DATA step but are not needed in the final data set, then using the WHERE= option on the created data set is necessary.

In addition to being valid in both DATA and PROC steps, another difference between the WHERE statement and the KEEP/DROP/RENAME statements is that the former applies to SAS data sets as their records are read into the PDV, while the latter apply as the PDV is output to the created data set. Thus, the WHERE statement acts as a filter that prevents records from entering the PDV, just as the WHERE= option does when it applies to an input data set. Program 4.4.10 demonstrates the effects of using both the WHERE statement and WHERE= options simultaneously on input and output data sets.

Program 4.4.10: Using the WHERE Statement and WHERE= Option Simultaneously

```
data work.Metro4(where=(metro eq 4)) ❶ work.AllMetro;
  set BookData.Ipums2001Basic(obs = 3)
      BookData.Ipums2005Basic(where=(countyfips eq 73) obs = 3) ❷;
  where MortgagePayment gt 0; ❸
run;
```

❶ The first WHERE= option still prevents records from being written to the Metro4 data set unless the value of Metro is 4.

❷ The 2005 data is still filtered to only allow three records with CountyFips = 73 to enter the PDV.

❸ The WHERE statement is applied to all incoming data sets that do not already have a WHERE= option attached. In this case, only the 2001 data is filtered to include records with mortgages over 0 in the PDV.

Using the same REPORT procedures from Program 4.4.9 on the data sets generated in Program 4.4.10 produces Outputs 4.4.10A and 4.4.10B. Note that the results in Output 4.4.10A match those in 4.4.9A, while the results in Output 4.4.10B differ from those in Output 4.4.9B. Although this may not be the intended result, this serves as a reminder to always check the SAS log after submitting a program.

Output 4.4.10A: Using the WHERE Statement and WHERE= Option Simultaneously

Household serial number	First mortgage monthly payment	Total household income	House value	state	MortgageStatus
2	0	12000	9999999	Alabama	N/A
4	900	185000	137500	Alabama	Yes, mortgaged/ deed of trust or similar debt
10	0	10200	9999999	Alabama	N/A

Output 4.4.10B: Using the WHERE Statement and WHERE= Option Simultaneously

Household serial number	First mortgage monthly payment	Total household income	House value	state	MortgageStatus
4	890	12000	95000	Alabama	Yes, mortgaged/ deed of trust or similar debt
5	170	20000	27500	Alabama	Yes, mortgaged/ deed of trust or similar debt
7	790	63670	137500	Alabama	Yes, mortgaged/ deed of trust or similar debt
2	0	12000	9999999	Alabama	N/A
4	900	185000	137500	Alabama	Yes, mortgaged/ deed of trust or similar debt
10	0	10200	9999999	Alabama	N/A

Log 4.4.10 shows that SAS issues a warning as a result of submitting Program 4.4.10 because a WHERE statement is used while a WHERE= option is also specified for at least one incoming data set. As a result, SAS warns that it could not apply the WHERE statement to all incoming data sets. However, the program runs successfully and the data set is created as requested using the following criteria.

1. work.AllMetro is created by concatenating records from 2001 with MortgagePayment > 0 with records from 2005 with CountyFips = 73.
2. work.Metro4 is created with the same records as AllMetro but limited to those with Metro = 4.

Log 4.4.10: Partial Log from Program 4.4.10

```
where MortgagePayment gt 0;
WARNING: The WHERE statement cannot be applied to the data set on the last
SET/MERGE/UPDATE/MODIFY statement.  Either the data set failed to open or it already
specifies a WHERE data set option.
```

Since the WHERE statement is applied to all incoming data sets, the variables in the WHERE statement must exist in each of the incoming data sets. If the variables do not exist, then the WHERE statement cannot check whether the current record should be loaded into the PDV. Program 4.4.11 demonstrates several common ways this issue is encountered, and Log 4.4.11 shows the errors generated as a result of the program.

Program 4.4.11: Attempting to Apply a WHERE Statement when Variables Do Not Exist

```
data work.Combined;
  set BookData.Ipums2001Basic BookData.Ipums2005Basic;
  Year = '2001 or 2005';
  where CityPop ne 0 and Year eq '2001 or 2005';
run;
```

Log 4.4.11: Partial Log after Submitting Program 4.4.11

```
ERROR: Variable CityPop is not on file BOOKDATA.IPUMS2001BASIC. ❶
ERROR: Variable Year is not on file BOOKDATA.IPUMS2005BASIC. ❷
```

❶ CityPop is one of the variables included in the 2005 basic data but not the 2001 data. Because the WHERE statement is applied to all incoming data sets, this results in an error when SAS cannot evaluate the WHERE condition against the Ipums2001Basic file.

❷ Year is a new variable created during the current DATA step. Thus, it is not in either of the incoming data sets, and again SAS generates an error since the WHERE condition cannot be checked against all incoming data sets.

Of course, checking the condition CityPop is not zero in Program 4.4.11 is possible by modifying the 2005 data with the appropriate WHERE= option. However, checking the value of Year cannot be handled in this manner since Year is not in any of the incoming data sets. In order to filter out records based on a variable that does not enter the PDV from a SAS data set, an alternative method is necessary. Program 4.4.12 demonstrates the use of this method, the subsetting IF statement.

Program 4.4.12: Using the Subsetting IF Statement

```
data work.SubIF;
  set BookData.Ipums2001Basic BookData.Ipums2005Basic;
  Year = '2001 or 2005';
  if CityPop ne 0 and Year eq '2001 or 2005';
run;
```

The results of a subsetting IF statement are similar to those of a WHERE statement: records that do not meet the condition are not present in the final data set. However, the process SAS uses to select the records is quite different. In Program 4.4.11, all records from the 2001 and 2005 files are read into the PDV and are available for use in the DATA step. However, only those that meet the conditions in the subsetting IF statement are sent from the PDV to the data set. As such, while the WHERE statement serves as a way to filter records on the way into the PDV, the subsetting IF statement is a way to filter records on the way out of the PDV. Figure 4.4.1 illustrates the location of the effects of the WHERE and subsetting IF statements relative to the creation of the PDV.

Figure 4.4.1: Illustrating Flow of Records in DATA Steps using WHERE and Subsetting IF Statements

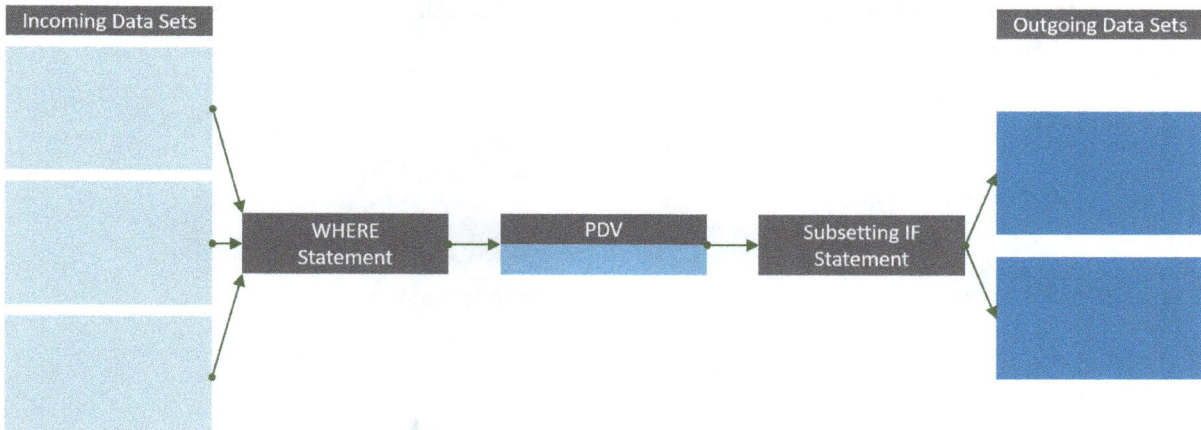

4.4.4 Identifying Record Sources with IN=

Notice that when the 2001 and 2005 IPUMS CPS data was combined using concatenation in Program 4.3.1 or using an interleave in Programs 4.3.4 and 4.3.5, once the new data sets are created there is no information in the data sets to identify the year corresponding those records. One common approach to include such information in the data sets is via the IN= data set option.

Unlike the DROP=/KEEP=/RENAME=/WHERE= data set options, the IN= option is only valid for data sets being read during a DATA step. Specifically, it can only be used in the SET, MERGE, UPDATE, and MODIFY statements. In this chapter, various applications of the SET statement are covered, while Chapter 5 discusses uses of the MERGE statement. For more information about the UPDATE and MODIFY statements, see the SAS Documentation.

The IN= option also differs in that it creates a temporary variable in the PDV. Temporary variables are similar to automatic variables in that they are present in the PDV, automatically dropped and cannot be kept, and are not written to the output data set. However, since they appear in the PDV, they are available for use during the current DATA step for programming purposes. Program 4.4.13 updates the interleave from Program 4.3.4 by including the IN= options.

Program 4.4.13: Using the IN= Option to Keep Track of Record Sources

```
data work.Ipums0105Basic;
  set work.Ipums2001Basic(in = a) ❶
      work.Ipums2005Basic(in = b) ❷;
  by serial;
run;
```

❶   The IN= option specifies the name of the temporary variable, which is given the name A here. Any record in the PDV that comes from Ipums2001Basic has A=1; all other records have A=0. This variable must follow SAS naming conventions and should be given a name distinct from any other variable used in the DATA step.

❷   Records from the 2005 data set are associated with the temporary variable B. All records from the Ipums2005Basic file have B=1 in the PDV, while all other records have B=0.

Program 4.4.13 added the temporary variables A and B to the PDV but did not use them. As a result, the final data set looks the same as the results from Program 4.3.4. Because variables created via the IN= option are temporary, they are not available in the output data set to identify the source of the records. To create a variable that tracks the source of these records, the temporary variables A and B are used in additional programming statements in Program 4.4.14. This assignment statement creates the variable Year, which results in either 2001 (when A=1) or 2005 (when B=1) as shown in Output 4.4.14. Since an interleave is a vertical method for combining the data sets, A and B never have the value 1 simultaneously.

Program 4.4.14: Using the IN= Tracking to Compute an Identifying Variable

```
data work.InDemo1;
  set BookData.Ipums2001Basic(in = a)
      BookData.Ipums2005Basic(in = b);
  by serial;
  Y05=a;
  Y10=b;
  Year = 2001*a + 2005*b;
run;

proc report data = work.InDemo1;
 column Y05 Y10 Year Serial HHIncome HomeValue MortgageStatus;
run;
```

Output 4.4.14: Using the IN= Tracking to Compute an Identifying Variable

Y05	Y10	Year	Household serial number	Total household income	House value	MortgageStatus
1	0	2001	1	6400	9999999	N/A
1	0	2001	2	30000	45000	No, owned free and clear
1	0	2001	3	37500	12500	No, owned free and clear
0	1	2005	2	12000	9999999	N/A
0	1	2005	3	17800	9999999	N/A
0	1	2005	4	185000	137500	Yes, mortgaged/ deed of trust or similar debt

4.4.5 Implicit Conversion

Section 3.9.2 introduces the INPUT function for explicitly converting character values into numeric values and a similar function, PUT, is available for carrying out the reverse process explicitly. Failure to use these functions (or an equivalent mechanism) forces SAS to interpret this type mismatch using an automatic process called automatic or implicit conversion. There are several reasons to avoid implicit conversion—in some cases it is not available and in other cases it does not perform as expected. In addition, implicit conversion uses substantially more computational power than explicit conversion to accomplish the same task, and as such, the loss in efficiency is an unavoidable pitfall.

The two common potential drawbacks are failed conversions and loss of data integrity. Not all programming statements in SAS support implicit conversion, naturally limiting its utility. Moreover, even when SAS does support implicit conversion, there is a potential for missing, truncated, or imprecise values. Log 4.4.15 demonstrates both pitfalls and includes the code that generates these notes, warnings, and errors.

Log 4.4.15: Notes and Errors Generated During Implicit Conversion (Partial Log Shown)

```
55    data work.Base;
56      length y $6; ❶
57      x=123456789;
58      y=x; ❶
59      put _all_ ; ❷
60    run;

NOTE: Numeric values have been converted to character values at the places given by:
      (Line):(Column).
        58:5 ❶
y=1.23E8 x=123456789 _ERROR_=0 _N_=1 ❷

61
62    data work.Subset1;
63      set Base;
64      where x = '123456789';  ❸
ERROR: WHERE clause operator requires compatible variables. ❸
65    run;
NOTE: The SAS System stopped processing this step because of errors. ❸
WARNING: The data set WORK.SUBSET1 may be incomplete.  When this step was stopped
there were 0 observations and 2 variables. ❸
66
67    data work.Subset2;
68      set Base;
69      if x = '123456789'; ❹
70    run;

NOTE: Character values have been converted to numeric values at the places given by:
      (Line):(Column).
        69:10 ❹
```

```
NOTE: There were 1 observations read from the data set WORK.BASE.
NOTE: The data set WORK.SUBSET2 has 1 observations and 2 variables. ❹
```

❶ The LENGTH statement declares Y as a character variable with a length of six bytes. As written, the assignment statement requests that the value of the numeric variable X be assigned to the character variable Y. Since this is not possible directly, it forces SAS to carry out an implicit conversion of the value of X to character, generating the note in the SAS log in the process. This note clearly indicates that column 5 in log line 58 (where SAS encounters the numeric value of X in the assignment statement) is the cause of the note.

❷ The PUT statement along with the keyword _ALL_ places the current PDV into the log, which shows the result of the implicit conversion in this case. The value of 123456789—which has 9 digits—is too wide to fit into the six-byte variable, Y. Using the character length of 6 bytes, SAS formats the numeric value using a format of a matching width before converting. Due to this limited format width, the value is expressed in scientific notation and then converted to character, resulting in the value of Y of 1.23E8.

❸ This WHERE statement compares the numeric variable, X, against a character string, causing a type mismatch. Unlike in the assignment statement in ❶, implicit conversion is not valid in a WHERE statement or WHERE= option. SAS generates the associated error message as a result of the type mismatch when implicit conversion is not available. The note and warning indicate the DATA step does not create a data set and a similar result occurs for procedures.

❹ The logical condition in this statement is identical to the WHERE statement in ❸, but now appears in a subsetting IF statement. Notice that instead of a warning or error, this only generates a note—the same note that appears in ❶. Unlike the WHERE statement and WHERE= option, the subsetting IF statement does support implicit conversion. This support for implicit conversion is also present in IF-THEN/ELSE statements.

While it is not reasonable for this section to present an exhaustive list of the programming statements that do or do not support implicit conversion, it is important to note the pitfalls—potential or guaranteed—that accompany implicit conversion. The potential for a program that fails to compile, or worse a program that compiles but leads to a loss of data integrity, along with the guaranteed loss of computational efficiency are all drawbacks to implicit conversion. As a result, it is considered a good programming practice to always use explicit conversion via PUT or INPUT functions or an equivalent mechanism.

4.5 Creating Variables Conditionally

The assignment statement used in Program 4.4.14 allows for programmatic determination of the year associated with each record. However, it has limited utility since it requires the desired identifying value to be the result of an arithmetic expression.

Conditional logic is used in the WHERE statement in this and previous chapters, as well as in the subsetting IF statement earlier in this chapter. The WHERE and subsetting IF statements are only used to select observations; however, additional conditional logic statements are available and offer a wide variety of applications.

4.5.1 IF-THEN Statements

As an alternative to creating variables using a single formula for all records as was done in Program 4.4.14, Program 4.5.1 uses conditional logic statements to create a new variable for which the values are determined based on the values of the temporary variables created via the IN= data set options.

Program 4.5.1: Using IF-THEN Statements to Create New Variables
```
data work.IfThenDemo1;
  set BookData.Ipums2001Basic(in = in2001❶ obs = 3)
      BookData.Ipums2005Basic(in = in2005❶ obs = 3);
  if❷ (in2001 eq 1) ❸ then❹ year = 2001❺;
  if (in2005 eq 1) then year = 2005; ❻
run;
```

❶   It is a good programming practice for the IN= variable names to relate meaningfully to the corresponding data set, improving the readability of the code in more complex DATA steps. As demonstrated in Program 4.4.13, they must be legal variable names and not the same as any other variables in the PDV during the current DATA step.

❷   The IF-THEN statement is broken into two clauses. The first clause begins with keyword IF.

❸   The IF clause must contain a logical expression. The parentheses are optional, but are included here to help distinguish the logical expression from the SAS keywords on either side. Any logical expression or an expression that resolves to a numeric value is valid. Recall from Section 2.6—missing and zero are interpreted as false; any other number is interpreted as true.

❹   Unlike the subsetting IF, the IF-THEN statement has a second clause that begins with the keyword THEN.

❺   The THEN clause must include a single SAS statement to execute when the IF clause resolves as true. In this case, an assignment statement is used to create a new variable, Year, and assign it a value of 2001 for records read from the Ipums2001Basic data set.

❻   A second IF-THEN statement is used to detect records from the Ipums2005Basic data set and assign the value 2005 to Year for those records.

The IF-THEN statement differs from the subsetting IF statement due to the requirement of the THEN clause in the IF-THEN statement. Even if the THEN clause contains only a semicolon, which is valid syntax, the IF-THEN statement is not equivalent to the subsetting IF statement. Other differences between the IF-THEN and subsetting IF statements exist and are investigated in Chapter 5.

Program 4.5.1 demonstrates the use of two IF-THEN statements to create the variable Year in a way that is easily programmed. However, it is an inefficient approach since the use of multiple IF-THEN statements requires each IF-THEN statement to be evaluated for every record. As a result, a record that has already been identified as coming from Ipums2001Basic and has already had Year set to 2001 is still checked to determine whether it comes from Ipums2005Basic, which is not possible. As a result, the 483,000 records from 2001 are each checked twice, and the 1.2 million records from 2005 are also checked twice. The loss in efficiency grows as the number of data sets, records, or conditions to check grows. Thus, the IF-THEN statements are typically used only when independent checks need to be carried out on each record.

4.5.2 IF-THEN/ELSE Statements

A more efficient solution to creating the Year variable in Program 4.5.1 is provided by modifying the IF-THEN statement slightly. This modified approach uses IF-THEN/ELSE statements, which are actually a pair of statements—an IF-THEN statement paired with an ELSE statement. The gain in efficiency is due to the fact that SAS only executes an ELSE statement if the condition in the preceding IF-THEN statement evaluates as false. Program 4.5.2 demonstrates the use of the IF-THEN/ELSE statement.

Program 4.5.2: Using IF-THEN/ELSE Statements to Create New Variables

```
data work.IfThenDemo2;
  set BookData.Ipums2001Basic(in = in2001 obs = 3)
      BookData.Ipums2005Basic(in = in2005 obs = 3);
  length MetFlag $ 3; ❶
  if in2001 eq 1 then year = 2001; ❷
    else if in2005 eq 1 then year = 2005; ❸
  if Metro in (1,2,3) then MetFlag = 'Yes'; ❹
    else❺ if Metro eq 4 then MetFlag = 'No'; ❻
      else MetFlag = 'N/A'; ❼
run;
```

```
proc report data = work.IfThenDemo2;
  column Year MetFlag Serial HHIncome HomeValue State MortgageStatus;
run;
```

❶ It is a good programming practice to define the length of user-defined character variables, helping to prevent unexpected truncation.

❷ The IF-THEN/ELSE statement begins with an IF-THEN statement.

❸ Instead of another IF-THEN statement, an ELSE statement is used and, like the THEN clause, it contains a single SAS statement. That statement executes only if the previous IF-THEN condition was false—in this case, if the data did not come from the Ipums2001Basic data set. This prevents rechecking the 483,000 records from 2001 as done with two IF-THEN statements in Program 4.5.1.

❹ A second IF-THEN/ELSE chain creates another new variable, MetFlag. The first statement checks if the value of Metro is 1 or 2 or 3, assigning Yes to MetFlag if true.

❺ For records where Metro is neither 1 nor 2 nor 3, the first ELSE statement is executed. Since the ELSE statement includes a single SAS statement, it can invoke another IF-THEN statement. IF-THEN/ELSE statements can thus be chained to check a sequence of conditions of practically any length, subject only to the amount of memory allocated to SAS.

❻ Recall variable length is defined based on the first encounter of the variable during DATA step compilation. If MetFlag first appeared in the assignment of the literal value 'No' in the IF-THEN/ELSE chain, SAS would define MetFlag as a character variable with length two. The LENGTH statement is thus placed before the IF-THEN/ELSE chain to ensure the length of MetFlag is set explicitly.

❼ The final ELSE statement serves as a "catch-all"—defining what action SAS takes for records with a value for Metro other than 1, 2, 3, or 4. Without this final ELSE statement, MetFlag is missing (blank) if Metro has a value other than 1, 2, 3, or 4—no assignment to MetFlag is made under that condition and its value is unchanged from being initialized to missing at the start of the DATA step iteration.

Output 4.5.2: Using IF-THEN/ELSE Statements to Create New Variables

year	MetFlag	Household serial number	Total household income	House value	state	MortgageStatus
2001	N/A	1	6400	9999999	Alabama	N/A
2001	N/A	2	30000	45000	Alabama	No, owned free and clear
2001	N/A	3	37500	12500	Alabama	No, owned free and clear
2005	No	2	12000	9999999	Alabama	N/A
2005	Yes	3	17800	9999999	Alabama	N/A
2005	No	4	185000	137500	Alabama	Yes, mortgaged/ deed of trust or similar debt

The IF-THEN/ELSE chains allow increased efficiency over IF-THEN statements by only evaluating the ELSE statements when all preceding conditions in the same IF-THEN/ELSE chain are false. However, it is possible to further improve the efficiency of Program 4.5.2 by checking the condition for Ipums2005Basic first since that data set has more than twice as many records as the Ipums2001Basic data. Thus, reversing the order would save nearly 700,000 additional executions of the conditional logic. Be sure to consider all aspects of efficiency when writing and maintaining programs.

Each THEN clause and ELSE statement concludes with a single statement; however, it is often necessary to execute several statements when a condition is true. It is possible, but inefficient, to write multiple IF-THEN statements with the same IF clause but different THEN clauses. Instead, SAS allows the use of a DO group to carry out multiple actions based on a single logical expression, as Program 4.5.3 demonstrates. DO groups greatly increase the flexibility of conditional logic statements as they allow for multiple actions as the result of a single condition.

Program 4.5.3: Using Do Groups to Create Multiple Variables

```
data work.IfThenDemo3;
  set BookData.Ipums2001Basic(in = in2001 obs = 3)
      BookData.Ipums2005Basic(in = in2005 obs = 3);
  length FirstFlag $ 3; ❶
  if in2001 eq 1 then do; ❷
                    year = 2001; ❸
                    FirstFlag = 'Yes'; ❸
                  end; ❹
    else if in2005❺ then year = 2005;
run;

proc report data = work.IfThenDemo3;
  column Year FirstFlag Serial HHIncome HomeValue State MortgageStatus;
run;
```

❶   As before, define the length of new character variables.

❷   DO statements contain only the keyword DO.

❸   All statements to be executed when the logical expression is true must appear in the DO group.

❹   A DO group ends with the END statement. As always, the indentation is for increased readability.

❺   It is not necessary to explicitly compare to 0 or 1 since SAS interprets 0 as FALSE and 1 as TRUE. As a result, when In2005 has the value 1 the IF clause resolves as TRUE. However, the explicit comparison is generally considered a good programming practice.

Output 4.5.3: Using Do Groups to Create Multiple Variables

year	FirstFlag	Household serial number	Total household income	House value	state	MortgageStatus
2001	Yes	1	6400	9999999	Alabama	N/A
2001	Yes	2	30000	45000	Alabama	No, owned free and clear
2001	Yes	3	37500	12500	Alabama	No, owned free and clear
2005		2	12000	9999999	Alabama	N/A
2005		3	17800	9999999	Alabama	N/A
2005		4	185000	137500	Alabama	Yes, mortgaged/ deed of trust or similar debt

4.5.3 SELECT Groups

SELECT groups are most similar to the IF-THEN/ELSE chains seen previously since the purpose of a SELECT group is to evaluate an expression based on one or more logical conditions. When that expression resolves to TRUE, SAS executes one or more statements. However, SELECT groups differ from IF-THEN/ELSE chains by more than just syntax—the execution also differs slightly. This section provides several examples of how SELECT groups operate and gives a brief comparison between SELECT groups and IF-THEN/ELSE chains.

SELECT Groups Without A Select-Expression

Program 4.5.2 used the code snippet below to create a new variable, MetFlag. In this snippet the values of Metro are grouped into three sets of values, each associated with a value for the new variable MetFlag.

```
if Metro in (1,2,3) then MetFlag = 'Yes';
  else if Metro eq 4 then MetFlag = 'No';
    else MetFlag = 'N/A';
```

The following snippet is logically equivalent to the previous snippet, but uses a SELECT group.

```
select; ❶
  when(Metro in (1,2,3)) ❷ MetFlag = 'Yes' ❸;
  when(Metro eq 4) ❷ MetFlag = 'No' ❸;
  otherwise MetFlag = 'N/A'; ❹
end; ❺
```

❶ Every SELECT group begins with a SELECT statement. SAS syntax allows for two versions of the SELECT statement. The version shown here includes only the keyword SELECT—the other version is shown in a later example.

❷ WHEN statements identify distinct cases exactly like the IF and ELSE IF clauses do. The conditions listed in the parentheses are collectively referred to as a *when-expression*.

❸ Like a THEN clause or ELSE statement, the WHEN statement includes a single statement to execute if the current record satisfies the *when-expression*. Similar to the THEN clause in an IF-THEN statement, the WHEN statement also supports use of the DO statement as well as the use of a single semicolon to take no action.

❹ The OTHERWISE statement is available in a SELECT group to indicate what action SAS should take when the current record does not satisfy any of the *when-expressions*. In this example, any values of Metro other than 1, 2, 3, or 4 result in the SELECT group assigning a value of N/A as the value of MetFlag. The OTHERWISE and WHEN statements support the same programming statements.

❺ Every SELECT group must close with an END statement.

The results of the two snippets above, one using IF-THEN/ELSE chains and one using a SELECT group, produce identical results. Furthermore, the execution of the two blocks of code is similar as well—the first condition is checked, and if true, then SAS performs an action. Just as ELSE IF statements are not executed unless all previous conditions in the current chain are not true, a WHEN statement is not executed unless all previous WHEN conditions in the same SELECT group are not met. As such, in both snippets a subsequent condition is only checked if preceding conditions are all false. Finally, if neither condition is met, both approaches assign a value of N/A to MetFlag.

One major difference between the two is that an OTHERWISE statement is required in a SELECT block if any record fails to satisfy one of the set of WHEN conditions. In other words, the SELECT block must have an explicit action to apply to every record; otherwise, it produces an error in the SAS log and stops execution. The IF-THEN/ELSE chain has no such requirement as shown in the following modified code snippet.

```
if Metro in (1,2,3) then MetFlag = 'Yes'; ❶
  else if Metro eq 4 then MetFlag = 'No'; ❶
    ❷
```

❶ Recall, variables created via an assignment statement are initialized to missing at the beginning of each iteration of the DATA step. Only for records where one of these conditions is met is the value of MetFlag assigned.

❷ Without any further statements in the IF-THEN/ELSE chain, records with Metro values other than 1, 2, 3, or 4 do not result in the assignment of a value to MetFlag, and the initial value of missing (blank) remains.

In the DATA step shown in Log 4.5.4, the OTHERWISE statement is omitted and the set of WHEN conditions is not exhaustive. The submission of this DATA step generates the following lines in the SAS log as soon as a value other than 1,2,3, or 4 is encountered. Note that the line numbers are only for reference within the log—they do not correspond to line numbers in the Editor.

Log 4.5.4: Results of Omitting a Necessary OTHERWISE Statement

```
80    data work.OtherwiseNeeded;
81      Metro = 5;
82      select(Metro);
83        when(Metro in (1,2,3)) MetFlag = 'Yes';
84        when(Metro eq 4) MetFlag = 'No';
85      end;
86    run;
```

```
ERROR: Unsatisfied WHEN clause and no OTHERWISE clause at line 85 column 3.
Metro=5 MetFlag=   _ERROR_=1 _N_=1
NOTE: The SAS System stopped processing this step because of errors.
WARNING: The data set WORK.OTHERWISENEEDED may be incomplete.  When this step was
stopped there were 0 observations and 2 variables.
```

As with many errors, SAS identifies the exact cause—no satisfied WHEN clause and no OTHERWISE statement—and indicates the location in the SAS log (line and column number) at which the error occurs. In addition, SAS prints the current PDV to the log following the error to assist in identification of the unexpected value. From one perspective, this requirement is an advantage—the SELECT group does not permit the conditional logic to accidentally ignore any cases present in the data.

SELECT Groups with a Select-Expression

As shown in above, the first statement in a SELECT group must be the SELECT statement. The version of the SELECT statement demonstrated so far uses only the keyword SELECT in this statement. Another version of the SELECT statement is shown in the snippet below.

```
select(Metro); ❶
  when(1,2,3) ❷ MetFlag = 'Yes';
  when(4) ❷ MetFlag = 'No';
  otherwise MetFlag = 'N/A';
end;
```

❶   The SELECT statement now contains a value, called a *select-expression*, in parentheses following the keyword SELECT, which can be any valid expression in SAS.

❷   WHEN statements still identify the distinct cases, a comma delimited list is used to indicate multiple values or a single value is valid as well. The comparison is for equality of the *select-expression* to values in the various *when-expressions*.

This SELECT group produces the same results as the SELECT snippet given at the opening of this section, and the use of a *select-expression* makes the conditioning slightly more compact than the equivalent IF-THEN/ELSE. If the number of conditions to check is large, and the *select-expression* is complex (and all comparisons are equalities), this can be a significant advantage in coding efficiency. Programs 4.5.5 and 4.5.6 consider conditioning on relatively complex *select-expressions*, each of which introduces a function useful in a variety of contexts. Program 4.5.5 creates a flag variable that is 1 if the property is mortgaged, 0 if it is not, and missing (.) if unknown. Conditioning on the full value of MortgageStatus is possible, but the full value is not required to achieve the conditional assignment, so the SCAN function is used to simplify this process.

Program 4.5.5: Using the SCAN Function to Simplify Conditioning

```
data work.one;
  set BookData.Ipums2005Basic;
  select(upcase(scan(MortgageStatus,1,',')));  ❶
    when('YES') ❷ MortFlag = 1;
    when('NO') ❷ MortFlag = 0;
    when('N/A') ❷ MortFlag = .;
  end;
run;
```

❶ The *select-expression*, uses both the SCAN and UPCASE functions—the UPCASE function is discussed in Section 3.9.2. The first argument of the SCAN function is an expression which is taken as a character value (numeric values are converted). The second argument is the piece or "word" to extract from the string—words are determined based on a set of default delimiters that are operating system dependent. The third argument allows for specification of the set of delimiters to use, much like the DLM= option in the INFILE statement. A fourth argument is possible to set special behaviors for the SCAN function. Here, the value of MortgageStatus up to the first comma is extracted, which is sufficient to determine each potential value of the flag being created.

❷ Even though the casing of the values of MortgageStatus is consistent throughout, defensive programming techniques do not assume this. Using the UPCASE function allows for a single condition to be checked irrespective of what casings are used in the data set.

In Program 4.5.6, the flag variable is created with a simplified set of rules: 1 if the property is mortgaged, 0 if it is not or the status is unknown. While extracting the first word with the SCAN function is still effective, the first letter is sufficient to properly make this assignment, which is achieved with the SUBSTR (substring) function.

Program 4.5.6: Using the SUBSTR Function to Simplify Conditioning

```
data work.two;
  set BookData.Ipums2005Basic;
  select(upcase(substr(MortgageStatus,1,1)));  ❶
    when('Y') ❷ MortFlag = 1;
    when('N') ❷ MortFlag = 0;
  end;
run;
```

❶ The first argument of the SUBSTR function is an expression which is taken as a character value (numeric values are converted). The second argument is the starting position that forms the substring, and the third argument is the number of characters to read. The third argument is optional—if omitted, SUBSTR reads from the starting position to the end of the string. Here, starting at the first character and reading one character means only the first character from the value of MortgageStatus is extracted.

❷ Using the UPCASE function again provides for some measure of defensive programming. Also, both here and in Program 4.5.5, not using the OTHERWISE statement is a form of defensive programming—if any values not accounted for in the conditioning actually exist in the data, the processing stops and an error is transmitted to the log.

It is possible to use a *select-expression* with logical expressions in the *when-expressions*, but this is typically a dangerous practice—this danger is highlighted in Program 4.5.7.

Program 4.5.7: Demonstrating Inappropriate Usage of a Select-Expression

```
data work.wrong;
  length MetFlag $ 3;
  Metro = 2;  ❶
  select(Metro);  ❷
    when(Metro in (1,2,3)) ❸ MetFlag = 'Yes';
    when(Metro eq 4) ❹ MetFlag = 'No';
    otherwise MetFlag = 'N/A';  ❺
  end;
run;
```

❶ For the current record, the value of Metro has been set to 2.

❷ The SELECT statement contains a *select-expression*, Metro. *When-expressions* are compared to its value, 2, until one is found to match.

❸ For the first WHEN statement, since Metro=2 the first *when-expression* resolves to TRUE, and SAS assigns the value 1 for this TRUE logical expression (it assigns 0 for those that are FALSE). As a result, SAS compares the *select-expression* value of 2 with the *when-expression* value of 1, giving a result of FALSE.

❹ SAS moves to the next *when-expression*, but with a similar result. Metro eq 4 evaluates to FALSE, which SAS assigns the value 0. Thus, the *select-expression* to *when-expression* comparison is 2 = 0, which also resolves as FALSE.

❺ Finally, since no matches occurred between the *select-expression* and any *when-expression*, SAS moves to the OTHERWISE statement and sets MetFlag equal to N/A.

Select-expressions are commonly used in cases where a single variable with distinct levels is of interest. In cases where intervals of values are of interest, such as household income or mortgage payment amounts, a SELECT group without a *select-expression* is more beneficial. In cases where multiple variables are of interest, either compound *when-expressions* or nested SELECT groups may be used exactly as is possible with IF-THEN/ELSE chains.

4.6 Working with Dates and Times

Section 3.6 introduced the concept of how SAS dates, times, and datetimes are numeric values: dates are the number of days since January 1, 1960; times are the number of seconds from midnight; and datetimes are the number of seconds since midnight on January 1, 1960. Informats provide an easy method for reading such values into numeric variables for future use, and formats provide a variety of ways to display date, time, and datetime values. SAS also provides a rich set of functions for working with dates, times, and datetimes. Program 4.6.1 demonstrates several techniques for creating and working with dates.

Program 4.6.1: Creating and Doing Arithmetic Operations on Dates
```
data work.DatesDemo01;
  set BookData.Ipums2001Basic(in = in2001 obs = 3)
      BookData.Ipums2005Basic(in = in2005 obs = 3);
  if in2001 eq 1 then year = 2001;
    else if in2005 eq 1 then year = 2005;
  Date = mdy(12,31,year); ❶
  DynamicDay = today(); ❷
  FixedDay = '14MAR2019'd; ❸
  Diff1 = DynamicDay - Date; ❹
  Diff2 = FixedDay - Date; ❹
  Diff3 = Time() - '13:56't; ❺
run;

proc report data = work.DatesDemo01;
  column Serial State Year Date DynamicDay FixedDay Diff1 Diff2 Diff3;
  label Serial = 'Serial';
run;
```

❶ The MDY function requires three numeric arguments that represent the month, day, and year for a date, respectively. As shown here, both numeric constants and DATA step expressions are valid.

❷ The TODAY function retrieves the current date and stores it as a SAS date in the DynamicDay variable. If the code is executed again on a different date, then the new value is retrieved and stored. The function has no arguments, but the parentheses are required—this is to distinguish the TODAY function from a DATA step variable named Today. The DATE function also exists and performs the same action.

❸ In the case where fixed dates are necessary, one option is to place the date in a literal string followed immediately by the letter d—this is called a date literal. Note that only dates in the SAS DATE format, two- or four-digit year, are valid. SAS interprets the value in the literal by converting it to a SAS date. This is similar to how SAS interprets the value '09'x as a hexadecimal value due to the presence of the hexadecimal literal modifier.

❹ A simple approach to calculating a duration is the subtraction of the two date variables.

❺ Like the TODAY() function, the TIME() function retrieves the current time. The time literal modifier, t, interprets the quoted string as a SAS time (operating like the hexadecimal and date literal modifiers). A datetime modifier, dt, is also available.

Output 4.6.1 shows the results of the calculations run on a specific date and demonstrates the need for formats when displaying most date, time, and datetime fields. Variables such as Date, DynamicDay, and FixedDay are all calendar dates – but the stored values are not interpretable by most readers. However, the difference variables simply represent the number of days between two dates—for Diff1 and Diff2—and seconds between two times—for Diff3. Date or time formats are inappropriate for these variables, though they would perhaps benefit from the use of a COMMA format to improve readability.

Output 4.6.1: Creating and Doing Arithmetic Operations on Dates

Serial	state	year	Date	DynamicDay	FixedDay	Diff1	Diff2	Diff3
1	Alabama	2001	15340	21691	21622	6351	6282	-3636.229
2	Alabama	2001	15340	21691	21622	6351	6282	-3636.229
3	Alabama	2001	15340	21691	21622	6351	6282	-3636.229
2	Alabama	2005	16801	21691	21622	4890	4821	-3636.229
3	Alabama	2005	16801	21691	21622	4890	4821	-3636.229
4	Alabama	2005	16801	21691	21622	4890	4821	-3636.229

SAS provides several functions for calculating durations at specific scales and extracting information from date, time, and datetime values. Program 4.6.2 demonstrates several useful date functions as well as the application of a few common date formats.

Program 4.6.2: Using Date Functions

```
data work.DatesDemo02;
  set BookData.Ipums2001Basic(in = in2001 obs = 3)
      BookData.Ipums2005Basic(in = in2005 obs = 3);
  if in2001 eq 1 then year = 2001;
    else if in2005 eq 1 then year = 2005;
  Date = mdy(12,31,year);
  FixedDay = '14MAR2019'd;
  Years = yrdif(mdy(12,31,year) ❶, '14MAR2019'd❶, 'Actual'); ❷
  Int1 = intck('year', mdy(12,31,year), '14MAR2019'd,'continuous'); ❸
  Int2 = intck('month'❹, mdy(12,31,year), '14MAR2019'd);
  Next1 = intnx('month', '14MAR2019'd, 3, 'beginning'); ❺
  Next2 = intnx('month', '14MAR2019'd, 3, 'sameday'); ❻
  Day1a = weekday(Next1); ❼
  Day1b = weekday(Next2); ❼
  Format Date date7.❽ FixedDay yymmdd8.❽ Day1b next2 downame. ❾ Next1 date7.;
run;
```

```
proc print data = work.DatesDemo02 noobs;
  var Serial Date FixedDay Years Int1 Int2 Next1 Day1a Day1b Next2;
run;
```

❶ The results of date functions and date literals are valid for direct input into other functions.

❷ Based on the dates in the first and second argument, the YRDIF function calculates the time between the dates, in years, using a predefined calculation method. Here, the ACTUAL method instructs SAS the actual number of days (365/non-leap year and 366/leap year). The DATDIF function is available for calculating differences, in days, between dates. For details about the various calculation methods available, see the SAS Documentation.

❸ The INTCK function calculates the number of time interval boundaries between two dates. Here, the INTCK function uses the CONTINUOUS method to calculate the number of complete years between the two dates. For example, when using the CONTINUOUS method, the number of years between 14MAR2019 and 13MAR2023 is three since only three full years have elapsed.

❹ Here, the INTCK function calculates the number of months between the two dates using the default DISCRETE method. For example, the number of months between 14MAR2019 and 01APR2019 is one since the dates include one month boundary—the first of the month. Custom intervals and boundaries are available. For more information, see the SAS Documentation.

❺ The INTNX function is similar to the INTCK function in that it is based on time intervals. However, it is used to determine the future date based on a time interval and a target date during the interval. Here, 14MAR2019 is the initial date, and the INTNX function calculates a date three months in the future. The BEGINNING method aligns the calculated date to the beginning of the interval—first day of the month in this case.

❻ This INTNX function also projects a date three months in the future, but sets the target date using the SAMEDAY method to ensure the day of the month is the same as the initial date.

❼ The WEEKDAY function extracts the day of the week from the provided date and stores it as a numeric value (1=Sunday, 2=Monday, ...). Other functions such as QTR, YEAR, MONTH, and more are also available. The same value is calculated twice to demonstrate the use of formats in ❽.

❽ As discussed in previous chapters, a wide variety of formats are available—including some specifically designed for use with dates and times. If the width value is too small to allow for a full representation of the date, SAS adjusts how it displays the dates. For more details about this, see the SAS Documentation and Chapter Note 2 in Section 4.10.

❾ It is important to understand that date formats treat the underlying value as a SAS date. Here, Day1b is no longer representing a date, it represents the day of the week, but the DOWNAME format interprets the current value—six—as a date value (Thursday, January 7, 1960) and thus Day1b is displayed as Thursday while Next2 appears as Friday.

Output 4.6.2: Using Date Functions

Serial	state	year	Date	DynamicDay	FixedDay	Diff1	Diff2	Diff3
1	Alabama	2001	15340	21696	21622	6356	6282	14776.169
2	Alabama	2001	15340	21696	21622	6356	6282	14776.169
3	Alabama	2001	15340	21696	21622	6356	6282	14776.169
2	Alabama	2005	16801	21696	21622	4895	4821	14776.169
3	Alabama	2005	16801	21696	21622	4895	4821	14776.169
4	Alabama	2005	16801	21696	21622	4895	4821	14776.169

4.7 Data Exploration with the UNIVARIATE Procedure

Demonstrations on using the MEANS and FREQ procedures as data exploration tools are given in Chapter 3; this section presents data exploration methods using PROC UNIVARIATE. The UNIVARIATE procedure produces several summary statistics by default, is capable of producing others, and can also produce a variety of plot types.

4.7.1 Summary Statistics in PROC UNIVARIATE

When PROC UNIVARIATE is executed with only the PROC UNIVARIATE statement and DATA= as its only option, the default behavior is to summarize all numeric variables. Since the IPUMS CPS data contains variables such as Serial and CountyFIPS that are numeric but not quantitative, many of the default summaries from PROC UNIVARIATE are undesirable. To choose variables for analysis, PROC UNIVARIATE uses a VAR statement that works in much the same manner as it does in PROC MEANS. Program 4.7.1 summarizes selected variables from the combination of the IPUMS CPS data from 2001 and 2005 (similar to Program 4.5.2) using PROC UNIVARIATE.

Program 4.7.1: Default Summaries Generated by PROC UNIVARIATE

```
data work.Ipums2001and2005;
  set BookData.Ipums2001Basic(in = in2001)
      BookData.Ipums2005Basic(in = in2005);
  if in2001 eq 1 then Year = 2001;
    else if in2005 eq 1 then Year = 2005;
run;

proc univariate data= work.Ipums2001and2005;
  var MortgagePayment HomeValue Citypop;
run;
```

By default, PROC UNIVARIATE creates five output tables for each variable listed. The resulting tables for the MortgagePayment variable are shown in Output 4.7.1

Output 4.7.1: Default Output from PROC UNIVARIATE on the MortgagePayment Variable

Moments			
N	1642156	Sum Weights	1642156
Mean	472.363975	Sum Observations	775695336
Std Deviation	707.232195	Variance	500177.377
Skewness	2.45508546	Kurtosis	10.4176149
Uncorrected SS	1.18778E12	Corrected SS	8.21369E11
Coeff Variation	149.721874	Std Error Mean	0.55189291

Basic Statistical Measures			
Location		Variability	
Mean	472.3640	Std Deviation	707.23219
Median	0.0000	Variance	500177
Mode	0.0000	Range	7900
		Interquartile Range	790.00000

Tests for Location: Mu0=0				
Test	**Statistic**		**p Value**	
Student's t	t	855.8979	Pr > |t|	<.0001
Sign	M	387391.5	Pr >= |M|	<.0001
Signed Rank	S	1.501E11	Pr >= |S|	<.0001

Quantiles (Definition 5)	
Level	**Quantile**
100% Max	7900
99%	3000
95%	1800
90%	1400
75% Q3	790
50% Median	0
25% Q1	0
10%	0
5%	0
1%	0
0% Min	0

Extreme Observations			
Lowest		**Highest**	
Value	**Obs**	**Value**	**Obs**
0	1.64E6	7900	694144
0	1.64E6	7900	694186
0	1.64E6	7900	694359
0	1.64E6	7900	695015
0	1.64E6	7900	695754

The default tables in Output 4.7.1 contain information about moments (basic and central moments of various orders), some measures of center and variability (a few of which are repeats from the moments table), tests for a mean of zero, a set of common quantiles, and an extreme observations table (denoting the five largest and smallest values for the variable and the record number in the data set for each). In several instances, a subset of these tables is desired and, to achieve this, the ODS SELECT or ODS EXCLUDE statements are available, as first shown in Program 1.5.2. (As noted in Program 1.5.1, table names can be determined via the ODS TRACE statement.) Program 4.7.2 modifies Program 4.7.1 to display only the table for basic measures of center and variability along with the quantiles table for all variables (output not shown).

Program 4.7.2: Using ODS SELECT to Subset Output Tables

```
proc univariate data= work.Ipums2001and2005;
  var MortgagePayment HomeValue Citypop;
  ods select Quantiles BasicMeasures;
run;
```

PROC UNIVARIATE also permits the use of a CLASS statement, and it behaves in much the same manner as seen with PROC MEANS in Section 2.3.2. Year is used as a class variable in Program 4.7.3, with ODS SELECT limiting results to the Basic Statistical Measures table, with the results shown in Output 4.7.3.

Program 4.7.3: Using a Class Variable in PROC UNIVARIATE

```
proc univariate data= work.Ipums2001and2005;
  class year;
  var HomeValue Citypop;
  ods select BasicMeasures;
run;
```

Output 4.7.3: Using a Class Variable in PROC UNIVARIATE

Variable: HomeValue (House value)
Year = 2001

Basic Statistical Measures			
Location		Variability	
Mean	2964943	Std Deviation	4444513
Median	162500	Variance	1.97537E13
Mode	9999999	Range	9994999
		Interquartile Range	9914999

Variable: HomeValue (House value)
Year = 2005

Basic Statistical Measures			
Location		Variability	
Mean	2793526	Std Deviation	4294777
Median	225000	Variance	1.84451E13
Mode	9999999	Range	9994999
		Interquartile Range	9887499

Variable: CITYPOP (City population)
Year = 2005

Basic Statistical Measures			
Location		Variability	
Mean	2916.656	Std Deviation	12316
Median	0.000	Variance	151690509
Mode	0.000	Range	79561
		Interquartile Range	0

Note that the only level for the CLASS variable Year is 2005 for the analysis variable CityPop; CityPop is not present in the IPUMS CPS 2001 data, so all values are missing for that class. By default, PROC UNIVARIATE produces a table noting the missing values when they occur, but this is suppressed by the selection of only the BasicMeasures table in ODS SELECT. Also note the mode of 9,999,999 for HomeValue, which is due to a special encoding in the IPUMS CPS data. If it is desired to restrict the analysis to records without that special encoding, care should be taken to achieve the desired result. Program 4.7.4 uses a WHERE statement to eliminate these values, with two of the output tables shown in Output 4.7.4A and 4.7.4B.

Program 4.7.4: Subsetting Records with the WHERE Statement in PROC UNIVARIATE

```
proc univariate data= work.Ipums2001and2005;
  class year;
  var HomeValue Citypop;
  ods select BasicMeasures;
  where HomeValue ne 9999999;
run;
```

Output 4.7.4A: Where Subsetting (Partial Output, HomeValue for 2005)

Variable: HomeValue (House value)
Year = 2005

Basic Statistical Measures			
Location		Variability	
Mean	238922.4	Std Deviation	219007
Median	162500.0	Variance	4.7964E10
Mode	225000.0	Range	995000
		Interquartile Range	255000

The WHERE subsetting has the desired effect on the HomeValue results. However, since CityPop is also part of the PROC UNIVARIATE analysis, those results are also different.

Output 4.7.4B: Where Subsetting (Partial Output, CityPop for 2005)

Variable: CITYPOP (City population)
Year = 2005

Basic Statistical Measures			
Location		**Variability**	
Mean	1799.078	**Std Deviation**	9196
Median	0.000	**Variance**	84572682
Mode	0.000	**Range**	79561
		Interquartile Range	0

Since the WHERE condition subsets the entire record based on the provided condition, a record having a HomeValue of 9,999,999 also has its Citypop value excluded from the analysis. If this is not what is desired, separate submissions of PROC UNIVARIATE must be made with the appropriate variables and conditions. Program 4.7.5 also produces the table shown in Output 4.7.4B.

Program 4.7.5: How WHERE Subsets Records

```
proc univariate data= work.Ipums2001and2005;
  class year;
  var CityPop;
  ods select BasicMeasures;
  where HomeValue ne 9999999;
run;
```

4.7.2 Graphing Statements in PROC UNIVARIATE

The UNIVARIATE procedure also allows for a variety of diagnostic plots to be constructed which typically include the ability to check the fit of certain theoretical distributions. PROC UNIVARIATE includes probability plots, cumulative distribution plots, probability-probability plots, quantile-quantile plots, and histograms. Only quantile-quantile plots and histograms are considered in this section. Plotting statements can be made for any variable listed in the VAR statement, as is shown in Program 4.7.6.

Program 4.7.6: Generating Plots in PROC UNIVARIATE

```
proc univariate data= work.Ipums2001and2005;
  class year;
  var MortgagePayment;
  histogram MortgagePayment ❶ / normal(mu=est sigma=est) ❷;
  qqplot MortgagePayment ❸ / weibull(c=est sigma=est theta=est) ❹;
  where MortgagePayment gt 0 ❺;
run;
```

❶ The HISTOGRAM statement creates a two-panel graph in Output 4.7.6A, one panel for each level of the class variable Year.

❷ Several different distributions can be requested with parameters specified or estimated (EST) from the data. See the SAS Documentation for more information about distribution choices and associated parameters.

❸ The QQPLOT statement also generates a two-panel graph in Output 4.7.6B.

❹ The quantiles from the data are plotted against the theoretical quantiles of the chosen distribution. If a sufficient set of parameters is chosen, a reference line is produced.

❺ MortgagePayment is subset in this case to remove the dominant number of zeros in the data and to make the parameter set of the Weibull distribution estimable.

Output 4.7.6A: Generating Histograms in PROC UNIVARIATE

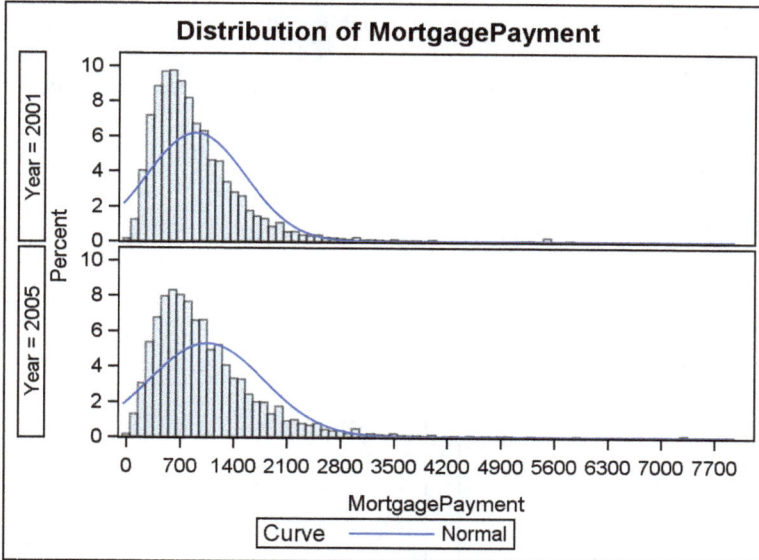

Distribution of MortgagePayment

Output 4.7.6B: Generating QQ Plots in PROC UNIVARIATE

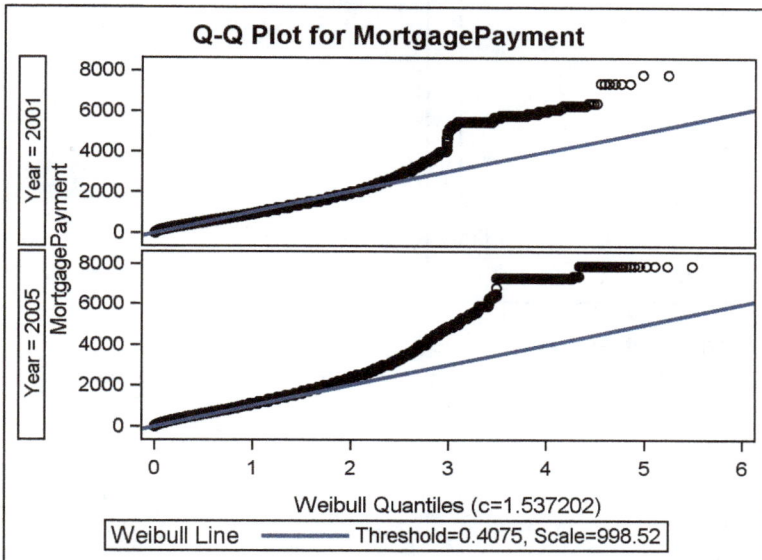

Q-Q Plot for MortgagePayment

The next section revisits histograms as part of the SGPLOT procedure; however, quantile, probability, and probability-probability plots are often useful diagnostic tools for assessing distributions of data, but are not directly implemented in PROC SGPLOT. As these plots and their use are beyond the scope of this book, refer to the SAS Documentation for more information about using them.

4.8 Data Distribution Plots

Histograms and boxplots are commonly used to display information about data distributions and are available through the SGPLOT procedure. Various statements and options for creating data distribution plots are covered in this section, drawing on similarities to the charting concepts covered in Sections 3.3 through 3.5.

4.8.1 Histograms

The HISTOGRAM statement in the SGPLOT procedure requires a numeric variable from a data set, producing a relative frequency histogram across a set of bins. Program 4.8.1 uses the combined 2001 and 2005 IPUMS CPS data created in Program 4.7.1 to create a histogram for nonzero values of MortgagePayment shown in Output 4.8.1.

Program 4.8.1: Creating a Histogram in PROC SGPLOT

```
proc sgplot data= work.Ipums2001and2005;
  histogram MortgagePayment;
  where MortgagePayment gt 0;
run;
```

Output 4.8.1: Histogram of Mortgage Payments

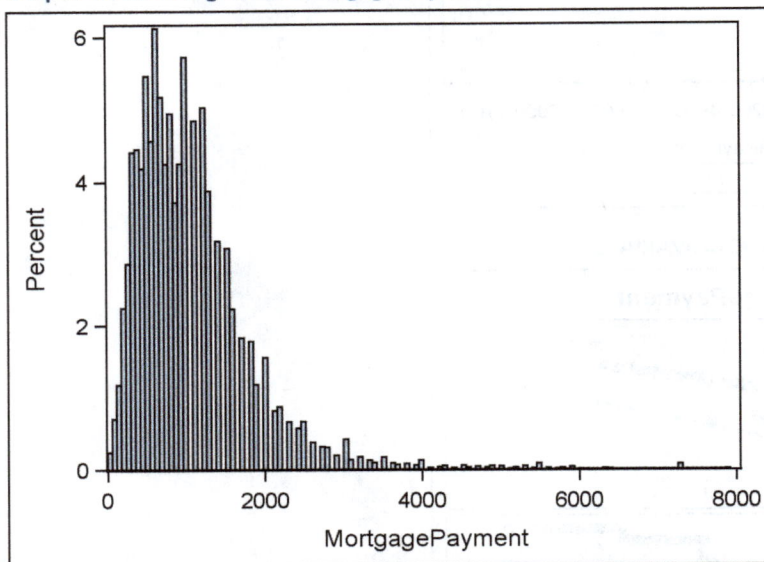

In this case, the default choice of bins is poor, and other aspects of the graph can be improved as well. Program 4.8.2 uses AXIS statements, initially discussed in Section 3.4.2, and some options specific to the HISTOGRAM statement to improve Output 4.8.1.

Program 4.8.2: Histogram Options and Axis Modifications

```
proc sgplot data= work.Ipums2001and2005;
  histogram MortgagePayment / binstart=250 binwidth=500❶ scale=proportion❷
                              dataskin=gloss❸;
  xaxis label='Mortgage Payment' valuesformat=dollar8.;❹
  yaxis display=(nolabel) valuesformat=percent7.;❹
  where MortgagePayment gt 0;
run;
```

❶ BINSTART= and BINWIDTH= set the starting point for the first bin and the width of all bins, respectively. The location of the bin is referenced by its midpoint, so making the first bin span 0 to 500 requires a starting value of 250 in conjunction with a width of 500.

❷ The SCALE= option allows for choices for summarizing frequency or relative frequency, here PROPORTION summarizes relative frequency with a decimal value. The VALUESFORMAT= option in the YAXIS statement alters the display of the values.

❸ Like bar charts, bar fills for a histogram accept the DATASKIN= option.

❹ XAXIS and YAXIS statements are available as discussed in Section 3.4.2. Refer to that section and the SAS Documentation for more details about these and other options.

Output 4.8.2: Setting Options for a Histogram

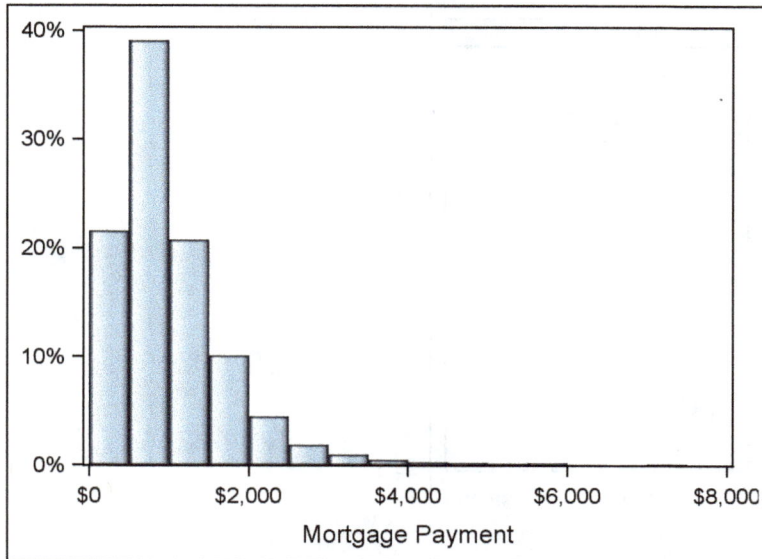

These histograms cannot be separated by year with a CLASS statement as PROC UNIVARIATE does in Section 4.7.2. Previous chapters show that many plotting calls in PROC SGPLOT allow for grouping; however, the HISTOGRAM statement does not. To produce a similar plot to the histogram panels in Output 4.7.6, the SGPANEL procedure can be used.

4.8.2 Histograms in SGPANEL

The SGPANEL procedure provides the same essential set of plotting statements as PROC SGPLOT with other statements and options to create multi-panel graphs. The panels are determined by the levels of the variables listed in the PANELBY statement, and the XAXIS and YAXIS statements are replaced by the COLAXIS and ROWAXIS statements, respectively. Program 4.8.3 modifies the previous histogram into two panels across the Year variable.

Program 4.8.3: Multi-Panel Histogram

```
proc sgpanel data= work.Ipums2001and2005;
  panelby Year; ❶
  histogram MortgagePayment / binstart=250 binwidth=500 scale=proportion
                              dataskin=gloss;
  colaxis label='Mortgage Payment' valuesformat=dollar8.; ❷
  rowaxis display=(nolabel) valuesformat=percent7.; ❷
  where MortgagePayment gt 0;
run;
```

❶ The PANELBY statement treats each variable as categorical, much like a CLASS statement in PROC MEANS or PROC UNIVARIATE. The specified plots are repeated in panels for each level of the variable or each combination of levels when multiple variables are given.

❷ The options listed in COLAXIS and ROWAXIS here are the same as those in Program 4.8.2. However, given the nature of the panel graph, these options now apply across axes on both histograms

Output 4.8.3: Multi-Panel Histogram

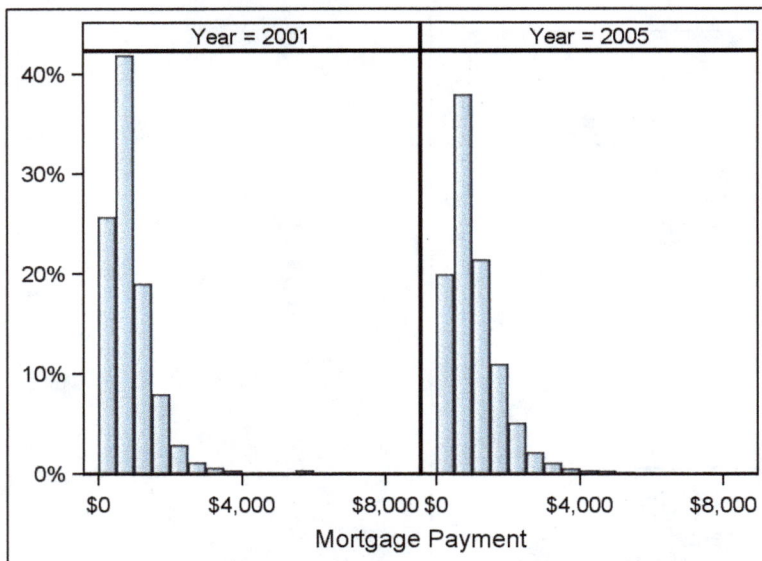

Several options are available as part of the PANELBY statement to control the arrangement and appearance of the panels. Program 4.8.4 illustrates some of these, see the SAS Documentation for information about more available options.

Program 4.8.4: Altering Panel Structure with PANELBY Statement Options

```
proc sgpanel data= work.Ipums2001and2005;
   panelby Year / columns=1  ❶  novarname ❷ headerattrs=(family='Georgia') ❸;
   histogram MortgagePayment / binstart=250 binwidth=500 scale=proportion
                               dataskin=gloss;
   colaxis label='Mortgage Payment' valuesformat=dollar8.;
   rowaxis display=(nolabel) valuesformat=percent7.;
   where MortgagePayment gt 0;
run;
```

❶ COLUMNS= or ROWS= can be used to control the size of the grid in either direction—only one of the two should be specified.

❷ NOVARNAME suppresses the variable name from appearing in the heading label—only the variable value appears.

❸ ATTRS are available for several elements, including the headers. As discussed in Section 3.4.2, FAMILY= sets the font face for text elements.

Output 4.8.4: Altering Panel Structure with PANELBY Statement Options

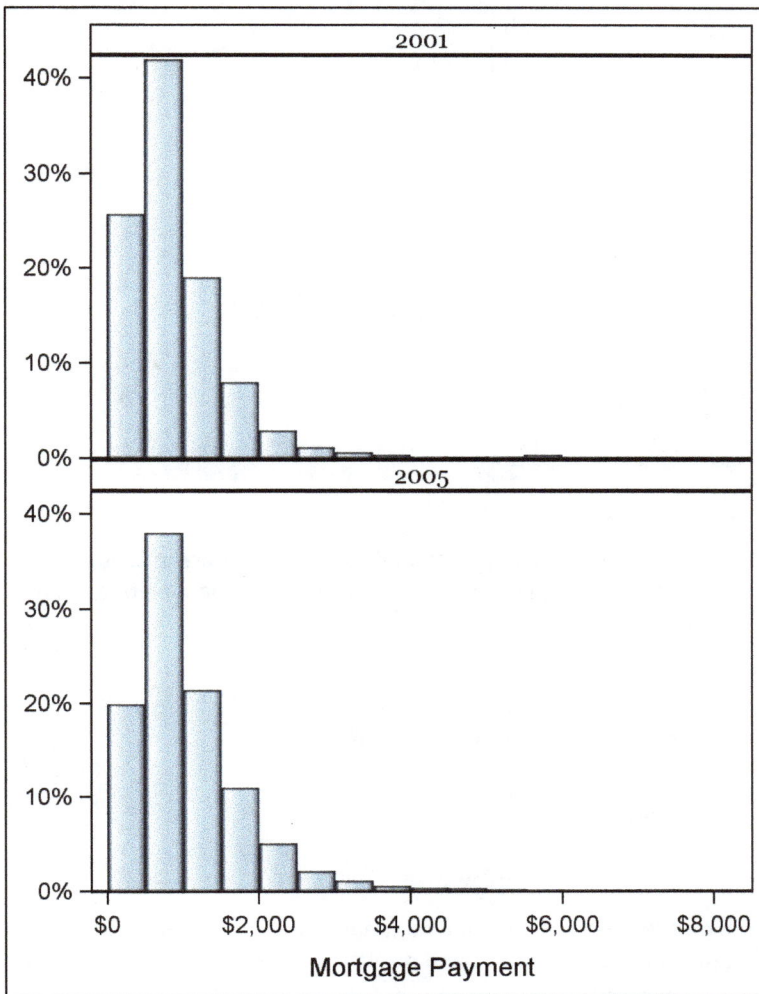

4.8.3 Boxplots

Boxplots are available in PROC SGPLOT via the VBOX or HBOX statement for vertical or horizontal boxplots, respectively. The default boxplot style is the typical outlier boxplot, though other styles are available. The VBOX and HBOX statements each permit the GROUP= option and the GROUPDISPLAY=CLUSTER setting, first used for bar charts in Section 3.3.3, as Program 4.8.5 shows.

Program 4.8.5: Grouped Boxplot

```
proc sgplot data= work.Ipums2001and2005;
  vbox MortgagePayment / group=year groupdisplay=cluster;
  where MortgagePayment gt 0;
run;
```

Output 4.8.5: Grouped Boxplot

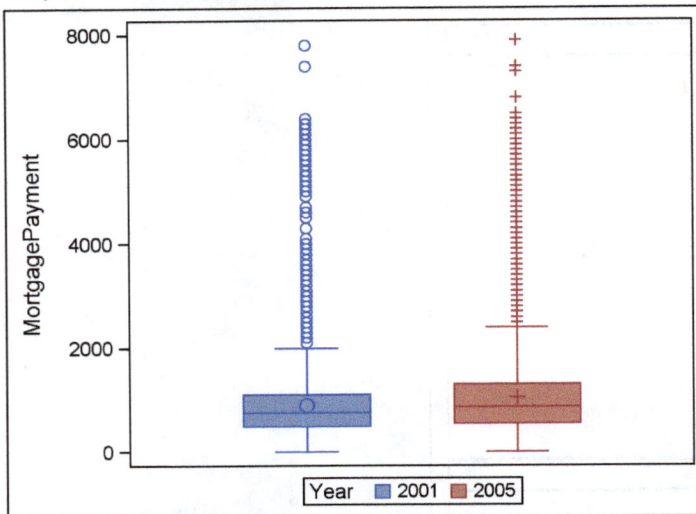

Various options are available for modifying the boxplot, and axis and legend modifications are available to set options similar to those discussed in previous plotting examples. Program 4.8.6 modifies the type of boxplot and several styles on the graph in Output 4.8.6.

Program 4.8.6: Boxplot Modifications

```
proc sgplot data= work.Ipums2001and2005;
  vbox MortgagePayment / group=year groupdisplay=cluster extreme ❶
                         whiskerattrs=(color=red) ❷;
  keylegend / position=topright location=inside title=''; ❸
  yaxis display=(nolabel) valuesformat=dollar8.;
  where MortgagePayment gt 0;
run;
```

❶ EXTREME draws the whiskers of the boxplot to the maximum and minimum values; thus, no outliers are displayed. Other options to control the positioning of whiskers and display of outliers are available.

❷ The boxplot is made up of several line elements, each of which has an ATTRS option. Here, both sets of whiskers are colored red in the WHISKERATTRS= option.

❸ Since this is a grouped chart, a legend is produced, and its attributes are changeable via options in the KEYLEGEND statement, as first shown in Section 3.4.2.

Output 4.8.6: Boxplot Modifications

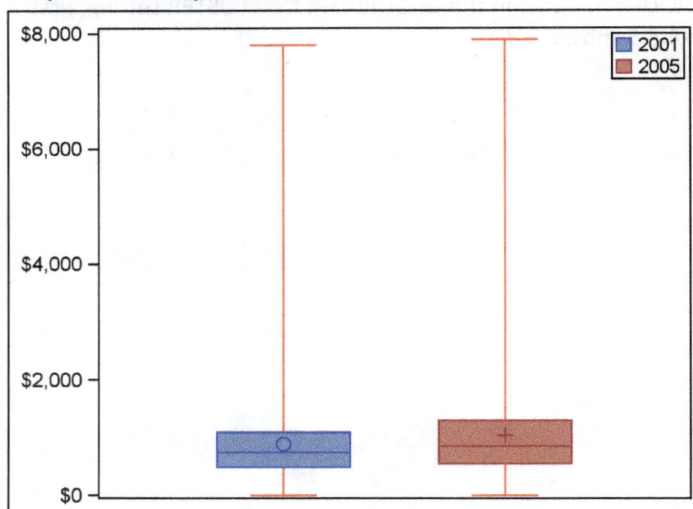

The number of graphing statements and options available in PROC SGPLOT is extensive; however, the general logic enables relatively easy associations between them. Reviewing the PROC SGPLOT documentation and examples from various sources is important for developing sound plotting skills. Section 4.8.4 introduces the high-low plot and some advanced variations on it to create a customized version of the boxplot.

4.8.4 High-Low Plots

High-low plots have their origins and most common use in financial time series; however, they are valuable in a variety of other scenarios. The HIGHLOW statement requires specification of three variables to generate the plot. Two of these are variables specified for each of LOW= and HIGH= with the requirement, as expected, that the value for the variable set as HIGH= is greater than or equal to that of LOW= for all records in the data set. The third variable is set as one of X= or Y= and controls the orientation of the graph. Choosing an X= variable provides a vertical axis for the high-low range; Y= orients this range horizontally.

The programs in this section, starting with Program 4.8.7, use data containing several percentiles on MortgagePayment, which is generated from the combined IPUMS CPS data for 2001 and 2005 using PROC MEANS and the ODS OUTPUT statement.

Program 4.8.7: High-Low Plot

```
proc means data= work.Ipums2001and2005 min p10 p25 median p75 p90 max;
   class year;
   var MortgagePayment;
   where MortgagePayment gt 0;
   ods output summary= work.MPQuantiles;
run;

proc sgplot data= work.MPQuantiles;
   highlow x=year low=MortgagePayment_p25 high=MortgagePayment_P75;
run;
```

Output 4.8.7: High-Low Plot

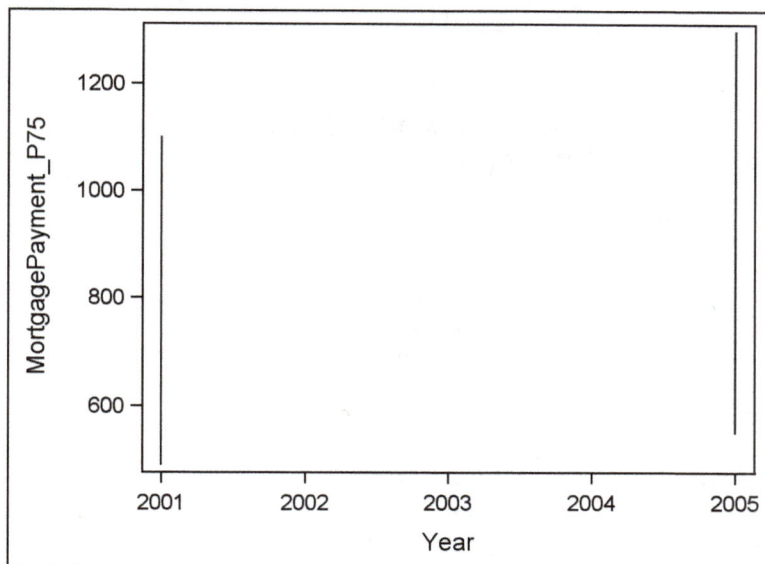

The default plot, as shown in Output 4.8.7, is a line between the high and low value at each distinct value of the third value chosen for X= or Y=. As with other statements, the HIGHLOW statement contains various options to alter the display. Other PROC SGPLOT statements are also of utility here, Program 4.8.8 uses some of these to enhance the graph.

Program 4.8.8: High-Low Chart Modifications

```
proc sgplot data= work.MPQuantiles;
  highlow x=year low=MortgagePayment_P25 high=MortgagePayment_P75
          / type=bar❶  fillattrs=(color=cx99FF99) dataskin=sheen❷;
  xaxis values=(2001 2005) ❸ display=(nolabel);
  yaxis label='Mortgage Payment' valuesformat=dollar8.;
run;
```

❶ The display between the high and low values can be set as a bar instead of a line.

❷ Since the plotting elements are changed to bars, styling options for bars and their fills are available.

❸ The Year variable is effectively treated as quantitative even though it only has two distinct values. This is an instance where specifying a list of individual values for that axis is desirable.

Output 4.8.8: High-Low Plot Modifications

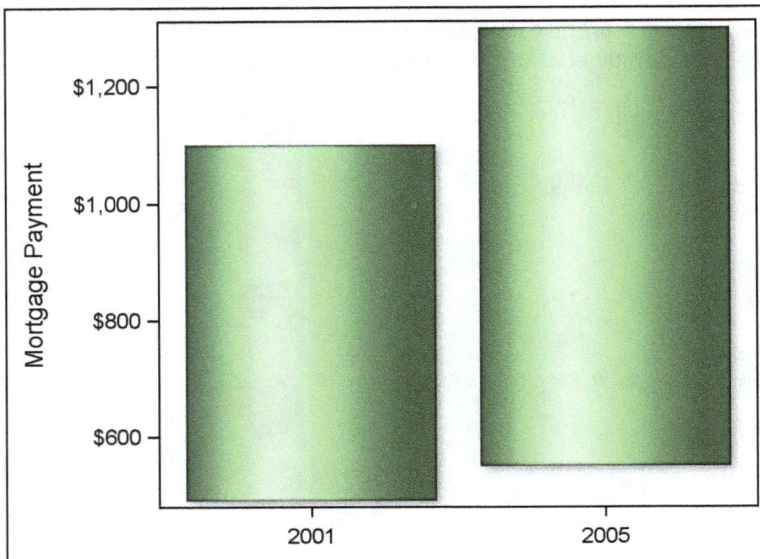

Using the ability to overlay several plots in a PROC SGPLOT call, Program 4.8.9 uses multiple HIGHLOW statements to create a custom variation of a boxplot in Output 4.8.9.

Program 4.8.9: Creating a Custom Boxplot from Multiple High-Low Plots

```
proc sgplot data= work.MPQuantiles;
  highlow x=year low=MortgagePayment_Min high=MortgagePayment_Max/
          legendlabel='Minimum to Maximum' lineattrs=(color=red) name='Line'; ❶
  highlow x=year low=MortgagePayment_P10 high=MortgagePayment_P90/
          legendlabel='10th to 90th Percentile' type=bar barwidth=.3
          fillattrs=(color=cx66AA66) name='Box1'; ❷
  highlow x=year low=MortgagePayment_P25 high=MortgagePayment_P75
          / type=bar legendlabel='Inter-Quartile Range' barwidth=.5
          fillattrs=(color=cx77FF77) name='Box2'; ❸
  xaxis values=(2001 2005) display=(nolabel);
  yaxis label='Mortgage Payment' valuesformat=dollar8.;
  keylegend 'Line' / position=topright location=inside noborder
                     valueattrs=(size=8pt);
  keylegend 'Box1' 'Box2' / across=1 position=topleft location=inside noborder
                            valueattrs=(size=8pt); ❹
run;
```

❶ The layer created by the first HIGHLOW statement is a red line from the maximum to minimum values. As a legend is generated for this overlay, the LEGENDLABEL= option is used to create a more useful description of this portion of the graph. NAME= uses a literal value to name the legend for use in a KEYLEGEND statement—this graph has two separate legends.

❷   The next plot, created by the second HIGHLOW statement, is from the 10th to 90th percentiles as a narrow bar in a green hue. A description for this portion of the graph is also transmitted to the legend via the LEGENDLABEL= option, and it is named with the NAME=option.

❸   The final plot spans the traditional first to third quartile span with a bar of somewhat more width in a different hue of green.

❹   As in previous examples, any legend generated can be altered with the KEYLEGEND statement. Unlike previous examples, two KEYLEGEND statements are used, with NAME= values from the plotting statements provided before the slash. Each is used to position and style the legend information— KEYLEGEND statements with multiple names include all legend information generated by all named plotting statements. A KEYLEGEND statement without a name thus includes all legend information from plots without a NAME= option.

Output 4.8.9: Custom Boxplot Created from Multiple High-Low Plots

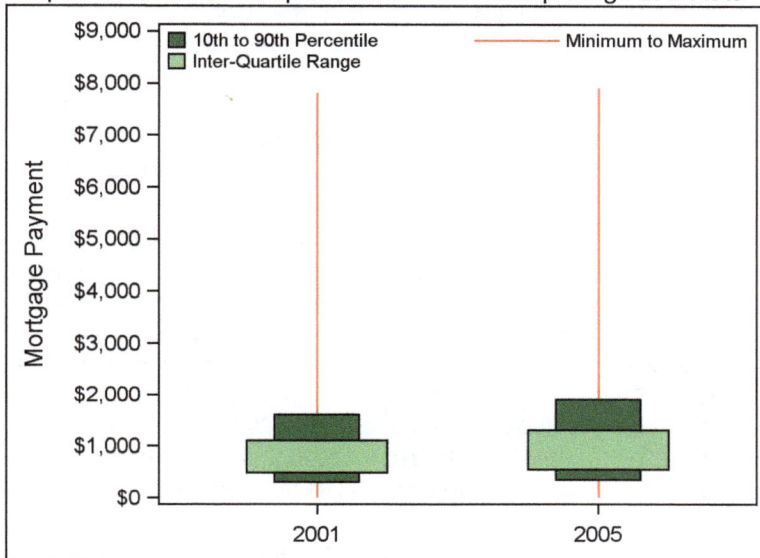

4.9 Wrap-Up Activity

Use the lessons and examples contained in this and previous chapters to complete the activity shown in Section 4.2.

Data

Various data sets are available for use to complete the activity. This data expands on the 2001 and 2005 data used for many of the in-chapter examples. Completing this activity requires the following files:

- lpums2005Basic.sas7bdat
- lpums2010Basic.sas7bdat
- lpums2015Basic.sas7bdat

Scenario

Use the skills mastered so far, including those from previous chapters, to assemble the listed files into a single data set and create the analyses shown in Section 4.2

4.10 Chapter Notes

1. *Attributes for a Common Variable Across Multiple Input Data Sets.* When multiple SAS data sets are referenced in a SET statement (or other DATA step statements, such as the MERGE statement introduced in Chapter 5), variables with the same name are established as a single variable in the PDV unless the types do not match, which generates an error and stops compilation of the DATA step. Mismatches for attributes like length, format, and label do not generate errors, and SAS sources these attributes from one of the data sets. Generally, these attributes are sourced from the first data set listed; however, this is not always the case. Default values for these attributes are often assigned, such as the default length of 8 bytes for character variables in list input, but default behaviors also exist for attributes that may be unassigned. The use of standard character or numeric formats or using variable names as labels in displayed output may be the result of a specific assignment of those attributes or a result default behavior when they are no format or label attribute is assigned—assignments of these attributes can be checked with PROC CONTENTS. If the attribute is not assigned in some of the referenced data sets, the first encounter of the attribute becomes the assignment for the final data set. Except in rare circumstances, the length attribute is always assigned and therefore is established by the first data set encountered. If that length is shorter than lengths assigned to that variable in other data sets, truncation can occur (and a warning is generated in the SAS log in such circumstances). It is a good programming practice to check attributes with PROC CONTENTS before attempting to combine data sets.

2. *YEARCUTOFF=.* When SAS displays a date using a format that limits the width so that full precision is not available, it generally begins by using only two-digit years, then by removing any delimiters—such as forward slashes—that appear in the date. Date, time, and datetime formats have a specific range of allowable widths, and the SAS Documentation demonstrates how values appear using a variety of widths. However, when reading date values—either from raw data or from a date literal—if only a two-digit year is present, then SAS uses the value of the system option YEARCUTOFF= to determine the century in which the two-digit year falls. For example, if YEARCUTOFF=1910 then SAS interprets the date 3/21/19 as March 21, 1919, while if YEARCUTOFF=1920, then 3/21/19 is interpreted as March 21, 2019. The two-digit year is compared to the last two digits in the YEARCUTOFF= value—if larger the century given is used; if smaller the next century is used. While two-digit years have become less common in modern databases, when dealing with historical data it is important to ensure date-related fields are properly represented. To determine the default value of the YEARCUTOFF= for a particular version of SAS, see the SAS Documentation.

4.11 Exercises

Concepts: Multiple Choice

1. Which of the following programs interleaves the data sets A and B according to the variable EmployeeID?

   a.
   ```
   data combined;
     set A B;
   run;
   ```

   b.
   ```
   data combined;
     set A B;
     by employeeID;
   run;
   ```

c.
```
proc sort data = A;
  by employeeID;
run;

proc sort data = B;
  by employeeID;
run;

data combined;
  set A B;
run;
```

d.
```
proc sort data = A;
  by employeeID;
run;

proc sort data = B;
  by employeeID;
run;

data combined;
  set A B;
  by employeeID;
run;
```

2. Which of the following variable attributes must match for a DATA step concatenation to combine columns from two contributing data sets?

   a. Type and Name

   b. Position and Name

   c. Position only

   d. Name only

3. Which of the following data set options can be used to identify which data sets contributed information to the current PDV?

   a. WHERE=

   b. KEEP=

   c. RENAME=

   d. IN=

4. Consider the following three items: 1) using a DROP= option on a data set in the SET statement, 2) using a DROP= option on the single data set listed in the DATA statement, and 3) using a DROP statement in the DATA step. Which of the following are equivalent?

   a. Item 1 and Item 2

   b. Item 1 and Item 3

   c. Item 2 and Item 3

   d. All three items are equivalent

5. Which of the following is false about DO groups?

   a. Must close with an END statement

   b. Can contain multiple SAS statements

   c. Cannot be used in an ELSE statement

   d. None of the above are false

6. Which of the following procedures can be used to generate histograms?
    a. UNIVARIATE
    b. SGPLOT
    c. SGPANEL
    d. All of the above

7. The GROUP= option is not available in PROC SGPLOT for which of the following plotting statements?
    a. VBAR
    b. VBOX
    c. HISTOGRAM
    d. HIGHLOW

8. For the SGPANEL procedure, which of the following is true regarding the XAXIS statement?
    a. It works the same as it does in PROC SGPLOT
    b. It is not available and is supplanted by the COLAXIS statement
    c. It is not available and is supplanted by the ROWAXIS statement
    d. None of the above

9. The orientation of a high-low plot in the SGPLOT procedure is determined by which of the following?
    a. The VERTICAL or HORIZONTAL option
    b. The choice of an X= or Y= variable
    c. The variable type (character or numeric)
    d. THE SGPLOT procedure itself

10. Which of the following programs could be used to interleave two data sets—Accounts1 and Accounts2—by the numeric key variables Date and Amount in order to produce the data set shown?

Combined

Date	Amount	Account
03JUL2019	$8,700	EK-1257
03JUL2019	$1,600	RJ-002X
18JUN2019	$3,200	JB-1977
18JUN2019	$425	JB-1941

a.
```
data combined;
  set Accounts1 Accounts2;
  by Date Amount;
run;
```

b.
```
data combined;
  set Accounts1 Accounts2;
  by descending Date Amount;
run;
```

c.
```
data combined;
  set Accounts1 Accounts2;
  by Date Amount descending;
run;
```

d.

```
data combined;
  set Accounts1 Accounts2;
  by descending Date descending Amount;
run;
```

Concepts: Short-Answer

1. For each of the following, what condition(s) must be satisfied?
   a. A successful concatenation of multiple data sets.
   b. A successful interleave of multiple data sets.
2. How is an IF-THEN/ELSE chain potentially more efficient than a series of IF-THEN statements?
3. Describe the difference between the WHERE and subsetting IF statements in the DATA step.
4. Both KEEP and DROP statements are available in the DATA step, and similarly both ODS SELECT and ODS EXCLUDE statements are available for use with any procedure. Why is it the case that both are available even though only one is used at a time?
5. Consider the following DATA step code.

```
data Logic;
  set InputData;
  select;
    when(1 <= status <= 5) rank = 'Assistant';
    when(3 <= status <=14) rank = 'Associate';
    when(status >= 10) rank = 'Full';
  end;
  if 0 <= SeqNum <= 100 then series = 3;
    else if 50 <= SeqNum <= 150 then series = 4;
run;
```

For each of the following, determine the results in each of the following scenarios and provide a justification.
   a. What is the value of Rank for a record with Status = 5?
   b. What is the value of Rank for a record with Status = 0?
   c. What is the value of Series for a record with SeqNum = 75?
   d. What is the value of Series for a record with SeqNum = 200?
6. Consider the following DATA step code, which is a modification of the DATA step from the previous question.

```
data Logic;
  set InputData;
  select;
    when(1 <= status <= 5) rank = 'Assistant';
    when(3 <= status <=14) ranks = 'Associate';
    when(status >= 10) rank = 'Full';
  end;
  if 0 <= SeqNum <= 100 then series = '3';
    else if 50 <= SeqNum <= 150 then series = '10';
      else series = 15;
run;
```

Use this DATA step to explain each of the following.
   a. In the SELECT block, both Rank and Ranks appear in the assignment statements. Describe the effects on the data set and list any notes, warnings, or errors that appear in the SAS log as a result of this discrepancy.
   b. In the IF-THEN/ELSE chain, Series appears in three assignment statements. Assuming each condition is satisfied at least once, what values does Series have in the data set?

7. For each of the following items, determine whether it correctly completes the DATA step below in order to write a subset of the data where Age is at least 55 or CholFlag is equal to Y.

```
data CheckUp;
   set PatientDemog(keep = Age Chol);
   if (Age <= 40 and Chol >= 130) or
      (Age >= 40 and Chol >= 100)
      then CholFlag = 'Y';
   <insert code here>
run;
```

   a. `if CholFlag = 'Y' or Age >= 55;`

   b. `where CholFlag = 'Y' or Age >= 55;`

   c. `if CholFlag = 'Y';`
      `where Age >= 55;`

   d. `where CholFlag = 'Y';`
      `if Age >= 55;`

8. The following code is a modified version of the code that generates Log 4.4.15. Execute the code below and use the results to help answer the following items.

```
data Base;
   length y $6;
   x=123456789;
   y=x;
   put _all_;
run;

data Subset3;
   set Base;
   if x = y;
run;
```

   a. Why does the Subset3 data set contain zero observations?

   b. Without introducing new statements or options, what modification of the code results in the implicit conversion successfully comparing X and Y to produce a non-empty Subset3 data set.

   c. Rewrite the second DATA step to correctly use explicit conversion when matching X and Y.

Programming Basics

1. The data sets BookData.Ratings2016 and BookData.Ratings2017 contain several reviews of restaurants. In both data sets, ID is used as a unique identifier for the restaurants. Str1 represents the rating and Str2 represents the rater's first name. Write a program that combines the data sets according to the following specifications. In each, ensure a variable exists that contains the year in which the restaurant was rated.

   a. Concatenates the data sets

   b. Interleaves the data sets based on Str2

   c. Interleaves the data sets based on Str2 and Str1

2. Using the data sets from the previous question, use the appropriate function(s) to create a new variable, Stars, that is numeric and contains the number of stars assigned to each restaurant. (For example, in the 2016 data the values should be 4.5, 4, 5, 3, and 5.) Interleave the resulting data sets by Stars and ID, and include the year in which the restaurant was rated.

3. Use the Sashelp.Stocks data set to do the following.

   a. Create a new data set that includes two new variables, with one being Year. The other is FiscalSeason, which has the values Federal1 during the months October, November, and December; Federal2 during the months January, February, and March; Federal3 during the months April, May, June; and Federal4 for the months of July, August, and September.

   b. Summarize the maximum and minimum of monthly high price for each combination of Year and FiscalSeason.

c. Create the following panel graph:

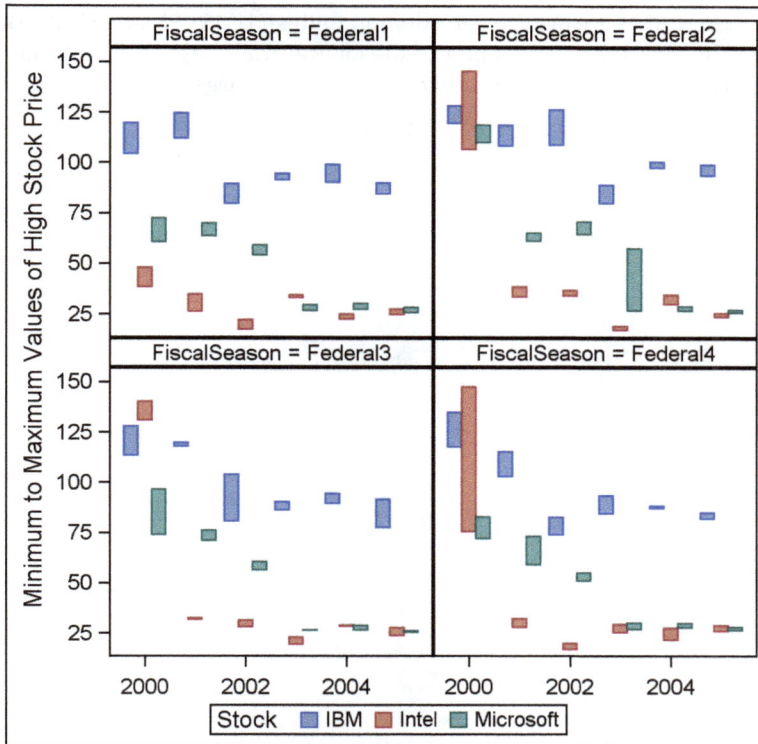

4. Using PROC UNIVARIATE, obtain the quantiles, extreme observations, and QQ-plots for the Sales, Inventory, and Returns variables in the Sashelp.Shoes data set. Ensure the procedure does not produce any additional results.

5. The previous question used the raw Sales, Inventory, and Returns values from Sashelp.Shoes. However, each record is based on a variable number of stores. Re-create the quantiles, extreme observations, and QQ-plots using a per-store average value of Sales, Inventory, and Returns.

6. Write a program that does all of the following:

   ○ Reads in the Flights.csv data set, which contains the same fields read by Program 2.9.1, and ensures data values are read correctly and in a manner that allows for future analyses.

   ○ Uses the PUTLOG statement to write the FlightNum and FirstClass values along with the associated variable names only when the value of FirstClass is missing.

   ○ Uses the LIST statement only when the flight is on a date prior to 12/14/2000 or is from DFW.

7. Write a program that does all of the following:

   ○ Reads in the Flights.txt data set, which contains the same fields read by Program 2.9.1.

   ○ Uses the PUTLOG statement to write the complete PDV if any variable is missing.

   ○ Uses the LIST statement to write the complete input buffer if any variable is missing.

   ○ Includes a comment detailing whether the PUTLOG or LIST statement is a better debugging tool in this scenario.

   ○ Validates the data set against the Flights data set created in the previous exercise.

8. Read in the raw data contained in Cars.datfile and compare it to the Cars data set in the Sashelp library. Modify the DATA step that reads the raw file to correct any data differences that arise so that the file generated validates against Sashelp.Cars.

Case Studies

For additional practice, multiple case studies are available in addition to the IPUMS CPS case study used in the chapters. See Section 8.4 to apply the skills from this chapter to the Clinical Trials Case Study. For additional case studies, including extensions to the IPUMS CPS case study, see the author pages.

Chapter 5: Joining Data Sets on Common Values and Measuring Association

5.1 Learning Objectives

At the conclusion of this chapter, mastery of the concepts covered in the narrative includes the ability to:

- Differentiate between the one-to-one reading, one-to-one merging, and the three types of match-merging and apply the appropriate technique in a given scenario

- Describe the process by which SAS carries out a match-merge

- Compare and contrast one-to-one, one-to-many, and many-to-many match-merges

- Formulate a strategy for selecting observations and variables during a join and for determining the data set into which the DATA step writes

- Apply the CORR procedure to numerically assess the association between numeric variables

- Apply the FREQ procedure to numerically assess the association between categorical variables

- Apply the SGPLOT procedure to graphically assess the association between variables

- Apply the TRANSPOSE procedure to exchange row and column information in a SAS data set

- Assess the advantages and disadvantages of working with either different analysis variables stored in several columns or a single analysis variable with one or more classification variables

Use the concepts of this chapter to solve the problems in the wrap-up activity. Additional exercises and case studies are also available to test these concepts.

5.2 Case Study Activity

Continuing the case study covered in the previous chapters, the tables and graphs shown in Outputs 5.2.1 through 5.2.3 are based on the IPUMS CPS Basic data sets for the years of 2005, 2010, and 2015, with the addition of information about select utility costs for those same years. The objective, as described in the Wrap-Up Activity in Section 5.8, is to assemble the data from all the individual files and produce the results shown below.

Output 5.2.1: Correlations Between Utility Costs and Household Income and Home Value

Year=2005

Pearson Correlation Coefficients Prob > |r| under H0: Rho=0 Number of Observations				
	electric	gas	water	fuel
HomeValue House value	0.19794 <.0001 849608	0.07862 <.0001 560777	0.18506 <.0001 672900	0.22044 <.0001 155903
HHINCOME Total household income	0.25157 <.0001 1114390	0.10719 <.0001 687656	0.14358 <.0001 789073	0.15827 <.0001 175434

Year=2010

Pearson Correlation Coefficients Prob > |r| under H0: Rho=0 Number of Observations				
	electric	gas	water	fuel
HomeValue House value	0.16522 <.0001 850656	0.10212 <.0001 545203	0.17250 <.0001 673226	0.19814 <.0001 144422
HHINCOME Total household income	-0.34537 <.0001 1234931	-0.32431 <.0001 764654	-0.32735 <.0001 891981	-0.53562 <.0001 244182

Year=2015

	electric	gas	water	fuel
Pearson Correlation Coefficients **Prob > |r| under H0: Rho=0** **Number of Observations**				
HomeValue House value	0.16246 <.0001 844377	0.07654 <.0001 511979	0.17698 <.0001 672683	0.17246 <.0001 107552
HHINCOME Total household income	-0.44558 <.0001 1322181	-0.40194 <.0001 801318	-0.40052 <.0001 980570	-0.59934 <.0001 271108

Output 5.2.2: Plot of Mean Electric and Gas Costs

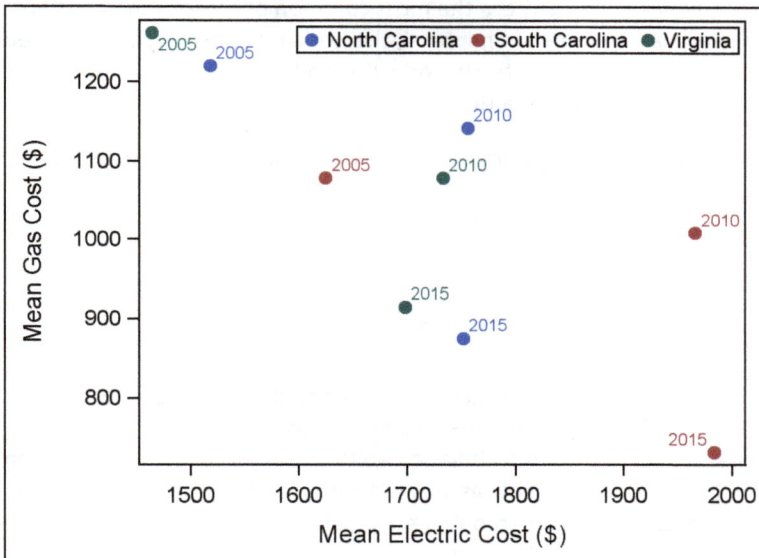

Output 5.2.3: Fitted Curves for Utility Costs Versus Home Values

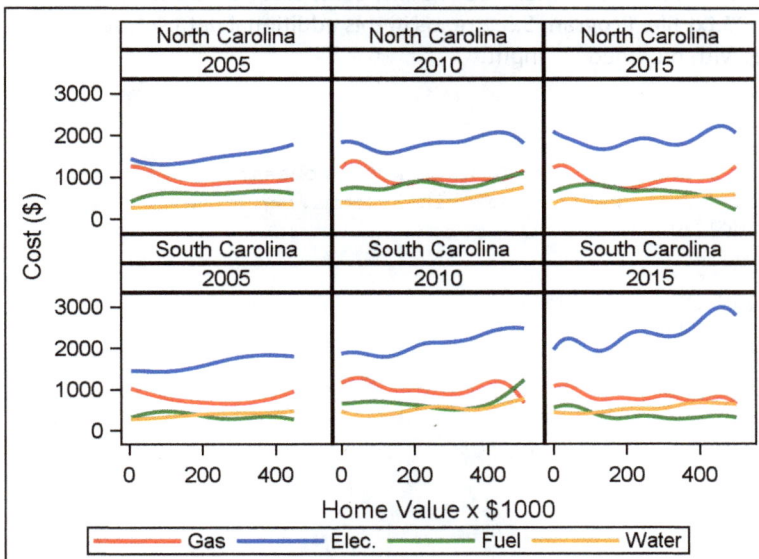

5.3 Horizontally Combining SAS Data Sets in the DATA Step

Recall from Chapter 4 that there are several ways to describe methods available in SAS for combining data sets. Table 5.3.1 below (a copy of Table 4.3.1) shows the classifications used here.

Table 5.3.1: Overview of the Four Methods for Combining Data Via the DATA Step

		Orientation of Combination	
		Vertical	**Horizontal**
Grouping Used?	**No**	Concatenate	One-to-One Read
			One-to-One Merge
	Yes	Interleave	Match-Merge

Chapter 4 presented vertical techniques—concatenating and interleaving—which are appropriate when the records are not matched on any key variables. However, it is not uncommon for these rows to contain key variables designed to link the records across data sets. In these cases, the new records are formed by combining information from one or more of the contributing records. As a result, these data combination methods can be described as horizontal techniques and in SAS are the most common of these techniques are typically referred to as merges. As with the vertical methods of Chapter 4, the horizontal techniques discussed here are applicable with two or more data sets, but to simplify the introduction to these methods, the examples in this chapter are limited to joining two data sets at a time.

The SAS DATA step provides three distinct ways to combine information from multiple records into a single record:

- One-to-one reading
- One-to-one merging
- Match-merging

Only match-merging incorporates key variables to help ensure data integrity during the joining process. In addition, there are different classifications of match-merges: one-to-one, one-to-many, and many-to-many. One-to-one reading and one-to-one merging are seldom-used techniques, so this chapter focuses primarily on match-merges. However, to help highlight some critical elements associated with match-merges, this section also includes a brief discussion of one-to-one reading and one-to-one merging.

To facilitate the join techniques shown in this chapter, this section introduces another data source related to the observations in the IPUMS 2005 Basic data set. Information about utility costs associated with these records is contained in the Utility Cost 2005.txt file. Program 5.3.1 reads in this additional data set, serving as a reminder on how to read delimited files with modified list input.

Program 5.3.1: Reading Utility Cost Data

```
data work.ipums2005Utility;
    infile RawData❶('Utility Cost 2005.txt'❷) dlm='09'x dsd firstobs=4;
    input serial electric:comma. gas:comma. water:comma. fuel:comma.;❸
    format electric gas water fuel dollar.;
run;

proc report data = work.ipums2005utility(obs = 5);
    columns serial electric gas water fuel;
    define Serial / display 'Serial';❹
    define Electric / display;❹
    define Gas / display;❹
    define Water / display;❹
    define Fuel / display;❹
run;
```

❶ Here RawData is a *fileref* that is assigned via a FILENAME statement pointing to the folder where the text file is stored. See Program 2.8.3 for details about such assignments.

❷ Details on the contents and layout of this file are contained in the top lines of the file itself. These details provide important information for the construction of the INFILE and INPUT statements, causing the first data record to appear in the fourth line of the raw text file; thus, FIRSTOBS=4 appears to skip these lines.

❸ When reading the raw file in the first DATA step, any legal variable names can be chosen. Here, the first column is given the name Serial as it is used to match records in Ipums2005Basic during the merge.

❹ Recall from Program 4.3.3 that reports that only include numeric variables can produce unwanted behavior by collapsing all included records into a one-line report. Including explicit values for the usage in each DEFINE statement is one way to prevent such unwanted results. While including the DISPLAY usage in a single DEFINE statement is sufficient, the defensive programming tactic of including it on all columns ensures the appearance of one column is unaffected by the removal of another column.

Output 5.3.1: Reading Utility Cost Data

Serial	electric	Gas	water	fuel
2	$1,800	$9,993	$700	$1,300
3	$1,320	$2,040	$120	$9,993
4	$1,440	$120	$9,997	$9,993
5	$9,997	$9,993	$9,997	$9,993
6	$1,320	$600	$240	$9,993

5.3.1 One-to-One Match-Merging

As mentioned in the opening of this section, it is often the case that records from different data sets are related based on the values of variables common to the input data sets. These variables that relate records across data sets are called *key variables* in many programming languages. A match-merge combines records from multiple data sets into a single record by matching the values of the key variables. As such, one of the first steps in carrying out a match-merge is to identify the key variables that link records across data sets and ensure they are the same type (character or numeric) and have the same variable name. Output 5.3.1 shows the variables included in the Ipums2005Utility data: Serial, Electric, Gas, Water, and Fuel. Input Data 5.3.2 shows a partial listing of the SAS data set Ipums2005Basic.

Input Data 5.3.2: Partial Listing of Ipums2005Basic (Only Four Variables for the 3rd through 6th Records)

Household serial number	state	Metropolitan status	Total household income
4	Alabama	4	185000
5	Alabama	1	2000
6	Alabama	3	72600
7	Alabama	1	42630

In order to join this data set to other data sets in a match-merge, one or more of these variables must appear in the other data sets. In conjunction with Output 5.3.1, an investigation of the Ipums2005Basic data set shows that Serial is the only variable common to both data sets. An investigation with PROC CONTENTS reveals this common variable is numeric in both data sets. Program 5.3.2 uses Serial as the key variable in a one-to-one match-merge.

Program 5.3.2: Carrying Out a One-to-One Match-Merge Using a Single Key Variable

```
data work.OneToOneMM;
  merge ❶ BookData.ipums2005basic(firstobs = 3 ❷ obs = 6)
          work.ipums2005Utility(obs = 5) ❸;
  by Serial; ❹
run;

proc print data = work.OneToOneMM; ❺
  var serial state metro hhincome electric gas water fuel; ❻
run;
```

❶ All match-merging in the DATA step is done with the MERGE statement. All data sets must appear in the same MERGE statement.

❷ The FIRSTOBS= data set option operates like the FIRSTOBS= option in the INFILE statement—it defines the first record read by the MERGE statement.

❸ The full data sets have the exact same set of values for Serial and are sorted on it. Here, the OBS= option selects a subset of records from this data set with some different values of Serial to help highlight how a one-to-one match-merge handles both records with and without matching values of the key variable. These options are local to each data set, so there is no need to reset FIRSTOBS=1.

❹ Match-merges require a BY statement that identifies the key variables. All input data sets must be compatibly sorted or indexed by the key variables or the DATA step prematurely exits and places an error in the SAS log. See Programs 2.3.5 and 4.3.5 for a review of sorting using multiple variables.

❺ Since the reporting is done here simply to view the results of the merge, PROC PRINT is a reasonable alternative to the REPORT procedure.

❻ The results of a match-merge include all columns and all rows from all input data sets. For brevity, only a subset of the variables is shown.

Due to the requirement of a BY statement to specify the key variables, the term BY variables is synonymous with key variables in the SAS Documentation. Similarly, the term BY group refers to the set of observations defined by a combination of the levels of the BY variable. A one-to-one match-merge occurs when each distinct BY group has at most one observation present for every data set in the MERGE statement. Note that the partial listings in Output 5.3.1 and Input Data 5.3.2 show that each Serial appears at most once per data set. Because this holds true for the full data sets, the use of FIRSTOBS= and OBS= in Program 5.3.2 are not necessary to ensure a one-to-one match-merge—they are used to simplify the output tables and force a mismatch on some records.

Output 5.3.2: Carrying Out a One-to-One Match-Merge Using a Single Key Variable

Obs	SERIAL	state	METRO	HHINCOME	electric	gas	water	fuel
1	2		.	.	$1,800	$9,993	$700	$1,300
2	3		.	.	$1,320	$2,040	$120	$9,993
3	4	Alabama	4	185000	$1,440	$120	$9,997	$9,993
4	5	Alabama	1	2000	$9,997	$9,993	$9,997	$9,993
5	6	Alabama	3	72600	$1,320	$600	$240	$9,993
6	7	Alabama	1	42630	.	.	.	.

Compare Output 5.3.1 and Input Data 5.3.2—representing the two input data sets for the match-merge in Program 5.3.2—with Output 5.3.2, the results of the one-to-one match-merge. When compared, they demonstrate that the match-merge links information from rows for which a match exists for the key variables. When identifying key variables, be careful not to look only for variables with the same name. Sometimes the same information appears with different names, such as EmpID versus EmployeeID or Temp versus Temperature. Similarly, variables with the same name may contain different information, such as two variables named Units, which contains measurement units (for example, kg and cm) in one data set and a count of the number of units (for example, 100 or 500) in the other data set. If necessary, use the RENAME= option or RENAME statement to standardize variable names across the data sets as shown in Section 4.4.4.

As mentioned, following Program 5.3.2, when using a match-merge, the resulting data set contains all variables from all input data sets. If the same variable appears in multiple data sets in the MERGE statement, the variable attributes are determined by the first encounter of the attribute for that variable. (See Chapter Note 1 in Section 4.10 for additional details.) Conversely, when a variable appears in multiple data sets in the MERGE statement, the variable values are determined by the last encounter of a variable. As with the SET statement, the MERGE statement moves left-to-right when multiple data sets are present.

5.3.2 Comparing One-to-One Reading, Merging, and Match-Merging

A match-merge refers to any merge process that uses one or more BY variables to determine which records to join. As discussed above, a one-to-one match-merge occurs when the BY variables create BY groups that contain at most one observation in every input data set. Despite the similarity in naming, the one-to-one reading and one-to-one merging behave quite differently—they combine information from multiple rows into a single row based on observation number instead of variable values. This means they combine information in the first record from each input data set, even if those records have no information in common. Program 5.3.3 demonstrates the syntax for a one-to-one reading using the same input files and records as Program 5.3.2.

Program 5.3.3: Carrying Out a One-to-One Reading

```
data work. OneToOneRead;
   set ❶ BookData.ipums2005basic(firstobs = 3 obs = 6);
   set ❶ work.ipums2005Utility(obs = 5);
   ❷
run;
```

❶ A one-to-one reading uses a separate SET statement for each data set.

❷ When carrying out a one-to-one reading, do not use a BY statement.

Applying the PROC REPORT code from Program 5.3.2 to the OneToOneRead data set produced in Program 5.3.3 produces Output 5.3.3.

Output 5.3.3: Carrying Out a One-to-One Reading

Household serial number	state	Metropolitan status	Total household income	electric	gas	water	fuel
2	Alabama	4	185000	$1,800	$9,993	$700	$1,300
3	Alabama	1	2000	$1,320	$2,040	$120	$9,993
4	Alabama	3	72600	$1,440	$120	$9,997	$9,993
5	Alabama	1	42630	$9,997	$9,993	$9,997	$9,993

At first glance, the results shown in Output 5.3.3 may not appear problematic. However, a close comparison of Output 5.3.3 with the source files (Output 5.3.1 and Input Data 5.3.2) reveals the following issues.

- Output 5.3.3 only contains four records, despite one input data set containing four records and the other having five records. This occurs because the one-to-one reading stops reading all input data sets as soon as it reaches the end of any input data set. As a result, the number of observations resulting from a one-to-one reading is always the minimum of the number of observations in the input data sets.

- The values of Serial in Output 5.3.3 correspond to the values from the second SET statement. This occurs because the values of any common variable are determined by the last value sent to the PDV.

- The first record in Output 5.3.3 contains values for State, Metro, and HHIncome when Serial=2, but no such record appears in the input data sets (Output 5.3.1 and Input Data 5.3.2). This occurs because a one-to-one read matches records by relative position instead of key variables. The first record from Output 5.3.1 (Serial=2) is combined with the first record from Input Data 5.3.2 (Serial=4).

Furthermore, comparing Output 5.3.3 to Output 5.3.1 and Input Data 5.3.2 shows that the label for Serial comes from BookData.Ipums2005Basic rather than Ipums2005Utility. This is an example of the MERGE statement using the first-encountered attribute—formats and lengths behave similarly. Due to the potential for data fidelity issues (for example, overwritten values and lost records), one-to-one reading is rarely used without additional programming statements that control the data reading process. Note that comparing altering Program 5.3.3 to use the same set records (for example, the first five rows) or the full data sets produces a reasonable result. However, that is because the incoming data sets are already sorted by Serial, have no other variables in common, and have records for the exact same set of values of Serial. A one-to-one match-merge is still superior in that case as it produces the same result while ensuring data fidelity.

A one-to-one merge has elements in common with both the one-to-one reading (they both use observation number to link records) and the one-to-one match-merge (they both use all records from all input data sets). However, due to the continued use of observation number to join records, the one-to-one merge still suffers from some of the data fidelity issues present in Output 5.3.3. Program 5.3.4 shows the syntax for a one-to-one merge with the same input data sets as Program 5.3.3.

Program 5.3.4: Carrying Out a One-to-One Merge

```
data work.OneToOneMerge;
  merge ❶ BookData.ipums2005basic(firstobs = 3 obs = 6)
          work.ipums2005Utility(obs = 5);
  ❷
run;
```

❶ A one-to-one merge uses a single MERGE statement for all input data sets just as a match-merge does.

❷ Unlike the match-merge, the one-to-one merge does not use a BY statement.

Applying the PROC REPORT code from Program 5.3.2 to the OneToOneMerge data set produces Output 5.3.4.

Output 5.3.4: Carrying Out a One-to-One Merge

Household serial number	state	Metropolitan status	Total household income	electric	gas	Water	fuel
2	Alabama	4	185000	$1,800	$9,993	$700	$1,300
3	Alabama	1	2000	$1,320	$2,040	$120	$9,993
4	Alabama	3	72600	$1,440	$120	$9,997	$9,993
5	Alabama	1	42630	$9,997	$9,993	$9,997	$9,993
6		.	.	$1,320	$600	$240	$9,993

Comparing Output 5.3.4 to the previous results, it is clear that using a one-to-one merge resolves one of the data-fidelity issues: records are no longer lost due to the DATA step stopping when it reaches the end of the shortest input data set. However, because the last encounter in the MERGE statement defines the values of Serial, and the one-to-one merge uses observation number, the record that appears with Serial=2 still contains information with Serial=2 from the Ipums2005Utility data and Serial=4 from Ipums2005Basic. As with the one-to-one reading, the one-to-one merge also defines formats, lengths, and other attributes of common variables based on the first time the MERGE statement encounters the attribute for that variable.

The results shown in Output 5.3.3 and 5.3.4 should provide a cautionary note that one-to-one reading and one-to-one merging are only appropriate in circumstances where the records are correctly structured for position-based matching. However, this is no small concern, and programmers should use these methods sparingly even when taking due precautions.

5.3.3 One-To-Many Match-Merge

The Ipums2005Basic and Ipums2005Utility data sets from Sections 5.3.1 and 5.3.2 included at most one record per value of Serial in each data set. To facilitate a discussion of one-to-many match-merges, Program 5.3.5 revisits the one-to-one match-merge of Program 5.3.2 to carry out some data cleaning before using PROC MEANS to calculate the mean of the HomeValue, HHIncome, and MortgagePayment variables and store the results in a data set for later use on a one-to-many match-merge.

Program 5.3.5: Data Cleaning and Generating Summary Statistics

```
data work.ipums2005cost;
  merge BookData.ipums2005basic work.ipums2005Utility;
  by serial;

  if homevalue eq 9999999 then homevalue=.; ❶
  if electric ge 9000 then electric=.; ❷
  if gas ge 9000 then gas=.; ❷
  if water ge 9000 then water=.; ❷
  if fuel ge 9000 then fuel=.; ❷
run;

proc means data=bookdata.ipums2005basic;
  where state in ('North Carolina', 'South Carolina'); ❸
  class state mortgageStatus;  ❹
  var homevalue hhincome mortgagepayment;
  output out= work.means mean=HVmean HHImean MPmean;  ❺
run;

proc format;
  value $MStatus
    'No'-'Nz'='No'
    'Yes, a'-'Yes, d'='Yes, Contract'
    'Yes, l'-'Yes, n'='Yes, Mortgaged'
  ;
run; ❻

proc report data = work.means;
  columns State MortgageStatus _type_ _freq_ HVmean HHImean MPmean;
  define State / display;
  define MortgageStatus / display 'Mortgage Status' format=$MStatus.;
  define _type_ / display 'Group Classification';
  define _freq_ / display 'Frequency';
  define hvmean / display format = dollar11.2 'Mean Home Value';
  define hhimean / display format = dollar10.2 'Mean Household Income';
  define mpmean / display format = dollar7.2 'Mean Mortgage Payment';
run;
```

❶ Values of 9,999,999 represent missing data for the HomeValue variable but should not be used to compute any statistics on this variable.

❷ Values over 9,000 represent different types of missing data for these variables, and likewise should not be used in statistical computations. For details about one way to handle different types of missing data, see Chapter Note 1 in Section 5.9.

❸ Because the numeric variables appear in separate columns but are included in the same MEANS procedure, it is not appropriate to merely include a condition such as HomeValue NE 9,999,999 in the WHERE statement. Omitting those records would also omit records with nonmissing values of HHIncome and MortgagePayment from the computation of their means. Since missing values are not used in the computation of summary statistics, using the . to represent missing numeric values ensures all statistics are computed on the appropriate value set.

❹ Request statistics separately based on State and MortgageStatus groupings.

❺ Use the OUTPUT statement to save only the named statistics to the Means data set.

⑥ Program 2.7.6 first introduced the possibility of using ranges when formatting character values.

The results of the REPORT procedure in Output 5.3.5 show the computed means for each combination of State and MortgageStatus as well as the means for each level of State and MortgageStatus separately and the overall means. Note that each combination of State and MortgageStatus appears exactly once in this data set.

Output 5.3.5: Data Cleaning and Generating Summary Statistics

state	Mortgage Status	Group	Frequency	Mean Home Value	Mean Household Income	Mean Mortgage Payment
		0	40187	$166,724.50	$63,426.91	$537.38
	No	1	14536	$143,377.30	$45,860.32	$0.00
	Yes, Contract	1	403	$87,698.51	$44,355.32	$522.88
	Yes, Mortgaged	1	25248	$181,427.54	$73,844.92	$847.00
North Carolina		2	26783	$169,739.20	$64,568.02	$554.02
South Carolina		2	13404	$160,700.72	$61,146.83	$504.13
North Carolina	No	3	9438	$145,771.35	$46,098.38	$0.00
North Carolina	Yes, Contract	3	231	$94,004.33	$45,326.50	$541.65
North Carolina	Yes, Mortgaged	3	17114	$183,979.20	$75,013.34	$859.72
South Carolina	No	3	5098	$138,945.17	$45,419.61	$0.00
South Carolina	Yes, Contract	3	172	$79,229.65	$43,051.01	$497.67
South Carolina	Yes, Mortgaged	3	8134	$176,058.83	$71,386.55	$820.23

A common use of summary statistics is for comparison with the individual data values. One way to carry this out in SAS is to combine the summary statistics with the original data in a merge. To ensure appropriate comparisons are made, use a match-merge to match a summary statistic with a data record if they come from the same State and have the same MortgageStatus. Program 5.3.6 carries out the one-to-many match-merge and then computes some comparison values.

Program 5.3.6: One-to-Many Match-Merge

```
proc sort data= work.ipums2005cost out= work.cost; ❶
  by state mortgagestatus;
  where state in ('North Carolina', 'South Carolina');
run;

proc sort data= work.means; ❶
  by state mortgagestatus;
run;

data work.OneToManyMM;
  merge work.cost(in=inCost) work.means(in=inMeans);
  by state mortgagestatus; ❷
  if inCost eq 1 and inMeans eq 1; ❸
```

```
   HVdiff=homevalue-HVmean; ❹
   HVratio=homevalue/HVmean; ❹
   HHIdiff=hhincome-HHImean; ❹
   HHIratio=hhincome/HHImean; ❹
   MPdiff=mortgagepayment-MPmean; ❹
   MPratio=mortgagepayment/MPmean; ❹
run;

proc report data = work.OneToManyMM(obs=4); ❺
   where mortgageStatus contains 'owned'; ❺
   columns Serial State MortgageStatus HomeValue HVMean HVRatio;
   define Serial / display 'Serial';
   define State / display 'State';
   define MortgageStatus / display 'Mortgage Status';
   define HomeValue / display 'Home Value';
   define HVMean / display format = dollar11.2 'Mean Home Value';
   define HVRatio / display format = 4.2 'Ratio';
run;

proc print data = work.OneToManyMM(obs=4); ❺
   where mortgageStatus contains 'contract'; ❺
   var Serial State MortgageStatus HomeValue HVMean HVRatio;
run;
```

❶ Remember to properly sort or index before carrying out a match-merge. The OUT= option is important to ensure PROC SORT does not overwrite the original data set since the use of a WHERE statement filters out rows, causing the Cost data set to be a subset of the Ipums2005Cost data set.

❷ The BY statement names the two key variables necessary for use during the match-merge. This is a one-to-many merge because exactly one of the data sets (Work.Cost) in the MERGE statement contains records that are not uniquely identified by the BY groups.

❸ By default, the merging process preserves information from all records from all data sets, even if no match occurs (also referred to as a full outer join). In this case, the Means data set includes extra summary statistics that are missing one or both BY variables and thus do not match any record in the Cost data set. Using the IN= variables with the subsetting IF statement ensures processing continues only if the current PDV has information from both data sets—that is, a match has occurred. This is also called an inner join—various forms of joins are discussed in Section 5.5.

❹ These calculations do not consider missing values or division by zero, both of which lead to undesirable messages in the SAS log. The exercises in Section 5.10 offer an opportunity to improve this program to prevent such messages from appearing in the log.

❺ For brevity, only four observations are displayed in each of Outputs 5.3.6A and 5.3.6B. However, to highlight the results clearly, the first four results are displayed from two of the groups to demonstrate that the same effects of the one-to-many match-merge are present throughout the resulting data set.

Output 5.3.6A shows a partial listing of the computed means and ratios from Program 5.3.6. Note that the HVMean variable contains the same value for every record in the same BY group—Section 5.4 discusses the details of the DATA step logic that achieves this. As with the previous techniques in this section, the one-to-many match-merge determines attributes of common variables based on first encounter of that attribute in the MERGE statement and it determines values based on the last encounter in the MERGE statement. Similarly, Output 5.3.6B demonstrates that the same behavior is present for other combinations of State and MortgageStatus.

Output 5.3.6A: One-to-Many Match-Merge – Group 1: Homeowners with No Mortgage

Serial	State	Mortgage Status	Home Value	Mean Home Value	Ratio
817019	North Carolina	No, owned free and clear	5000	$145,771.35	0.03
817020	North Carolina	No, owned free and clear	162500	$145,771.35	1.11

Serial	State	Mortgage Status	Home Value	Mean Home Value	Ratio
817031	North Carolina	No, owned free and clear	45000	$145,771.35	0.31
817032	North Carolina	No, owned free and clear	32500	$145,771.35	0.22

Output 5.3.6B: One-to-Many Match-Merge – Group 2: Homeowners with a Contract to Purchase

Obs	SERIAL	state	MortgageStatus	HomeValue	HVmean	HVratio
9439	817029	North Carolina	Yes, contract to purchase	95000	94004.33	1.01059
9440	817053	North Carolina	Yes, contract to purchase	112500	94004.33	1.19675
9441	817109	North Carolina	Yes, contract to purchase	85000	94004.33	0.90421
9442	817665	North Carolina	Yes, contract to purchase	55000	94004.33	0.58508

As an aside from the merging process, note that the programs in this section also juxtapose the PRINT and REPORT procedures to emphasize the advantages, and disadvantages, of each. Earlier chapters used PROC PRINT for its simplicity, while future chapters focus more on PROC REPORT for its flexibility. However, when a quick inspection of the data is needed, either procedure is acceptable and the simplicity of PROC PRINT is a likely advantage. In some cases, it may even be sufficient to simply use the VIEWTABLE window (or equivalent, depending on whether the data is viewed in SAS Studio, SAS University Edition, or other SAS product) if no printout is necessary. Be aware that the VIEWTABLE in the windowing environment has a potentially high resource overhead (see Chapter Note 4 in Section 1.7), so that is not a recommended approach for large data sets. Compared to PROC PRINT, the versatility of the REPORT procedure makes creating professional-quality output not only possible, but relatively simple as well. In the remainder of the text, both procedures appear—not because they are interchangeable in general—but because it is a good idea to be well-versed in both procedures and know when it is advantageous to choose one over another.

5.3.4 Many-to-Many Match-Merge

As noted in Section 5.3.1, a one-to-one match-merge occurs when each record in every data set in the MERGE statement is uniquely identified via its BY-group value, while a one-to-many match-merge occurs when the BY-group values uniquely identify records in all but exactly one of the data sets in the MERGE statement. The third possibility, a many-to-many match-merge, occurs when multiple data sets contain repeated observations within the same BY-group value. To explore the potential pitfalls when carrying out a many-to-many match-merge, Program 5.3.7 generates some additional summary statistics for use during a match-merge.

Program 5.3.7: Many-to-Many Match-Merge

```
proc means data= work.ipums2005cost(where = (state in ('North Carolina',
                                                      'South Carolina')));
  class state metro; ❶
  var homevalue hhincome mortgagepayment;
  output out= work.medians median=HVmed HHImed MPmed; ❷
run;

proc sort data = work.medians;
  by state metro; ❸
run;
```

```
data work.ManyToManyMM;
  merge work.means work.medians; ➍
  by state; ➎
run;

proc report data = work.ManyToManyMM(firstobs = 7);
  columns State MortgageStatus Metro _FREQ_ ➏ HVMean HVMed;
  define State / display;
  define MortgageStatus / display 'Mortgage Status';
  define Metro / display 'Metro Classification';
  define _freq_ / display 'Frequency'; ➏
  define HVMean / display format = dollar11.2 'Mean Home Value';
  define HVMed / display format = dollar11.2 'Median Home Value';
run;
```

➊ This PROC MEANS uses State and Metro as key variables for computing the medians. Note this differs from previous programs in this section that use State and MortgageStatus.

➋ Save the medians to a new data set. The Medians data set has the same structure as the Means data set from Program 5.3.5, but with some different columns.

➌ Sort the Medians data set by both classification variables.

➍ As with the one-to-one and one-to-many match-merges, the many-to-many match-merge simply uses a single MERGE statement which includes all data sets.

➎ Note that only State appears as a key variable for this match-merge. Referring to Output 5.3.5, it is clear that each value of State occurs more than once for both data sets in the MERGE statement. Thus, this is a many-to-many match-merge.

➏ As with all other techniques in this section, many-to-many match-merges overwrite common variable values based on last encounter in the MERGE statement. Thus, this _FREQ_ variable contains the frequencies only for the computation of medians based on State and Metro.

Output 5.3.7 shows a partial listing of the results of this many-to-many merge. Note that the last value of MortgageStatus is repeated for every distinct value of State. In particular, notice that the value "Yes, mortgaged/ deed of trust or similar debt" appears three times per state even though the other two values of MortgageStatus only appear once each. This demonstrates the need for caution when carrying out a many-to-many match-merge via the DATA step.

Output 5.3.7: Many-to-Many Match-Merge

state	Mortgage Status	Metro Classification	Frequency	Mean Home Value	Median Home Value
North Carolina		.	36063	$169,739.20	$137,500.00
North Carolina	No, owned free and clear	0	621	$145,771.35	$162,500.00
North Carolina	Yes, contract to purchase	1	11657	$94,004.33	$112,500.00
North Carolina	Yes, mortgaged/ deed of trust or similar debt	2	5571	$183,979.20	$137,500.00
North Carolina	Yes, mortgaged/ deed of trust or similar debt	3	4323	$183,979.20	$162,500.00
North Carolina	Yes, mortgaged/ deed of trust or similar debt	4	13891	$183,979.20	$137,500.00
South Carolina		.	17508	$160,700.72	$112,500.00

state	Mortgage Status	Metro Classification	Frequency	Mean Home Value	Median Home Value
South Carolina	No, owned free and clear	0	3170	$138,945.17	$95,000.00
South Carolina	Yes, contract to purchase	1	3682	$79,229.65	$85,000.00
South Carolina	Yes, mortgaged/ deed of trust or similar debt	2	458	$176,058.83	$137,500.00
South Carolina	Yes, mortgaged/ deed of trust or similar debt	3	3698	$176,058.83	$112,500.00
South Carolina	Yes, mortgaged/ deed of trust or similar debt	4	6500	$176,058.83	$137,500.00

Section 5.4 discusses the details of how this process unfolds in the DATA step and why it typically produces an undesirable result. In fact, when the DATA step identifies a many-to-many match-merge, SAS prints the following note to the log.

Log 5.3.7: Partial Log After Submitting Program 5.3.7

```
NOTE: MERGE statement has more than one data set with repeats of BY values.
```

As discussed in Section 1.4, some notes are indications of potential problems during execution and are not inherently benign. Just like errors and warnings, notes must be carefully reviewed. It is a good programming practice to investigate key variables first, in order to ensure that the DATA step uses the correct data-handling techniques. Several common options include using PROC FREQ to determine combinations of classification variables with more than one record or using a PROC SORT with the NODUPKEY and DUPOUT= options to create a data set of duplicate values based on the key variables.

Section 4.3 and this section introduce the DATA step techniques for combining data sets vertically (concatenating and interleaving) and joining data sets horizontally (one-to-one reading, one-to-one merging, and match-merging). In addition, the DATA step provides two additional techniques—updating and modifying—for combining information from multiple SAS data sets. More details about these two techniques can be found in Chapter Note 2 in Section 5.9 and in the SAS Documentation.

5.4 Match-Merge Details

To demonstrate the way the DATA step match-merges data sets, this section moves step-by-step through the process. The flowchart in Figure 5.4.1 presents a brief overview of the process.

Figure 5.4.1: Flowchart of the DATA Step Match-Merge Process

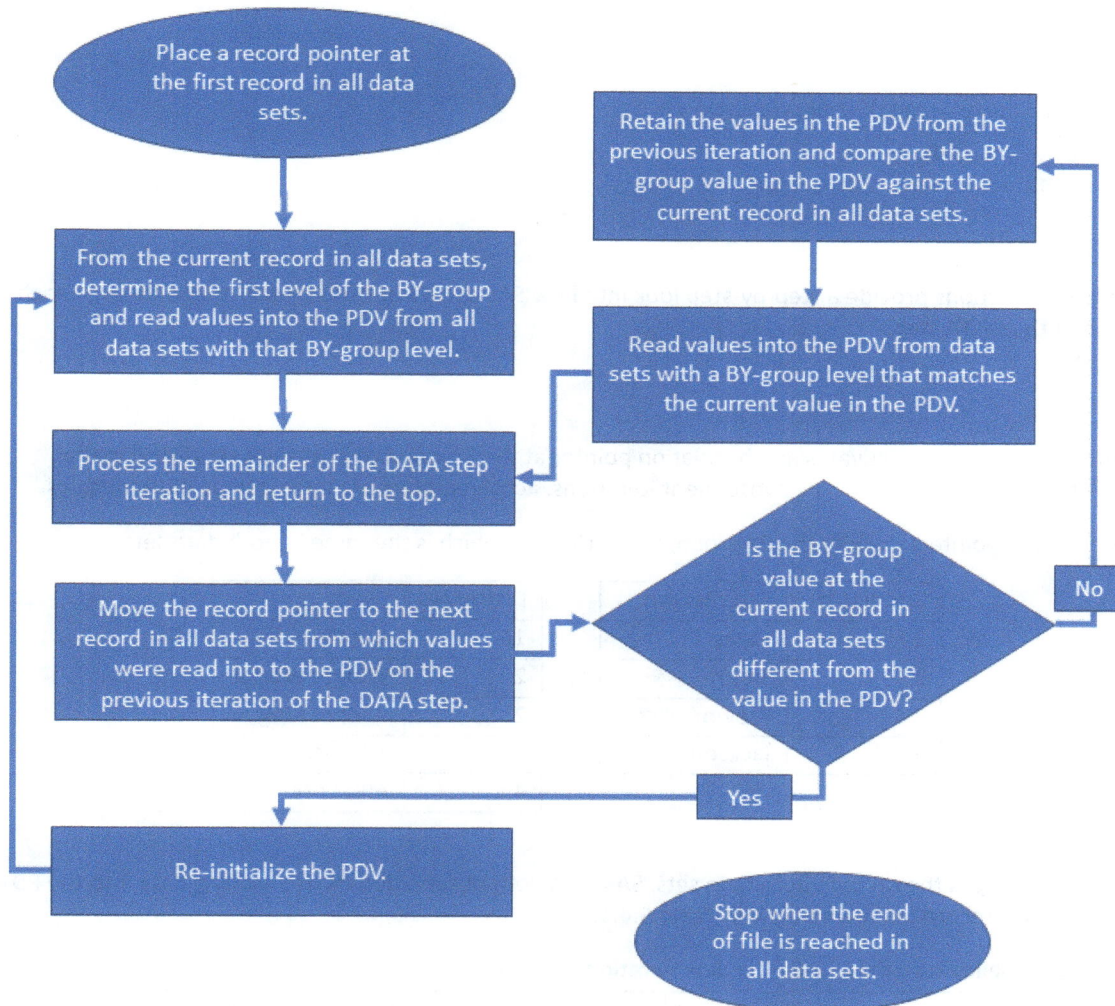

When a BY statement is present in a DATA step, SAS begins by activating an observation pointer in each data set located in a MERGE or SET statement. These pointers are used to identify the records currently available to the PDV. As Figure 5.4.1 indicates, the PDV can only use records from one BY group at a time. The first BY group is defined based on the sort sequencing done for the BY variable(s).

To demonstrate the process, Program 5.4.1 merges the data sets shown in Input Data 5.4.1 by the variable DistrictNo. Note the values of DistrictNo are designed to highlight the salient details: DistrictNo = 1 appears in both data sets exactly once, DistrictNo = 2 appears multiple times in both data sets, and DistrictNo =4 and DistrictNo = 5 appear in only one of the data sets.

Input Data 5.4.1: Staff and Clients Data Sets

DistrictNo	SalesPerson
1	Jones
2	Smith
2	Brown
5	Jackson

+

DistrictNo	Client
1	ACME
2	Widget World
2	Widget King
2	XYZ Inc.
4	ABC Corp.

Program 5.4.1: Investigating a Match-Merge

```
data work.Investigate;
  merge BookData.Staff BookData.Clients;
  by DistrictNo;
run;

proc report data = work.Investigate;
  columns DistrictNo SalesPerson Client;
  define DistrictNo / display;
  define SalesPerson / display;
  define Client / display;
run;
```

The following sections provide a step-by-step look into how SAS carries out this match-merge through each iteration of the DATA step.

Iteration 1

As execution begins, SAS activates an observation pointer at the first record in all data sets listed in the MERGE statement. Throughout the subsequent iterations, active pointers are indicated by blue triangles.

In this case, the pointers identify the BY group, DistrictNo = 1, which is the same in both data sets.

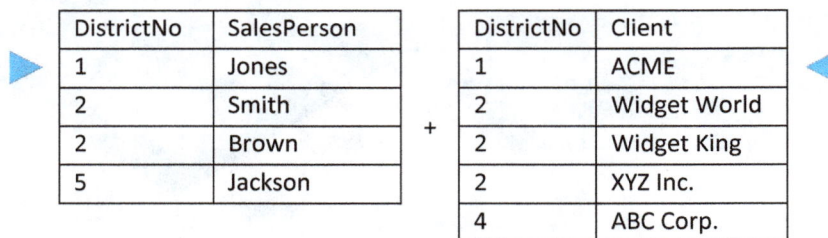

DistrictNo	SalesPerson
1	Jones
2	Smith
2	Brown
5	Jackson

+

DistrictNo	Client
1	ACME
2	Widget World
2	Widget King
2	XYZ Inc.
4	ABC Corp.

Since the BY group is the same in both data sets, SAS uses both observations when reading data into the PDV. The PDV for this record is shown below in Table 5.4.1.

Table 5.4.1: Visualization of the PDV After One Iteration of the DATA Step

_N_	_ERROR_	DistrictNo	SalesPerson	Client
1	0	1	Jones	ACME

Since this BY group only has one record in each data set, the DATA step joins the two records as expected and subsequently outputs this record to the data set.

Iteration 2

Values are read from both data sets into the PDV, so SAS advances the observation pointers in each. Table 5.4.2 shows the updated location of the pointers. SAS checks the value of the BY groups in each data set, comparing them to the value in the current PDV. Since the BY group has changed in all incoming data sets, the PDV is reset, causing DistrictNo, SalesPerson, and Client to be reinitialized.

Table 5.4.2: Location of the Pointers During the Second Iteration of the DATA Step

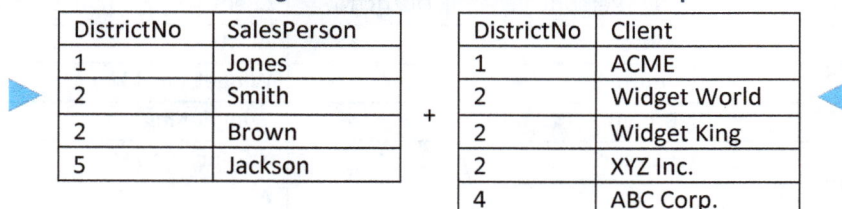

DistrictNo	SalesPerson
1	Jones
2	Smith
2	Brown
5	Jackson

+

DistrictNo	Client
1	ACME
2	Widget World
2	Widget King
2	XYZ Inc.
4	ABC Corp.

The observation pointers in both data sets now point to a new BY group and, since the BY values are equal, values are read from both data sets into to the PDV shown below in Table 5.4.3. The DATA step then outputs this to the data set.

Table 5.4.3: Visualization of the PDV After Two Iterations of the DATA Step

_N_	_ERROR_	DistrictNo	SalesPerson	Client
2	0	2	Smith	Widget World

Iteration 3

Since values are read into the PDV from both data sets, SAS again advances the observation pointers in each data set and Table 5.4.4 shows the updated locations. SAS checks the value of the BY groups in each data set, comparing them to the value in the current PDV. The BY value has not changed in all data sets (in fact, it has not changed in either), so the PDV is not reset. Since the BY group is the same in both data sets as it is in the current PDV, SAS reads values into the PDV from both data sets, overwriting the values of DistrictNo, SalesPerson, and Client.

Table 5.4.4: Location of the Pointers During the Third Iteration of the DATA Step

DistrictNo	SalesPerson
1	Jones
2	Smith
2	Brown
5	Jackson

+

DistrictNo	Client
1	ACME
2	Widget World
2	Widget King
2	XYZ Inc.
4	ABC Corp.

Table 5.4.5 shows the PDV at this point which is output to the data set as the third record (except for the automatic variables, _N_ and _ERROR_).

Table 5.4.5: Visualization of the PDV After Three Iterations of the DATA Step

_N_	_ERROR_	DistrictNo	SalesPerson	Client
3	0	2	Brown	Widget King

Iteration 4

Since values are read into the PDV from both data sets, SAS again advances the observation pointers in each data set and Table 5.4.6 shows the updated locations. The BY value has not changed in all data sets, so the PDV is not reset. The pointer in the Staff data set is not associated with an observation in this BY group, so SAS does not read from it (this is indicated visually with the red octagon). The BY group value of DistrictNo = 2 in the Clients data set does match the value in the current PDV, so values are read from that data set overwriting the value of Client, and Table 5.4.7 shows the result.

Table 5.4.6: Location of the Pointers During the Fourth Iteration of the DATA Step

DistrictNo	SalesPerson
1	Jones
2	Smith
2	Brown
5	Jackson

+

DistrictNo	Client
1	ACME
2	Widget World
2	Widget King
2	XYZ Inc.
4	ABC Corp.

Since the PDV was not reset, the fourth observation retains the value of SalesPerson from the previous record but reads a new for Client from the current record in the Clients data set.

Table 5.4.7: Visualization of the PDV After Four Iterations of the DATA Step

_N_	_ERROR_	DistrictNo	SalesPerson	Client
4	0	2	Brown	XYZ Inc.

Iteration 5

On iteration 4, new information is read into the PDV from the Clients data set only, so SAS advances the observation pointer in the Clients data set and leaves the observation pointer in the Staff data set in the same position. SAS checks the value of the BY groups in each data set, comparing them to the value in the current PDV. Since the BY value is different in all incoming data sets, the PDV is reset, causing DistrictNo, SalesPerson, and Client to be reinitialized.

Table 5.4.8: Location of the Pointers During the Fifth Iteration of the DATA Step

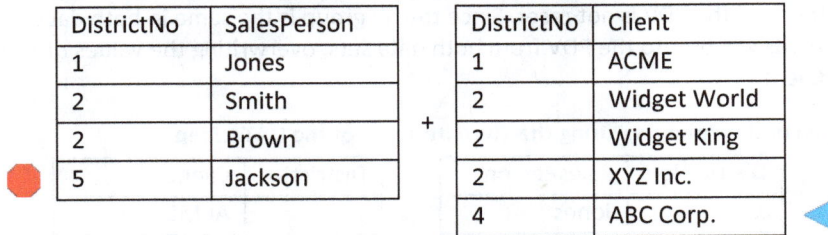

DistrictNo	SalesPerson		DistrictNo	Client
1	Jones		1	ACME
2	Smith		2	Widget World
2	Brown	+	2	Widget King
5	Jackson		2	XYZ Inc.
			4	ABC Corp.

SAS determines the next BY group has DistrictNo = 4 and reads only from the data sets with that value—limiting the read to the Clients data set in this case. At this iteration, no information is read from the Staff data, again indicated visually with the red octagon.

Figure 5.4.9 shows the PDV for the fifth observation. The new BY value of DistrictNo = 4 and the value of Client = ABC Corp. are read into the PDV from Clients, but no value of SalesPerson is populated for this BY group—it remains missing from the reinitialization of the PDV.

Table 5.4.9: Visualization of the PDV After Five Iterations of the DATA Step

_N_	_ERROR_	DistrictNo	SalesPerson	Client
5	0	4		ABC Corp.

Iteration 6

As new information is read into the PDV from the Clients data set only, SAS advances the observation pointer in the Clients data set and leaves the observation pointer in the Staff data set in the same position. The observation pointer has now reached the end-of-file marker in the Clients data; however, the Staff data set still has an active pointer. SAS checks the value of the BY groups in each active data set, comparing them to the value in the current PDV. Since the BY value is different in all active data sets, the PDV is reset, causing DistrictNo, SalesPerson, and Client to be reinitialized.

Table 5.4.10: Location of the Pointers During the Sixth Iteration of the DATA Step

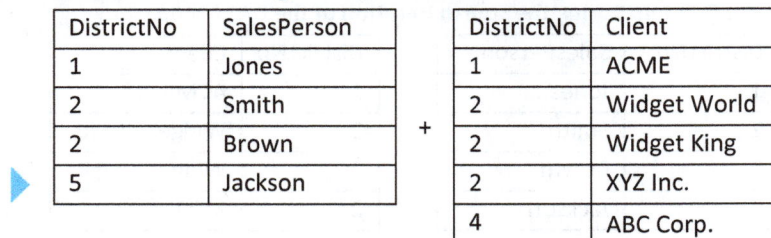

DistrictNo	SalesPerson		DistrictNo	Client
1	Jones		1	ACME
2	Smith		2	Widget World
2	Brown	+	2	Widget King
5	Jackson		2	XYZ Inc.
			4	ABC Corp.

Table 5.4.11 shows the PDV for the sixth observation, which reads DistrictNo and SalesPerson from the Staff data set, but Client is not read from any active data set.

Table 5.4.11: Visualization of the PDV After Six Iterations of the DATA Step

_N_	_ERROR_	DistrictNo	SalesPerson	Client
6	0	5	Jackson	

The final data set is shown in Output 5.4.1. While this data set is likely not the intended goal of Program 5.4.1, it is the result when following the logic summarized by the diagram in Figure 5.4.1. In addition to providing an example of the way the match-merge combines records, Output 5.4.1 also demonstrates that the default is for the match-merge to include information from all rows and all columns from all contributing tables. Section 5.5 discusses this result, and ways to modify it, in more detail.

Output 5.4.1: Investigating a Match-Merge

DistrictNo	SalesPerson	Client
1	Jones	ACME
2	Smith	Widget World
2	Brown	Widget King
2	Brown	XYZ Inc.
4		ABC Corp.
5	Jackson	

This example demonstrates several key behaviors created by the default logic of a match-merge in a DATA step. The action of resetting or not resetting the PDV based on checks of the BY-group value in the incoming data to the current value is critical for ensuring that a one-to-many match-merge gives the expected result and that records without full matching information in all data sets have missing values for the appropriate variables. The reset/non-reset behavior is irrelevant to the one-to-one match-merge and is insufficient to attain a reasonable result for the many-to-many match-merge. The example merge of Sales and Clients allows for a review of each of these cases.

If the Sales and Clients data sets are each limited to their first two records, the match-merge is one-to-one. In any case such as this, the PDV always resets at every iteration because the BY group always changes—one-to-one match-merges occur when all records in all data sets are uniquely identified via their BY-group values. If all records involving DistrictNo = 2 are eliminated from the incoming data sets, it is still a one-to-one match-merge, but not all records have matches. Since the PDV resets each time the BY value changes and values are read only from data sets containing the first value among the current BY values, any variables without a record having a matching by value remain missing—as happens with DistrictNo =4 and DistrictNo = 5 in Output 5.4.1.

If the second record is removed from both of Sales and Clients, and the second row of Output 5.4.1 is also removed, the match-merge is one-to-many. Iterations 3 and 4 of Program 5.4.1 illustrate how the retention of information when the BY value remains the same in a data set(s) when it is unique allows that information to be matched to multiple records from another data set having that BY value. If the second record were removed from only the Staff data (as done in Program 5.4.2), retracing iterations 1 through 6 reveals that the resulting data set is what is shown in Output 5.4.2, and the retention of the SalesPerson value from Staff when DistrictNo = 2 allows for matching with all Client values having that same DistrictNo.

Program 5.4.2: Match-Merge with the Second Observation Removed From the Staff Data Set

```
data Investigate2;
  merge BookData.Staff(where=(salesperson ne 'Smith')) BookData.Clients;
  by DistrictNo;
run;
```

```
proc report data = Investigate2;
  columns DistrictNo SalesPerson Client;
  define DistrictNo / Display;
  define SalesPerson / display;
  define Client / display;
run;
```

Output 5.4.2: Match-Merge with the Second Observation Removed From the Staff Data Set

DistrictNo	SalesPerson	Client
1	Jones	ACME
2	Brown	Widget World
2	Brown	Widget King
2	Brown	XYZ Inc.
4		ABC Corp.
5	Jackson	

With the records given in Sales and Clients, the merge process is a one-to-many merge. In this setting, with no other information available, a reasonable expectation would be to pair each SalesPerson with each Client for DistrictNo = 2, resulting in six records for that value of DistrictNo. Of course, Output 5.4.1 and the trace of the DATA step logic that generates it show that is not the case for the default match-merge process. Effectively, the many-to-many merge in the DATA step acts like one-to-one merging possibly combined with one-to-many merging. Consider iterations 2 and 3 from the logic previously shown: the retention of the values in the PDV from record 2 is irrelevant since the PDV is populated from both data sets and the retained values are overwritten. Thus, this looks like a one-to-one match, but it is dependent on the row order. Since the values are matched only on the BY variable DistrictNo, any sort on either data set that uses DistrictNo as its first sorting variable can be used prior to this merge. If Staff is sorted: BY DistrictNo SalesPerson; and the match-merge is conducted, the matching in records 2 and 3 is different. If the number of repeats for the BY value is the same across all data sets, this order-dependent one-to-one matching occurs.

In the Sales and Clients data, the number of repeats of the BY value DistrictNo = 2 is not the same across all data sets. Looking at iterations 3 and 4, the result is a matching of the last record with DistrictNo = 2 from Staff to all remaining records with DistrictNo = 2 from Clients—a one-to-many merge on these records. Neither this result, nor the one-to-one matching based on row order, is a reasonable result for such a scenario. Often, attempting a many-to-many merge is based on erroneous assumptions about the required matching variables. However, in cases where a many-to-many join is necessary, it can be done via the DATA step with significant interventions made to work around its default logic. A simpler and more standard approach is available through Structured Query Language (SQL), which can form all combinations of records matching on a set of key variables (a Cartesian product). For details about PROC SQL, see the SAS Documentation.

5.5 Controlling Output

By default, the match-merge process results in a data set that contains information from all records loaded into the PDV, even if full matching values are not found in all source data sets. This type of combination is referred to as a full outer join and it is one of several ways to select records that are joined based on key variables. Chapter 4 introduced the IN= option to create a temporary variable, identifying when a data set provides information to the current PDV. Chapter 4 also introduced the subsetting IF statement to select records for output and IF/THEN-ELSE statements for conditional creation of new variables.

Program 5.3.6 applies these concepts to demonstrate one way to override the default outer join and produce an inner join—that is, a match-merge that only includes records in the output data set if the BY group is present in all data sets in the MERGE statement. Program 5.5.1 revisits these concepts and provides further discussion of the options used to create the inner join, which are also used in later programs in this section to demonstrate additional techniques.

Program 5.5.1: Using a Subsetting IF Statement to Carry Out an Inner Join

```
data work.InnerJoin01;
  merge work.cost(in=inCost) ❶  work.means(in=inMeans) ❶;
  by State MortgageStatus;
  if (inCost eq 1) ❷ and inMeans❸; ❹
run;
```

❶ Recall the IN= option creates temporary variables in the PDV. These variables take on a value of 1 or 0 based on whether their associated data set contributes to the current PDV.

❷ The first condition in the expression checks whether the temporary variable inCost is equal to 1. This is true when the current PDV contains information from the Cost data set.

❸ The second condition does not compare the current value of inMeans to the necessary value, 1. Instead, it takes advantage of the fact that SAS interprets numeric values other than missing and zero as TRUE. SAS interprets missing or zero as FALSE.

❹ The subsetting IF statement checks the conditions in the expression and outputs the current record when the expression resolves as TRUE. Here the expression is true only if the PDV reads values from both data sets; that is, a match occurs.

In addition to full outer joins and inner joins, several other types of joins are possible, including:

- right outer joins
- left outer joins
- semijoins
- antijoins

Right and left outer joins occur when combining two data sets where records are included in the resulting data set based on two criteria: either the BY group is present in both data sets (as with an inner join) or the BY group is present in exactly one data set. If the BY group is only present in the first data set listed in the MERGE statement, then it is a left outer join. Similarly, if the BY group is only present in the second data set, then it is a right outer join. Program 5.5.2 demonstrates how to carry out either join. Output 5.5.2A and 5.5.2B show the results of the left and right joins, respectively, based on a REPORT procedure similar to the one in Program 5.5.1.

Program 5.5.2: Carrying out Left and Right Outer Joins

```
data work.LeftJoin01 work.RightJoin01; ❶
  merge work.cost(in=inCost) work.means(in=inMeans); ❷
  by State MortgageStatus;
  if inCost eq 1 then output work.LeftJoin01; ❸
  if inMeans eq 1 then output work.RightJoin01; ❹
run;
```

❶ As introduced in Program 4.4.3, listing multiple data set names in the DATA statement instructs SAS to create multiple data sets.

❷ Other than the inclusion of the IN= options, the MERGE and BY statements remain unchanged from carrying out a full outer join or an inner join. The match-merge process is the same for each of these joins, only the records selected for the final data sets differ.

❸ Rather than using a subsetting IF statement, an IF-THEN statement is necessary to instruct SAS where to send the record currently in the PDV. Only data sets listed in the DATA statement are valid.

❹ No ELSE statement is used for either IF-THEN statement since the records selected for the RightJoin01 data set are not mutually exclusive from those in the LeftJoin01 data set.

Output 5.5.2A: LeftJoin01 Data Set Created by Program 5.5.2

Household serial number	state	Mortgage Status	Home Value	Mean Home Value
817019	North Carolina	No, owned free and clear	$5,000.00	$145,771.35
817020	North Carolina	No, owned free and clear	$162,500.00	$145,771.35
817031	North Carolina	No, owned free and clear	$45,000.00	$145,771.35
817032	North Carolina	No, owned free and clear	$32,500.00	$145,771.35
817038	North Carolina	No, owned free and clear	$95,000.00	$145,771.35
817040	North Carolina	No, owned free and clear	$225,000.00	$145,771.35
817042	North Carolina	No, owned free and clear	$137,500.00	$145,771.35
817049	North Carolina	No, owned free and clear	$75,000.00	$145,771.35

Output 5.5.2B: RightJoin01 Data Set Created by Program 5.5.2.

Household serial number	state	Mortgage Status	Home Value	Mean Home Value
.			.	$166,724.50
.		No, owned free and clear	.	$143,377.30
.		Yes, contract to purchase	.	$87,698.51
.		Yes, mortgaged/ deed of trust or similar debt	.	$181,427.54
.	North Carolina		.	$169,739.20
817019	North Carolina	No, owned free and clear	$5,000.00	$145,771.35
817020	North Carolina	No, owned free and clear	$162,500.00	$145,771.35
817031	North Carolina	No, owned free and clear	$45,000.00	$145,771.35

Note that the observations shown in Output 5.5.2A match the observations shown in Output 5.3.6A since each BY value set in the left data set, Cost, also appears in the Means data set. Whenever every record in the left data set has a matching value in the right data set, a left outer join is equivalent to an inner join (right outer joins are equivalent to inner joins when the right data set has a match for all records). However, the first five records of RightJoin01 contain missing values on Serial and HomeValue, variables present in the Cost data set. This is because these combinations of State and MortgageStatus appear in Means (which is the second (right) data set in the MERGE statement) but not in Cost, so the right join preserves those mismatches. While it is not typically necessary to create both a right and a left join simultaneously, the approach presented in

Program 5.5.2 provides one possible template for creating any necessary joins in a single DATA step. Two additional types of joins, semijoins and antijoins, are discussed in Chapter Note 3 in Section 5.9.

In both IF-THEN statements in Program 5.5.2, the OUTPUT keyword is used to direct the current PDV to the named data set. If no data set is provided, then the record is written to all data sets named in the DATA statement. The use of the OUTPUT keyword overrides the usual process by which the DATA step writes records from the PDV to the data set. By default, the DATA step outputs the current PDV using an implicit output process triggered when SAS encounters the step boundary. (Recall, this text uses the RUN statement as the explicit step boundary in all DATA steps.) When using a subsetting IF, a false condition stops processing the current iteration, including the implicit output, and immediately returns control to the top of the DATA step.

Because the implicit process occurs at the conclusion of the DATA step, the resulting data set includes the results from operations that occur during the current iteration. However, the OUTPUT statement instructs the DATA step to immediately write the contents of the PDV to the resulting data set. Thus, any operations that occur after SAS encounters the OUTPUT keyword are executed but not included in the resulting data set because there is no implicit output. Program 5.5.2 demonstrates the good programming practice of only including the OUTPUT keyword in the final statement(s) of a DATA step. It is possible to use the DELETE keyword to select observations without affecting the execution of subsequent statements. For a discussion of using DELETE, see Chapter Note 4 in Section 5.9.

5.6 Procedures for Investigating Association

The MEANS, FREQ, and UNIVARIATE procedures were used in Chapters 2, 3, and 4 for various purposes. In this section, procedures for investigating pairwise association between variables are reviewed. While many methods for measuring association are available and are provided in several SAS procedures, this section focuses on the CORR procedure and gives further consideration to the FREQ procedure, along with extending concepts in the SGPLOT procedure to include some plots for associations between variables.

5.6.1 Investigating Associations with the CORR and FREQ Procedures

When PROC CORR is executed using only the CORR statement with DATA= as its only option, the default behavior is similar to the MEANS and UNIVARIATE procedures in that the summary is provided for all numeric variables in the data set. In particular, the set of Pearson correlations among all possible pairs of these variables is summarized in a table with some additional statistics. As noted previously in Section 2.4.1, given that the IPUMS CPS data contains variables such as Serial and CountyFIPS that are numeric but not quantitative, many of the default correlations provided by PROC CORR are of no utility. To choose variables for analysis, the CORR procedure employs a VAR statement that works in much the same manner as it does in the MEANS and UNIVARIATE procedures. Program 5.6.1 provides correlations among four variables from the IPUMS CPS data from 2005 using PROC CORR.

Program 5.6.1: Basic Summaries Generated by the CORR Procedure
```
proc corr data=BookData.ipums2005basic;
  var CityPop MortgagePayment HHincome HomeValue; ❶
  where HomeValue ne 9999999❷ and MortgageStatus contains 'Yes'❸;
run;
```

❶ A correlation is produced between each possible pair of variables listed in the VAR statement, including the variable with itself. These are Pearson correlations by default and are displayed in a matrix that also includes a p-value for a test for zero correlation. Other types of correlation measures, such as Spearman rank correlation, are available—see the SAS Documentation and its references for more details.

❷ This condition in the WHERE statement removes observations with HomeValue equivalent to 9999999 as introduced in Program 4.7.4. Look back to Section 4.7 to review why this is an appropriate subsetting here.

❸ As in any other procedure (and as shown in Program 2.6.1), the conditioning done in the WHERE statement is not limited to the analysis variables; here the analysis is limited to those cases with an active mortgage.

By default, PROC CORR creates three output tables: one containing the correlation matrix, one listing variables used in the analysis, and another providing set of simple statistics on each individual variable. The tables resulting from Program 5.6.1 are shown in Output 5.6.1.

Output 5.6.1: Basic Summaries Generated by the CORR Procedure

4 Variables:	CITYPOP HomeValue	MortgagePayment HHINCOME

Simple Statistics						
Variable	N	Mean	Std Dev	Sum	Minimum	Maximum
CITYPOP	555371	1821	9049	1011091529	0	79561
MortgagePayment	555371	1044	754.34069	579767754	4.00000	7900
HHINCOME	555371	83622	72741	4.6441E10	-29997	1407000
HomeValue	555371	262962	222021	1.46042E11	5000	1000000

Simple Statistics	
Variable	Label
CITYPOP	City population
MortgagePayment	First mortgage monthly payment
HHINCOME	Total household income
HomeValue	House value

Pearson Correlation Coefficients, N = 555371 Prob > |r| under H0: Rho=0				
	CITYPOP	MortgagePayment	HHINCOME	HomeValue
CITYPOP City population	1.00000	0.09370 <.0001	0.04275 <.0001	0.14248 <.0001
MortgagePayment First mortgage monthly payment	0.09370 <.0001	1.00000	0.51595 <.0001	0.69260 <.0001
HHINCOME Total household income	0.04275 <.0001	0.51595 <.0001	1.00000	0.50006 <.0001
HomeValue House value	0.14248 <.0001	0.69260 <.0001	0.50006 <.0001	1.00000

The WITH statement is available in the CORR procedure to limit the set of correlations generated. When provided, each of the variables listed in the WITH statement are correlated with each variable listed in the VAR statement, and no other pairings are constructed. Program 5.6.2 illustrates this behavior.

Program 5.6.2: Using the WITH Statement in the CORR Procedure

```
proc corr data=BookData.ipums2005basic;
  var HHIncome HomeValue; ❶
  with MortgagePayment; ❶
```

```
   where HomeValue ne 9999999 and MortgageStatus contains 'Yes';
   ods exclude SimpleStats; ❷
run;
```

❶ Even though HHIncome and HomeValue are each listed in the VAR statement, they are no longer paired
    with each other—they are only paired to MortgagePayment. Also, the table listing variables used by the
    procedure is separated by these two roles. (See Output 5.6.2.)

❷ As with other procedures, finding ODS table names allows for simplification of the output via an ODS
    SELECT or ODS EXCLUDE statement.

Output 5.6.2: Using the WITH Statement in the CORR Procedure

1 With Variables:	MortgagePayment	
2 Variables:	HHINCOME	HomeValue

Pearson Correlation Coefficients, N = 555371 Prob > |r| under H0: Rho=0		
	HHINCOME	**HomeValue**
MortgagePayment First mortgage monthly payment	0.51595 <.0001	0.69260 <.0001

Further information about utility costs associated with these observations is contained in the file Utility Cost
2005.txt. To explore associations between these cost variables and those included in Ipums2005Basic, the raw
text file containing the utility costs must be converted to a SAS data set and merged with the Ipums2005Basic
data. Program 5.6.3 completes these steps and constructs a set of correlations on several of those variables.

Program 5.6.3: Adding Utility Costs and Computing Further Correlations
```
data work.ipums2005Utility;
   infile RawData('Utility Cost 2005.txt') dlm='09'x dsd firstobs=4;
   input serial electric:comma. gas:comma. water:comma. fuel:comma.; ❶
   format electric gas water fuel dollar.;
run;

data work.ipums2005cost;
   merge BookData.ipums2005basic work.ipums2005Utility;
   by serial; ❷
run;

proc corr data= work.ipums2005cost;
   var electric gas water fuel; ❸
   with mortgagePayment hhincome homevalue;
   where homevalue ne 9999999 and mortgageStatus contains 'Yes';
   ods select PearsonCorr;
run;
```

❶ When reading the raw file in the first DATA step, any legal variable names can be chosen. Here the first
    column is given the name Serial as it is used to match records in Ipums2005Basic during the merge.

❷ Both data sets are ordered on the variable Serial, so no sorting is necessary for this example. For this
    example, unlike Program 5.3.5, values with special encodings are not reset to missing. This is to revisit the
    dangers and difficulties of having values encoded in this manner.

❸ These variable names are those chosen in the INPUT statement in the first DATA step.

Output 5.6.3: Computing Further Correlations with Utility Costs

Pearson Correlation Coefficients, N = 555371 Prob > |r| under H0: Rho=0				
	electric	gas	water	fuel
MortgagePayment First mortgage monthly payment	0.19071 <.0001	-0.07199 <.0001	-0.05763 <.0001	0.01549 <.0001
HHINCOME Total household income	0.18629 <.0001	-0.05584 <.0001	-0.02987 <.0001	0.00890 <.0001
HomeValue House value	0.15874 <.0001	-0.07995 <.0001	-0.03048 <.0001	0.00148 0.2712

Some of the correlations in Output 5.6.3 appear a bit strange, and an exploration of each of the utility variables following the techniques demonstrated in Section 3.9 shows why. Program 5.6.4 addresses the issue for the Electric variable.

Program 5.6.4: Subsetting Values for the Electric Variable

```
proc corr data= work.ipums2005cost;
  var electric;
  with mortgagePayment hhincome homevalue;
  where homevalue ne 9999999 and mortgageStatus contains 'Yes'
        and electric lt 9000;
  ods select PearsonCorr;
run;
```

Output 5.6.4: Computing Further Correlations with Utility Costs

Pearson Correlation Coefficients, N = 552600 Prob > |r| under H0: Rho=0	
	electric
MortgagePayment First mortgage monthly payment	0.22431 <.0001
HHINCOME Total household income	0.22240 <.0001
HomeValue House value	0.18481 <.0001

At this point, a separate procedure would need to be run for each variable, as subsetting the full set of utility variables considered in Program 5.6.3 to those values of interest is not possible. Any time the WHERE condition evaluates to false, the entire record is removed; so, when some variables have valid values and others do not, none of them are included in the analysis. This is an example of why it is best to encode missing or unknown values in a manner that SAS recognizes internally as a missing value. Program 5.6.5 converts values to missing and does the correlation matrix of Program 5.6.3 again. Chapter Note 1 in Section 5.9 introduces a method for encoding special missing values in SAS.

Program 5.6.5: Setting Missing Utility Costs and Re-Computing Correlations

```
data work.ipums2005cost; ❶
  set work.ipums2005cost; ❶

  if electric ge 9000 then electric=.; ❷
  if gas ge 9000 then gas=.; ❷
  if water ge 9000 then water=.; ❷
  if fuel ge 9000 then fuel=.; ❷
run;
```

```
proc corr data= work.ipums2005cost;
  var electric gas water fuel;
  with mortgagePayment hhincome homevalue;
  where homevalue ne 9999999 and mortgageStatus contains 'Yes'; ❸
  ods select PearsonCorr;
run;
```

❶   This DATA step clearly replaces the Work.Ipums2005Cost data set since it appears in both the SET and
    DATA statements. In general, this is a poor programming practice due to the potential for data loss.

❷   Each utility variable is reset to missing by an IF-THEN statement for the appropriate condition.

❸   Now none of these variables require subsetting—check the number of observations used for each
    correlation shown in Output 5.6.5. Also, compare the correlations on Electric in Output 5.6.5 to those in
    Output 5.6.4.

Output 5.6.5: Re-Computing Correlations for Missing Utility Costs

Pearson Correlation Coefficients Prob > |r| under H0: Rho=0 Number of Observations				
	electric	gas	water	fuel
MortgagePayment First mortgage monthly payment	0.22431 <.0001 552600	0.10947 <.0001 368714	0.16347 <.0001 448952	0.18342 <.0001 92646
HHINCOME Total household income	0.22240 <.0001 552600	0.10397 <.0001 368714	0.12547 <.0001 448952	0.16220 <.0001 92646
HomeValue House value	0.18481 <.0001 552600	0.07382 <.0001 368714	0.17611 <.0001 448952	0.20998 <.0001 92646

The CORR procedure is limited to working with numeric variables, but some measures of association are still
valid for ordinal data, which can be non-numeric. Consider a binning of each of the HHIncome and
MortgagePayment variables given at the top of Program 5.6.6, with a subsequent association analysis using
PROC FREQ.

Program 5.6.6: Creating Simplified, Ordinal Values for Household Income and Mortgage Payment

```
proc format;
  value HHInc
    low-40000='$40,000 and Below'
    40000-70000='$40,000 to $70,000'
    70000-100000='$70,000 to $100,000'
    100000-high='Above $100,000';
  value MPay
    low-500='$500 and Below'
    500-900='$500 to $900'
    900-1300='$900 to $1,300'
    1300-high='Above $1,300';
run; ❶

proc freq data=BookData.ipums2005basic;
  table MortgagePayment*HHincome / measures❷ norow nocol format=comma8.;
  where HomeValue ne 9999999 and MortgageStatus contains 'Yes';
  format HHincome HHInc. MortgagePayment MPay.; ❶
run;
```

❶   Formats are created to bin these values; the bins are based roughly on quartiles of the data.

❷   The MEASURES option produces a variety of ordinal association measures. As a thorough review of
    association metrics is beyond the scope of this book, see the SAS Documentation and its references for
    more details.

Output 5.6.6: Association Measures (Partial Listing) for Ordinal Categories

Table of MortgagePayment by HHINCOME					
MortgagePayment(First mortgage monthly payment)	HHINCOME(Total household income)				
Frequency Percent	$40,000 and Below	$40,000 to $70,000	$70,000 to $100,000	Above $100,000	Total
$500 and Below	57,985 10.44	38,683 6.97	15,934 2.87	9,693 1.75	122,295 22.02
$500 to $900	47,487 8.55	65,134 11.73	37,879 6.82	23,883 4.30	174,383 31.40
$900 to $1,300	17,214 3.10	37,206 6.70	35,475 6.39	35,219 6.34	125,114 22.53
Above $1,300	10,165 1.83	20,059 3.61	28,932 5.21	74,423 13.40	133,579 24.05
Total	132,851 23.92	161,082 29.00	118,220 21.29	143,218 25.79	555,371 100.00

Statistics (Partial) for Table of MortgagePayment by HHINCOME

Statistic	Value	ASE
Gamma	0.5228	0.0012
Kendall's Tau-b	0.4007	0.0010
Stuart's Tau-c	0.3984	0.0010
Pearson Correlation	0.4481	0.0011
Spearman Correlation	0.4669	0.0011

Though the original, quantitative values for HHIncome and MortgagePayment are present in the data set for the analysis shown in Program 5.6.6, this is not required. Program 5.6.7 builds a new version of the data set with the formats used to establish character variables that define the bins.

Program 5.6.7: Ordinal Values for Household Income and Mortgage Payment

```
data work.ipums2005Modified;
  set BookData.ipums2005basic;
  MPay=put(MortgagePayment,MPay.); ❶
  HHInc=put(HHIncome,HHInc.); ❶
  where HomeValue ne 9999999 and MortgageStatus contains 'Yes'; ❷
  keep MPay HHInc; ❸
run;

proc freq data= work.ipums2005Modified;
  table MPay*HHInc / measures norow nocol format=comma8.;
run;
```

❶ Using the PUT function, a character value is assigned for each record corresponding to the bin each value falls into based on the formats given in Program 5.6.6.

❷ The records are chosen to match those in the analyses for the previous two examples.

❸ This data is constructed to only include the character values created, which PROC CORR cannot use.

The output for Program 5.6.7 matches that of Output 5.6.6, which is as much a matter of good fortune as proper planning in this case. By default, the levels of a categorical variable are ordered by alphanumeric sequencing which, for the bins created by the formats in Program 5.6.6, matches the low to high ordering for each of those variables. Consider a case where Likert scale responses are encoded as: Strongly Disagree, Disagree, Neutral, Agree, Strongly Agree. The alphanumeric ordering for those categories does not match their natural ranking and therefore any of the association measures generated by PROC FREQ as shown from Output 5.6.6 are not correct for this case. In general, if a particular ordering of values is important, an intervention may be required for SAS to recognize that special ordering (which may include formatting or re-encoding the data).

5.6.2 Plots for Investigating Association

Associations are often investigated through visualizations, several of which are provided as part of the SGPLOT procedure. One of the most common visualizations, the scatterplot, is shown in Program 5.6.8.

Program 5.6.8: Creating Scatterplot for Utility Costs Versus Home Values

```
ods listing close;
proc means data= work.ipums2005cost median; ❶
  var gas electric fuel water;
  class homevalue; ❷
  where state eq 'Vermont' and homevalue ne 9999999; ❸
  ods output summary= work.medians;
run;

ods listing;
proc sgplot data= work.medians; ❹
  scatter y=gas_median x=homevalue; ❺
  scatter y=electric_median x=homevalue;
  scatter y=fuel_median x=homevalue;
  scatter y=water_median x=homevalue; ❻
run;
```

❶ The plot is created from summary data, median values for each of the utility variables are generated.

❷ HomeValue would generally be thought of as a quantitative variable; however, it actually forms a set of categories in this data and is well-suited for use as a class variable.

❸ The WHERE statement removes certain values of HomeValue as before and limits the data to a single state to make the plot easier to read (limiting the number of data points).

❹ The data set created from the ODS OUTPUT statement is used for the plot, inspect this data set to determine the variable names.

❺ The SCATTER statement requires choices for an X= and Y= variable, with the X= variable placed on the horizontal axis at the bottom (the X2AXIS option is available to move it to the top) and the Y= variable vertical on the left (or right if the Y2AXIS option is employed).

❻ As before with the SGPLOT Procedure, a series of compatible plot calls create an overlay of the plots requested, as shown in Output 5.6.8.

Output 5.6.8: Scatterplot for Utility Costs Versus Home Values

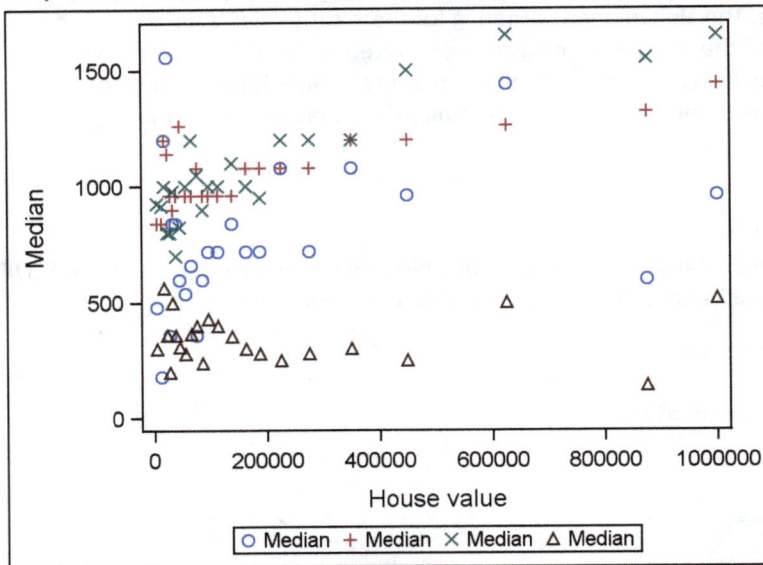

As with plots seen previously, a variety of options and statements are available to enhance the display, including the XAXIS, YAXIS, and KEYLEGEND statements. Program 5.6.9 illustrates the use of these statements and some other options.

Program 5.6.9: Enhancing Scatterplots for Utility Costs Versus Home Values

```
proc sgplot data= work.medians;
  scatter y=gas_median x=homevalue / legendlabel='Gas' ❶
          markerattrs=(color=red symbol=squarefilled) ❷;
  scatter y=electric_median x=homevalue / legendlabel='Elec.'
          markerattrs=(color=blue symbol=square);
  scatter y=fuel_median x=homevalue / legendlabel='Fuel'
          markerattrs=(color=green symbol=circlefilled);
  scatter y=water_median x=homevalue / legendlabel='Water'
          markerattrs=(color=orange symbol=circle);
  yaxis label='Cost ($)' values=(0 to 1600 by 200); ❸
  xaxis label='Value of Home' valuesformat=dollar12.;
  keylegend / position=topright; ❹
run;
```

❶ In Output 5.6.8, the Y= variable labels are used in the legend to name the differing symbols, but the labels are all unfortunately the same since they are generated from the same statistic in PROC MEANS. Use the LEGENDLABEL= option to specify values for display.

❷ The objects plotted for each data point are referred to as *markers* and have changeable attributes through the MARKERATTRS= option. Attributes that can be set include SIZE=, COLOR=, and SYMBOL=. See the SAS Documentation for a list of available symbol names and illustrations.

❸ The XAXIS and YAXIS statements are available for any plot that generates those axes and options available are consistent across plot types.

❹ As in Program 3.5.2, the overlaying of plots generates a legend and the KEYLEGEND statement is available to modify it. Recall that the KEYLEGEND statement requires the / character, separating a list of legend names (which may be null, as in this case) from the requested options.

Output 5.6.9: Enhanced Scatterplots for Utility Costs Versus Home Values

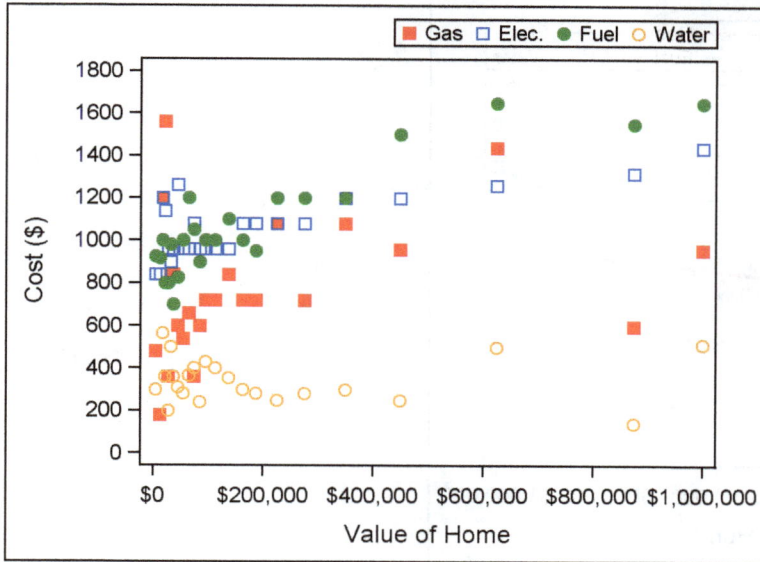

Plotting statements that create and display curve fits are also available as part of the SGPLOT procedure. Program 5.6.10 includes two polynomial regression curves (REG statement), a LOESS curve (LOESS statement), and a penalized B-spline (PBSPLINE statement). The statistical details of these curve fits, and others available in PROC SGPLOT, are beyond the scope of this book. However, the SAS Documentation contains further information about each of these methods and references other statistical publications.

Program 5.6.10: Fitting Curves for Utility Costs Versus Home Values

```
proc sgplot data= work.medians;
  reg y=gas_median x=homevalue❶ / degree=1❷ legendlabel='Gas'
      lineattrs=(color=red) ❸ markerattrs=(color=red symbol=squarefilled);
  reg y=electric_median x=homevalue / degree=5 legendlabel='Elec.'
      lineattrs=(color=blue) markerattrs=(color=blue symbol=square);
  loess❹ y=fuel_median x=homevalue / legendlabel='Fuel'
        lineattrs=(color=green) markerattrs=(color=green symbol=circlefilled);
  pbspline❹ y=water_median x=homevalue / legendlabel='Water'
          lineattrs=(color=orange) markerattrs=(color=orange symbol=circle);
  yaxis label='Cost ($)' values=(0 to 1600 by 200);
  xaxis label='Value of Home' valuesformat=dollar12.;
  keylegend / position=topright;
run;
```

❶   Like the SCATTER statement, each of these curve-fitting statements requires an X= and Y= variable. The curve is fit taking the Y= variable as a function of the X= variable.

❷   The REG statement fits a curve that is polynomial in the X= variable to the Y= variable. Though the available degrees of the polynomial fit vary across different maintenance releases of SAS, typically degrees from 1 to 10 are available.

❸   No matter which type of curve is fit, the curve itself is referred to as a line. Its attributes are set with LINEATTRS= and include COLOR=, THICKNESS=, and PATTERN=. PATTERN= includes both names and numbers as valid values. See the SAS Documentation for a list of names and numbers associated with various patterns.

❹   The LOESS and PBSPLINE statements fit a "smooth" curve to the data values. Both statements include the DEGREE= and SMOOTH= options to control the smoothness; however, they work differently for the two methods. See the SAS Documentation for more information and options.

Output 5.6.10: Fitted Curves for Utility Costs Versus Home Values

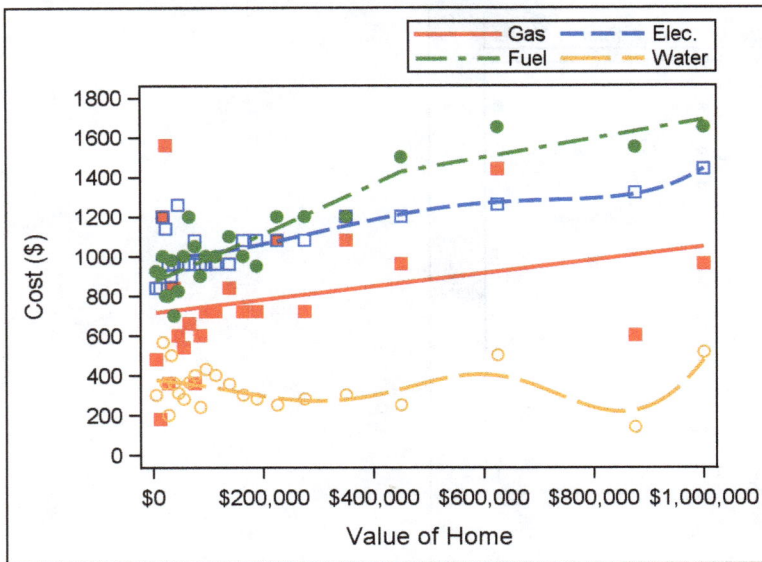

The previous examples used numeric variables exclusively for the X= and Y= variables, but these variables are not restricted to only the numeric type. Program 5.6.11 uses a character variable as one of the plotting variables and includes some options that are often helpful under that condition.

Program 5.6.11: Using a Categorical Variable with SCATTER

```
proc sgplot data=sashelp.cars;
  scatter x=origin❶ y=mpg_highway / jitter❷ jitterwidth=0.8❸;
run;
```

❶   The variable Origin is not only categorical, it is stored as character—which is legal for use in either (or both) of the X= or Y= variable positions.

❷   When either variable is categorical, several data points can occupy the same position. The JITTER option moves observations slightly around the original position; balanced left-to-right when moving horizontally, or up-and-down when moving vertically.

❸   The amount of space available to jitter the points is controlled by the JITTERWIDTH= option. Much like the BARWIDTH= option introduced in Program 3.5.2, it sets the maximum area occupied by the jittered points as the proportion of available space.

Output 5.6.11: SCATTER Using a Categorical Variable

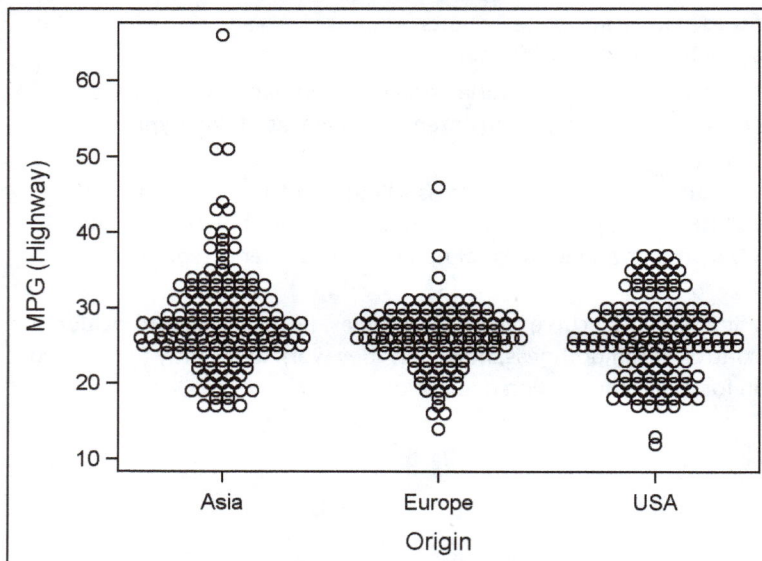

For these examples, plotting data for each of the utilities on the same graph requires multiple plotting statements which are subsequently overlaid. Section 5.7 shows a way to restructure the utility data and revisits these types of plots using the GROUP= option to also produce figures with distinct markers and/or curves for each utility type.

5.7 Restructuring Data with the TRANSPOSE Procedure

In the previous section, data for utility costs included one variable for each of four utilities. Under certain circumstances, it is advantageous (or even required) to restructure the data with one variable for the type of utility and another for its cost. This section shows how to perform this type of restructuring using the TRANSPOSE procedure and revisits some of the examples from Section 5.6 using this new data structure.

5.7.1 The TRANSPOSE Procedure

As a simple, preliminary example, Program 5.7.1 uses PROC TRANSPOSE with only a DATA= and OUT= option on a simplified data set.

Program 5.7.1: Transposing a Data Set

```
proc print data=sashelp.cars(keep=make model msrp mpg: obs=6) ❶ noobs;
run;

proc transpose data=sashelp.cars(keep=make model msrp mpg: obs=6) ❶
               out= work.carsTr❷;
run;

proc print data= work.carsTr noobs;
run;
```

❶ To make results easier to display, the Cars data is limited to a subset of the variables and observations available.

❷ Technically, OUT= is optional; however, PROC TRANSPOSE must produce a new data set to store the transposed data. If no name is supplied with the OUT= option, then SAS names the resulting data set automatically. It is considered a good programming practice to always explicitly name these output data sets.

Output 5.7.1 shows both the original data subset and its transpose. First, note that the make and model data do not appear in the transposed data set—only the numeric variables are transposed by default. However, a VAR statement is available to choose the variable values to be transposed, and character variables are only transposed when they are listed in it. Care should be taken with this as a record often contains a combination of character and numeric variables, but its transpose is to a single column. A column cannot contain both character and numeric types, so all values are converted to character.

Output 5.7.1: Transposing a Data Set

Make	Model	MSRP	MPG_City	MPG_Highway
Acura	MDX	$36,945	17	23
Acura	RSX Type S 2dr	$23,820	24	31
Acura	TSX 4dr	$26,990	22	29
Acura	TL 4dr	$33,195	20	28
Acura	3.5 RL 4dr	$43,755	18	24
Acura	3.5 RL w/Navigation 4dr	$46,100	18	24

NAME	_LABEL_	COL1	COL2	COL3	COL4	COL5	COL6
MSRP		36945	23820	26990	33195	43755	46100
MPG_City	MPG (City)	17	24	22	20	18	18
MPG_Highway	MPG (Highway)	23	31	29	28	24	24

Default behavior also includes naming the transposed variables with the prefix "col" and a numerical suffix and including a variable (_NAME_ by default) with values corresponding to the original variable names. In most cases a full transpose of rows and columns is not the objective; instead, like the example stated at the outset of this section, BY-group processing is exploited to transpose smaller structures throughout the data set. Program 5.7.2 turns each record from the utility cost data into four records, maintaining the Serial variable and replacing the four utility variables with one for utility type and another for the cost.

Program 5.7.2: Transposing Portions of Records Using BY-Group Processing
```
proc transpose data= work.ipums2005Utility out= work.Utility2;
   by serial; ❶
   var electric--fuel; ❷
run;

proc print data= work.Utility2 (obs=8) ❸ noobs;
run;
```

❶ The requested transpose is now applied to each level of Serial. Since Serial is unique to each record, the transpose is applied on each record.

❷ The values of the four utility variables are transposed into three variables and four records. By default, the values of the transposed variables are assigned to the variable Col1 and the original variable names are assigned to the _NAME_ variable. The BY-group variable, Serial, is also preserved in the transposed data set.

❸ A subset of the transposed data is displayed in Output 5.7.2 to show the new structure.

Output 5.7.2: Transposing Portions of Records Using BY-Group Processing

serial	_NAME_	COL1
2	electric	$1,800
2	gas	$9,993
2	water	$700
2	fuel	$1,300
3	electric	$1,320
3	gas	$2,040
3	water	$120
3	fuel	$9,993

Options are given to set names for the transposed variables, though the RENAME= option is also available. Program 5.7.3 illustrates some strategies.

Program 5.7.3: Transposing Portions of Records Using BY-Group Processing
```
proc transpose data= work.ipums2005Utility name= work.Utility❶
          prefix=Cost❷ out= work.Utility2;
   by serial;
   var electric--fuel;
run;
```

```
proc transpose data= work.ipums2005Utility
               out= work.Utility3(rename=(col1=Cost _name_=Utility) ❸);
   by serial;
   var electric--fuel;
run;
```

❶   The NAME= option replaces the default name of _NAME_ for the column containing the original variable name with the value specified.

❷   The PREFIX= option replaces the default prefix of COL for the names of the transposed variables. Whole number values are applied to each column resulting from the transpose. A SUFFIX= option is also available, with its value added to the end of the variable name following the column number.

❸   The RENAME= data set option can be applied to the output data set to directly set names of any columns, including those that result from the transpose operation.

If structure of the given data set is like that of Utility3 (four utility records on each value of Serial) and a structure like that of Ipums2005Utility (one record with four variables for each value of Serial) is desired, PROC TRANSPOSE can also perform this operation. Program 5.7.4 takes Utility3 as the input data set and returns it to the structure of the original Ipums2005Utility data.

Program 5.7.4: Transposing Multiple Records into a Set of Variables

```
proc transpose data= work.Utility3 out= work.Revert;
   by serial;
   var Cost;
run;

proc print data= work.Revert(obs=5) noobs;
run;
```

Output 5.7.4: Transposing Multiple Records into a Set of Variables

serial	_NAME_	COL1	COL2	COL3	COL4
2	Cost	$1,800	$9,993	$700	$1,300
3	Cost	$1,320	$2,040	$120	$9,993
4	Cost	$1,440	$120	$9,997	$9,993
5	Cost	$9,997	$9,993	$9,997	$9,993
6	Cost	$1,320	$600	$240	$9,993

Using BY-group processing results in the four records on the variable cost within each BY group for Serial being transformed into four variables on a single record for each unique Serial value. Here the _NAME_ variable is uniformly set to "Cost" and may not be particularly useful. The new variables containing utility costs have the default form of COL#, with # corresponding to the row order in the BY group.

A more useful set of variable names can be set using the RENAME= data set option; however, the TRANSPOSE procedure allows for another method via its ID statement. The ID statement uses the value of the selected variable as the name of the new variable, modifying those values to legal variable names if they do not satisfy proper naming conventions. (This modification produces variable names using the same process as PROC IMPORT, as shown in Program 3.8.1.) Program 5.7.5 uses the ID statement to select the values of the Utility variable as names for transposed cost values.

Program 5.7.5: Transposing Multiple Records into a Set of Variables

```
proc transpose data= work.Utility3 out= work.Revert2(drop=_name_);
   by serial;
   id Utility;
   var Cost;
run;
```

```
proc print data= work.Revert2(obs=5) noobs;
run;
```

Output 5.7.5: Transposing Multiple Records into a Set of Variables

serial	electric	gas	water	fuel
2	$1,800	$9,993	$700	$1,300
3	$1,320	$2,040	$120	$9,993
4	$1,440	$120	$9,997	$9,993
5	$9,997	$9,993	$9,997	$9,993
6	$1,320	$600	$240	$9,993

If the ID statement uses more than one variable, their values are concatenated to form the variable name. The formation of these variable names also respects the PREFIX= and SUFFIX= options. Any records with missing values for the ID variable(s) are not included in the transpose. It is important to ensure that the ID variables are unique within each BY group and that the values align across BY groups—unintended or unexpected differences in spelling, casing, or spacing in text values can create unwanted results. For example, if Utility values for electricity are given the value Electric in some records and Electricity in others, two separate variables using these values as names are created in the transposed data, resulting in a misalignment. See Section 3.9 for data diagnostics and cleaning methods that are applicable in situations like these.

5.7.2 Revisiting Section 5.6 Examples Using Transposed Data

In Section 5.6.1, the utility cost data was originally read with a structure having each of the four utility costs in separate columns for each record corresponding to a unique value of Serial. In section 5.7.1, the TRANSPOSE procedure was used to restructure the data into four records for each unique value of Serial, having one variable for utility type and another for its corresponding cost. Each data structure has its advantages, and for certain operations or analyses only one of the two works. However, it is often true that either structure can be used to achieve the same result using slightly different syntax. This section revisits some of the examples of Section 5.6 using the transposed data, Utility3, created as part of Program 5.7.3.

Program 5.7.6 uses the transposed utility data (Utility3) to revisit the objectives of Programs 5.6.3 through 5.6.5, merging the utility and Ipums2005Basic data together and producing correlations between utility costs and other variables.

Program 5.7.6: Building Correlations Between Utility Costs and Other Variables from Transposed Data
```
data work.ipums2005cost2;
  merge BookData.ipums2005basic work.Utility3;
  by serial; ❶
  utility=propcase(utility); ❷
  label utility='Utility';
run;

proc sort data= work.ipums2005cost2;
  by utility; ❸
run;

proc corr data= work.ipums2005cost2;
  by utility; ❸
  var cost; ❹
  with mortgagePayment hhincome homevalue;
  where homevalue ne 9999999 and mortgageStatus contains 'Yes' and cost lt 9000; ❹
  ods select PearsonCorr;
run;
```

❶ The fundamental syntax of the merge does not change from Program 5.6.3, only the data set names are updated, as the matching still occurs on the Serial variable. However, it should be noted that this is now a one-to-many match rather than the one-to-one in Program 5.6.3.

❷ The values of the Utility variable are the original variable names chosen in the DATA step in Program 5.6.3 that read the Utility Cost 2005.txt file. To improve the display of these for later use, they are converted to proper case via the PROPCASE function.

❸ The CORR procedure does not allow for a CLASS statement; therefore, separating the analysis across utility types requires BY-group processing. Since the original merge has both data sets sorted on Serial, a subsequent sort on Utility is required prior to invoking PROC CORR.

❹ All the utility costs are now contained in the Cost variable, and the desired conditioning is achieved on this variable only—no resetting values to missing, as in Program 5.6.5, is required. However, a more defensive programming strategy of setting the values to missing is advised.

Output 5.7.6: Correlations Using Transposed Data (Partial Output)

Utility=Electric

Pearson Correlation Coefficients, N = 552600 Prob > |r| under H0: Rho=0	
	Cost
MortgagePayment First mortgage monthly payment	0.22431 <.0001
HHINCOME Total household income	0.22240 <.0001
HomeValue House value	0.18481 <.0001

Utility=Fuel

Pearson Correlation Coefficients, N = 92646 Prob > |r| under H0: Rho=0	
	Cost
MortgagePayment First mortgage monthly payment	0.18342 <.0001
HHINCOME Total household income	0.16220 <.0001
HomeValue House value	0.20998 <.0001

Program 5.7.7 also uses the transposed utility data (Utility3) to revisit the objectives of Programs 5.6.9, using the GROUP= option in the SCATTER statement to make a graph like that of Output 5.6.9.

Program 5.7.7: Making Scatterplots Based on Transposed Utility Data

```
ods listing close;
proc means data= work.ipums2005cost2 median;
  var cost;
  class homevalue utility; ❶
  where state eq 'Vermont' and homevalue ne 9999999 and cost lt 9000; ❷
  ods output summary= work.medians2;
run;
```

```
ods listing;
proc sgplot data= work.medians2;
   scatter y=cost_median x=homevalue / group=utility; ❸
   yaxis label='Cost ($)' values=(0 to 1600 by 200);
   xaxis label='Value of Home' valuesformat=dollar12.;
   keylegend / position=topright; ❹
run;
```

❶ Adding Utility as a class variable produces the summary statistic for each of the four utility types, ensuring the Utility variable is available in the output data set.

❷ Again, the single analysis variable is Cost and it can be effectively conditioned for all records on all utility types without special intervention.

❸ The GROUP= option distinguishes the observations on the plot across the four utility types. Looking at the legend in Output 5.7.7, note the value of using the PROPCASE function on the original variable names.

❹ The XAXIS, YAXIS, and KEYLEGEND statements are unaltered from those used in Program 5.6.9.

Output 5.7.7: Scatterplots Based on Transposed Utility Data

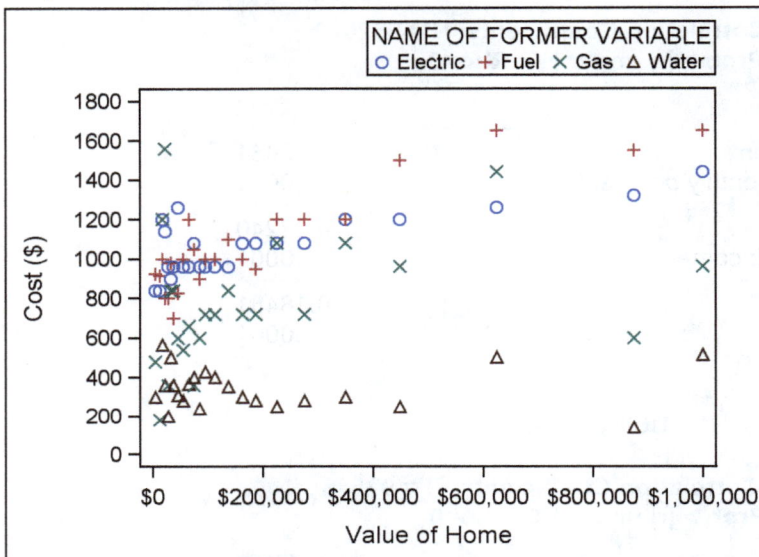

Since Program 5.7.7 does not include any options to modify the scatterplot markers, the colors and symbols used for them are the same as those shown in Output 5.6.8. When using the GROUP= option, control of markers and lines is handled somewhat differently. Program 5.7.8 uses the MARKERATTRS= option to make a limited modification to the markers.

Program 5.7.8: Setting Attributes for Markers in Grouped Scatter Plots

```
proc sgplot data= work.medians2;
   scatter y=cost_median x=homevalue / group=utility markerattrs=(symbol=square);
   yaxis label='Cost ($)' values=(0 to 1600 by 200);
   xaxis label='Value of Home' valuesformat=dollar12.;
   keylegend / position=topright;
run;
```

Output 5.7.8: Setting Attributes for Markers in Grouped Scatter Plots

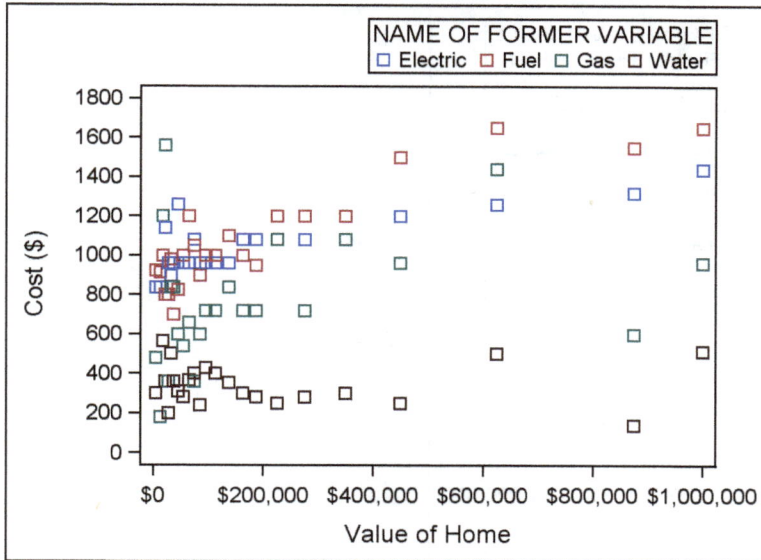

The SYMBOL= attribute makes the symbol for each group a square, leaving the color cycle as the only distinguishing feature among the levels of Utility. Setting both COLOR= and SYMBOL= attributes in the MARKERATTRS= option makes the groups indistinguishable. Since the ability to control the full set of colors and symbols (and line types) is desirable for these plots as well, Program 5.7.9 shows one method to take control of these sets.

Program 5.7.9: Using the STYLEATTRS Statement to Set Attributes in Grouped Plots

```
proc sgplot data= work.medians2;
   styleattrs❶ datacontrastcolors=(red blue green orange) ❷
              datasymbols=(circle square triangle diamond) ❸
              datalinepatterns=(solid) ❹;
   pbspline y=cost_median x=homevalue / group=utility;
   yaxis label='Cost ($)' values=(0 to 1600 by 200);
   xaxis label='Value of Home' valuesformat=dollar12.;
   keylegend / position=topright;
run;
```

❶ The STYLEATTRS statement allows for setting style attributes broadly, across all plotting statements included in the PROC SGPLOT call.

❷ The DATACONTRASTCOLORS= option specifies a list of colors to be applied to markers and lines. This modifies the list of colors SAS cycles through as it moves through levels of a GROUP= variable or as it moves through different plotting statements to be overlaid. DATACOLORS= is also available to modify fill colors.

❸ The DATASYMBOLS= option modifies the list of marker symbol shapes SAS cycles through.

❹ The DATALINEPATTERNS= option sets the list of line styles SAS cycles through. For any of these lists, it is important to ensure the number of elements of the list is sufficient to properly distinguish the different categories on the graph. In Output 5.7.9, the four levels of Utility each have different colors for lines and markers along with different marker shapes; however, each uses the same line type.

Output 5.7.9: Using the STYLEATTRS Statement to Set Attributes in Grouped Plots

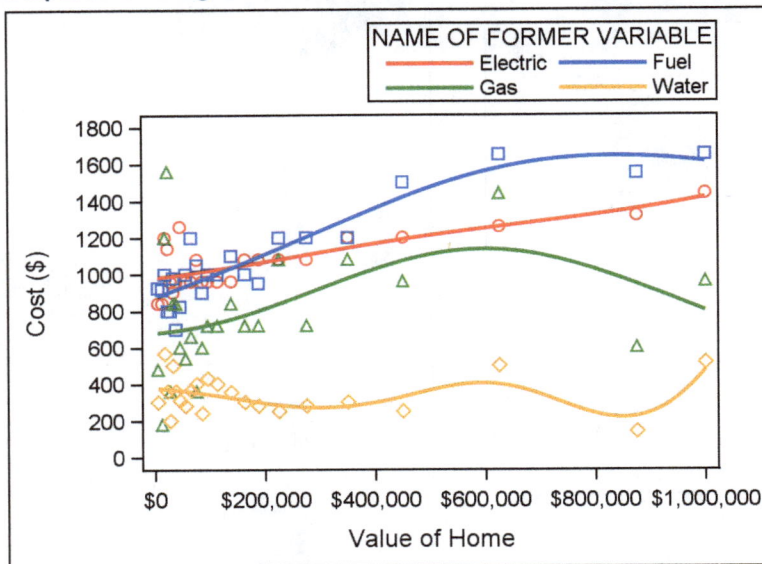

As an exercise, and to see that this behavior is not limited to the GROUP= option, remove the MARKERATTS= and LINEATTRS= option from each of the plotting statements in Program 5.6.10, add the STYLEATTRS statement used in Program 5.7.9, and submit the new version—the cycling of colors, symbols, line styles matches that of Output 5.7.9. However, note that Output 5.6.10 cannot be reproduced using a GROUP= option on the transposed data—Program 5.6.10 overlays plots of different types, while GROUP= splits across levels of the specified variable in a single plot type.

5.8 Wrap-Up Activity

Use the lessons and examples contained in this and previous chapters to complete the activity shown in Section 5.2.

Data

Various data sets are available for use to complete the activity. This data expands on the data used in the Wrap-Up Activity for Chapter 4, bringing in utility data for 2005 used in the in-chapter examples, plus similar data for 2010 and 2015. Completing this activity requires the following files:

- Ipums2005Basic.sas7bdat
- Ipums2010Basic.sas7bdat
- Ipums2015Basic.sas7bdat
- Utility Cost 2005.txt
- Utility Costs 2010.csv
- 2015 Utility Cost.dat

Scenario

Use the skills mastered so far, including those from previous chapters, to assemble the listed files into a single data set and create the analyses shown in Section 5.2. To get the maximum benefit from this activity, build the graphs shown in 5.2.2 and 5.2.3 from the data assembled in its original form and also from appropriately transposed versions of that data.

5.9 Chapter Notes

1. *Special Missing Numeric Values.* Program 5.3.5 demonstrates using conditional logic to assign values to missing using the standard numeric missing value—a single period. However, SAS provides 27 distinct special missing values for a numeric variable in addition to the standard missing value. The special missing values are denoted by a period followed by either an underscore or a letter (._, .A, .B, …, .Z). While a lowercase letter is valid during assignment, SAS displays these values using capital letters. Because the values are distinct, SAS assigns them different sort priorities with ._ being the smallest special missing value, the standard missing value is the second smallest, then .A through .Z with .Z being the largest missing value. The sort order is important when using numeric missing values in ranges as using the standard missing value as the lower bound, such as in a format, would exclude missing values encoded with ._, but would include all other missing values. To exclude all missing values, a range must use .Z as the excluded lower bound of the range.

   Special missing values are useful when a variable, such as the utility cost variables in the IPUMS CPS data sets, have missing values for multiple reasons. Program 5.9.1 demonstrates assigning two special missing values, .A and .B, based on the fact that in the IPUMS 2005 data a value of 9993 and 9997 had specific meanings. Outputs 5.9.1A and 5.9.1B show the results of displaying these special missing values with and without the application of a format.

Program 5.9.1: Assigning and Using Special Missing Values

```
data ipums2005UtilityB;
  infile RawData('Utility Cost 2005.txt') dlm='09'x dsd firstobs=4;
  input serial electric:comma. gas:comma. water:comma. fuel:comma.;
  format electric gas water fuel dollar.;
  if electric eq 9993 then electric=.A;
    else if electric eq 9997 then electric=.B;
run;

proc freq data = ipums2005UtilityB;
  table electric / missing;
  where electric in (.A, .B);
run;

proc format;
  value elecmiss .A = 'No Charge or None Used'
                 .B = 'Included in Rent or Condo Fee'
  ;
run;

proc freq data = ipums2005UtilityB;
  table electric / missing;
  where electric in (.A, .B);
  format electric elecmiss.;
run;
```

Output 5.9.1A: Using Special Missing Values without Applying a Format

electric	Frequency	Percent	Cumulative Frequency	Cumulative Percent
A	9428	21.10	9428	21.10
B	35244	78.90	44672	100.00

Output 5.9.1B: Using Special Missing Values with a Format

electric	Frequency	Percent	Cumulative Frequency	Cumulative Percent
Included in Rent or Condo Fee	35244	78.90	35244	78.90
No Charge or None Used	9428	21.10	44672	100.00

2. *Updating and Modifying.* The DATA step offers two additional techniques for combining information from multiple SAS data sets: updating and modifying. These techniques are most similar to the horizontal techniques discussed in Section 5.3. The UPDATE statement, as the name implies, is useful for adding new information into an existing data set. It requires a BY statement to join records between a master data set and a transaction data set. Records with matching values of the key variables in both data sets are updated so that any new information from the transaction data set is copied into the master data set. UPDATE is very similar to MERGE, with the notable exception that MERGE overwrites nonmissing data with missing data unless conditional logic is present to prevent it from occurring. The UPDATE statement can have the same behavior, but its default is to not overwrite nonmissing values with missing data. The MODIFY statement offers some of the same functionality, but with an expanded set of options and risks. For more information about the MODIFY statement, see the SAS Documentation.

3. *Semijoins and Antijoins.* Semijoins are directional, just like left and right joins. The effect of a semijoin is to filter a table based on whether its records have matching key variable values in another table; however, no information from the second table is included in the resulting semijoin. For example, a left semijoin of two tables, A and B, returns all the rows in A that have matching key variable values in B. This is very similar to a left join, but a left semijoin only returns the columns from A, while a left join returns the columns of both A and B. An antijoin is the row-complement of a semijoin; a left antijoin of the tables A and B returns all the rows in A for which the key variables do not have a match in table B. Like left semijoins, a left antijoin of tables A and B does not include any columns from table B. While the examples here are for left semijoins and antijoins, right semijoins and antijoins work in an equivalent manner. In the DATA step, use the IN= data set options and conditional logic along with KEEP and DROP lists to produce semijoins or antijoins.

4. *The DELETE Statement.* In addition to providing the OUTPUT statement discussed in Section 5.5, SAS provides the closely related DELETE statement. Recall, using OUTPUT in a program forces the DATA step to immediately write the current PDV to the data set but continues executing any remaining statements in the DATA step. Conversely, using DELETE prevents the DATA step from writing the current PDV to the resulting DATA set and immediately returns control to the top of the DATA step, foregoing the execution of any remaining programming statements during the current iteration of the DATA step. As a result, the following two statements are equivalent.

```
IF expression;
IF NOT (expression) THEN DELETE;
```

Due to their ability to avoid complex operations on unwanted observations, the subsetting IF and DELETE statements are both important tools when computational efficiency is a priority.

5.10 Exercises

Concepts: Multiple Choice

1. If the SAS data set Mayflies exists in the work library, what output is produced by the following code?

```
proc corr data = Mayflies;
run;
```

   a. None—it generates an error in the log.
   b. None—it generates a warning in the log.
   c. Correlations and associated statistics on all variables in the data set.
   d. Correlations and associated statistics on all numeric variables in the data set.

2. If the SAS data set DataDivas exists in the work library, what result is produced by the following code?

```
proc transpose data = DataDivas;
run;
```

   a.   None—the OUT= option must be used to specify where to store the transposed results.

   b.   Transposes of all variables in the data set.

   c.   Transposes of all numeric variables in the data set.

   d.   Transposes of all character variables in the data set.

3.   Assuming the SAS data set DataDivas exists in the work library, which of the following measures of association are produced by the following code?

```
proc freq data = DataDivas;
   table Size*FileType / norow format = comma8.;
run;
```

   a.   None—the CORR option is required.

   b.   None—the MEASURES option is required.

   c.   Only Pearson correlations between Size and FileType.

   d.   Pearson and Spearman correlations between Size and FileType.

4.   Consider the data set, Diagnosis, shown in the following table.

Diagnosis

Subject	Visit	DiagnosisCode
001	1	450
001	2	430
001	3	410
002	1	250
002	2	240
003	1	410
003	2	250
003	3	500
004	1	240

After running the following TRANSPOSE procedure, which of the answer choices below shows the resulting data set? (Assuming none of the variables have associated labels.)

```
proc transpose data = Diagnosis prefix = Dx
               out = Rotated(drop = _NAME_);
   by Subject;
   id Visit;
   var DiagnosisCode;
run;
```

   a.

Subject	Dx1	Dx2	Dx3
001	450	430	410
002	250	240	.
003	410	250	500
004	240	.	.

b.

Subject	001	001	001	002	002	003	003	003	004
Visit	1	2	3	1	2	1	2	3	1
Diagnosis	450	430	410	250	240	410	250	500	240

c.

Subject	Visit	Dx1	Dx2	Dx3
001	1	450	430	410
001	2	450	430	410
001	3	450	430	410
002	1	250	240	.
002	2	250	240	.
003	1	410	250	500
003	2	410	250	500
003	3	410	250	500
004	1	240	.	.

d. None of the above

5. Consider the data set, Height, shown in the following table.

Height

Height1	Height2	Height3
1	11	101

Height1	Height2	Height3
2	12	102
3	13	103
4	14	104
5	15	105

After running the following TRANSPOSE procedure, which of the answer choices below shows the resulting data set if none of the variables have associated labels?

```
proc transpose data = Height out = HeightRows;
  var Height3 Height2 Height1;
run;
```

a.

NAME	COL1	COL2	COL3	COL4	COL5
Height1	1	2	3	4	5
Height2	11	12	13	14	15
Height3	101	102	103	104	105

b.

COL1	COL2	COL3	COL4	COL5
1	2	3	4	5
11	12	13	14	15
101	102	103	104	105

c.

Height1	Height2	Height3	Height4	Height5
1	2	3	4	5
11	12	13	14	15
101	102	103	104	105

d. None of the above

6. Consider the Weight and Visit data sets shown below.

Weight (left) and Visit (right)

Subject	Visit	Weight
1	1	101
1	2	102
2	1	103
3	1	104
3	2	105

Subject	Visit	Date
1	1	18JAN2019
1	2	19JAN2019
1	3	20JAN2019
2	1	01FEB2019
2	2	03FEB2019

What type of join does the following DATA step carry out?

```
data combined;
  merge weight visit;
  by subject;
run;
```

   a. One-to-one merge

   b. One-to-one match-merge

   c. One-to-many match-merge

   d. Many-to-many match-merge

7. Using the data sets and code from the previous question, how many records appear in the data set Combined.

   a. 0—no data set is produced due to an error

   b. 5

   c. 7

   d. 10

8. The following DATA step is submitted.

```
data work.combined;
  merge work.Employee work.Demog;
run;
```

A character variable named EmpID is contained in both the Work.Employee and Work.Demog data sets. The variable EmpID has a length of six in Work.Employee and a length of eight in Work.Demog. What is the length of the variable EmpID in the Work.Combined data set?

   a. 6—because it is the first length from the two input data sets
   b. 8—because it is the larger length from the two input data sets
   c. 8—because it is the default length for character variables
   d. 8—because it is the last length from the two input data sets

9. Selected components of the descriptor portions of the Employee and Demog data sets are shown below.

Data Set	Variable	Type	Length	Format	Label
Employee	EmpID	Char	6	$6.	
	HireDate	Num	8	date9.	Date of Hire
	Salary	Num	8	dollar13.2	Current Salary
Demog	EmpID	Char	8	$6.	Employee ID
	HireDate	Num	8	mmddyy10.	Hire Date

If the following DATA step is submitted using the data sets described above, which of the answer choices below correctly identifies the format for HireDate?

```
data example;
  merge employee demog;
  by EmpID;
run;
```

   a. DATE9.
   b. MMDDYY10.
   c. DATE10.
   d. MMDDYY9.

10. Using the descriptor information and DATA step code from the previous question, which of the answer choices below correctly identifies the label for EmpID in the Example data set?

   a. EmpID
   b. Employee ID
   c. None—there is no label for this variable in the Employee data set.
   d. Employ

11. Consider the data set, OutputControl, shown below.

OutputControl

Y	Z
1	11
2	12
3	13
2	14
5	15
2	16

If the following program is submitted, how many observations are in each data set?

```
data A B;
  set OutputControl;
  if Y = 1 then output A;
    else if Y = 2 then output B;
      else output;
run;
```

    a.   A has 1 observation, B has 3 observations

    b.   A has 1 observation, B has 5 observations

    c.   A has 3 observations, B has 3 observations

    d.   A has 3 observations, B has 5 observations

12. Consider the data set, Quarks, shown below.

Quarks

Flavor	Sequence
Up	1
Down	2
Charm	3
Up	4
Strange	5
Charm	6

If the following program is submitted, how many observations are in each data set?

```
data Up Down Others;
  set Quarks;
  if Flavor = 'Up' then output Up;
    else if Flavor = 'Down' then output Down;
run;
```

   a.   Up has 2 observations, Down has 1 observation, Others has 3 observations

   b.   Up has 2 observations, Down has 1 observation, Others has 0 observations

   c.   Up has 5 observations, Down has 4 observations, Others has 0 observations

   d.   Up has 5 observations, Down has 4 observations, Others has 3 observations

13. The Start data set is shown in the table below.

Start

Var1	Var2
A	3
B	5
C	8

How many records appear in each of the data sets created by the following DATA step?

```
data ID Count;
  set start;
  if var1 = 'D' then output ID;
run;
```

   a.   ID has 3 observations, Count has 3 observations

   b.   ID has 0 observations, Count has 3 observations

   c.   ID has 3 observations, Count has 0 observations

   d.   ID has 0 observations, Count has 0 observations

14. Consider the data sets Model and Retail shown in the following tables.

Model

Brand	ModelID
Sunny	SN-657-NK
Acre	TG-983-FP

Brand

Brand	Retail
Sunny	$400
Sunny	$350
Dull	$370
Dull	.

If the following DATA step is submitted, which of the answer choices shows the resulting data set?

```
data work.all;
  merge work.model work.retail;
  by descending brand;
run;
```

a.

Brand	ModelID	Retail
Sunny	SN-657-NK	$400
Sunny	TG-983-FP	$350

b.

Brand	ModelID	Retail
Sunny	SN-657-NK	$400
Sunny	SN-657-NK	$350
Dull		$370
Dull		.
Acre	TG-983-FP	.

c.

Brand	ModelID	Retail
Sunny	SN-657-NK	$400
Sunny	TG-983-FP	$350
Dull		$370
Dull		.

d.

Brand	ModelID	Retail
Sunny	SN-657-NK	$400
Sunny		$350
Dull		$370

Brand	ModelID	Retail
Dull		.
Acre	TG-983-FP	.

Concepts: Short-Answer

1. Consider the different methods for combining three data sets—A, B, and C—in the DATA step. If the data sets contain a common variable named Sequence, and each data set is sorted by Sequence in ascending order, briefly describe the process SAS uses to join the records from the three input data sets in the following scenarios:
   a. Each data set appears in its own SET statement.
   b. All data sets appear in the same MERGE statement without a BY statement.
   c. All data sets appear in the same MERGE statement, and Sequence is used as the only BY variable.

2. In PROC SORT, the OUT= option prevents SAS from replacing the original data set with the new, sorted, data set. In contrast, the TRANSPOSE procedure creates a new data set even if the OUT= is absent. Why is the default in PROC TRANSPOSE to *not* overwrite the original data?

3. In PROC TRANSPOSE, the PREFIX= and SUFFIX= options allow for customization of the variable names for any newly-created columns that result from rows in the input data set. Given the existence of these options, justify the existence of the ID statement by explaining how it differs from the PREFIX= and SUFFIX= options.

4. When investigating the association between two numeric variables, why is it typically not useful to apply a user-defined format (or even certain formats provided by SAS) to either of those numeric variables in PROC CORR when it is often quite useful to do so in PROC FREQ?

5. Consider the three data sets shown below: Contacts, Sales, and FilingDate.

Contracts (left), Sales (middle), and FilingDate (right)

CustomerID	Contact	CustomerID	QTR	Sales	CustomerID	Date
1	Fulanah AlFulaniyyah	1	1	18000	1	01JAN2019
2	John Doe	1	2	23000	1	01APR2019
3	Tran Thi B	1	3	15000	1	01JUL2019
4	Yossi Cohen	1	4	35000	1	01OCT2019
		2	1	22000	3	01JAN2017
		2	2	10000	4	01JAN2018
		2	3	8000	4	01APR2018
		2	4	40000	4	05APR2018
		4	1	20000	4	01JUL2018

CustomerID	Contact	CustomerID	QTR	Sales	CustomerID	Date
		4	2	25000	4	09OCT2018
		4	3	22000	5	14APR2019
		4	4	24000	5	31DEC2019
		5	1	90000		
		5	2	120000		
		5	3	85000		
		5	4	140000		

a. Without executing the program below in a SAS session, write out the values of the user-defined variables in the PDV for iterations 2, 6, 9, and 10 of the following DATA step.

```
data Accounting;
  merge contacts sales filingdate;
  by customerID;
run;
```

b. If the variable Qtr is added to the FilingDate data set, is it possible to update the program in (a) so that it produces a one-to-many match-merge of the three provided data sets? Assume Qtr only takes on the values 1, 2, 3, or 4 and is not missing for any record.

6. Consider two data sets, A and B, that are connected through a single key variable, X. Describe the relationship between: the left antijoin of A with B, the right antijoin of A with B, the inner join of A with B, and the full outer join of A with B.

7. Explain why no observations appear in either of the created data sets when using the following DATA step.

```
data CompanyA CompanyB;
  set employee(where = (Age < 25));
  if Age eq 30 then output;
run;
```

Programming Basics

1. The following graph was generated using three variables from the Sashelp.Stocks data set: Stock (the name of the stock), High (the maximum value at which the stock was traded on a given date), and Date (the date on which the value of High was recorded).

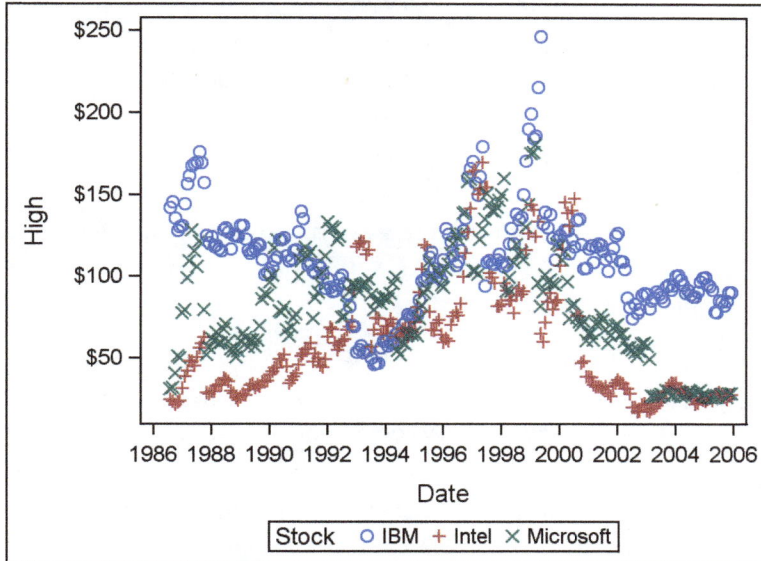

a. Use the Sashelp.Stocks data set to generate the graph shown above.

b. After submitting the following code, use the StockTr data set to create the graph shown above.

```
proc sort data=sashelp.stocks out=stockSort;
  by date;
run;

proc transpose data=stockSort out=stockTr;
  var high;
  by date;
  id stock;
run;
```

c. Compare the two SGPLOT procedures from part (a) and part (b). Which of the two techniques would scale up more easily to scenarios that needed to graph 500 different stocks instead of three?

2. Use the following code to create a subset of the Sashelp.Prdsale data set and exchange some row and column information from the subset.

```
proc sort data = sashelp.prdsale out = prdSubset(drop = year);
  by country region division prodtype product quarter;
  where year eq 1993;
run;

proc transpose data = PrdSubset out = salesWide prefix = month;
  by country region division prodtype product quarter;
  var actual predict;
run;
```

a. Write a program that reverses the transposition and recovers the PrdSubset data set.

b. Validate that the PrdSubset data generated by reversing the transposition matches the PrdSubset data set generated by the SORT procedure above.

3. Write a program that uses the Sashelp.Baseball data set and produces only the following output.

Pearson Correlation Coefficients, N = 322 Prob > |r| under H0: Rho=0			
	nHome	**nRuns**	**nRBI**
nHits Hits in 1986	0.54165 <.0001	0.91167 <.0001	0.79311 <.0001
nBB Walks in 1986	0.46835 <.0001	0.71288 <.0001	0.59070 <.0001

4. Use the BookData.Visit and BookData.PayStatus data sets to write a program that does the following in a single DATA step.

   ○ Joins the data sets by PatientID

   ○ Creates three data sets: AllRecords, Match, and Mismatch

   ○ Ensure every record goes into the AllRecords data set; only records that come from both data sets go into the Match data set; records that only come from one data set go into the Mismatch data set

   ○ Writes the PDV to the log any time a mismatched observation is written to the Mismatch data set

5. Use the Sashelp.Citimon and Sashelp.Citiday data sets to answer the following questions.

   a. What variables are available for use as key variables during a match-merge? Are the values useful for carrying out a match-merge?

   b. Write a DATA step that carries out a one-to-many match-merge so that the monthly indicators from Sashelp.Citimon are joined with each record from the same month in Sashelp.Citiday. Ensure that only records from Sashelp.Citiday remain in the final data set.

6. Program 5.3.6 created several difference and ratio variables. In that program, operations on missing values result in unnecessary—and in some industries, unallowed—notes in the log. Similarly, division by zero results in the DATA step setting _ERROR_ to 1 and writing the associated messages to the log. Write a DATA step that produces equivalent results but without generating any notes, warnings, or errors as a result of operations on missing values or division by zero.

Case Studies

For additional practice, multiple case studies are available in addition to the IPUMS CPS case study used in the chapters. See Section 8.5 to apply the skills from this chapter to the Clinical Trials Case Study. For additional case studies, including extensions to the IPUMS CPS case study, see the author pages.

Chapter 6: Restructuring Data and Introduction to Advanced Reporting

6.1 Learning Objectives

At the conclusion of this chapter, mastery of the concepts covered in the narrative includes the ability to:

- Apply DO loops to generate new records

- Differentiate between iterative, conditional, and iterative-conditional DO loops and apply the best choice in a given scenario

- Compare and contrast DO WHILE and DO UNTIL loops

- Apply one or more DO loops and arrays to carry out the same operations on multiple variables

- Employ the _TEMPORARY_ option to create elements only for use during the current DATA step

- Apply arrays to exchange row and column information in a SAS data set

- Summarize data with PROC REPORT using various statistics and stratifying the results with the GROUP usage option

- Create additional summaries using BREAK statements and the RBREAK statement in PROC REPORT

- Create groups and summary statistics in columns via the ACROSS usage option in PROC REPORT

- Generate a data set from a PROC REPORT submission

Use the concepts of this chapter to solve the problems in the wrap-up activity. Additional exercises and case-studies are also available to test these concepts.

6.2 Case Study Activity

Continuing the case study covered in the previous chapters, the tables and graphs shown below are based on the data assembled for the Wrap-Up Activity in Chapter 5. The objective, as described in the Wrap-Up Activity in Section 6.7, is to create the reports shown in Output 6.2.1 and 6.2.2, plus generate new data to create the graphs shown in Output 6.2.3, 6.2.4, and 6.2.5.

Output 6.2.1: Electricity Cost Summary Statistics

State	Year	Electricity Cost		
		Mean	**Median**	**Std. Dev.**
Connecticut	2005	$1,423.00	$1,200.00	$979.81
	2010	$1,746.45	$1,560.00	$1,215.27
	2015	$1,717.85	$1,440.00	$1,384.61
		$1,638.04	$1,440.00	$1,223.86
Massachusetts	2005	$1,173.31	$960.00	$862.60
	2010	$1,242.80	$1,080.00	$999.99
	2015	$1,369.06	$1,080.00	$1,275.12
		$1,269.12	$1,080.00	$1,078.57
New Jersey	2005	$1,477.86	$1,200.00	$1,137.00
	2010	$1,888.13	$1,560.00	$1,424.46
	2015	$1,749.70	$1,440.00	$1,428.80
		$1,711.41	$1,320.00	$1,353.26
New York	2005	$1,454.35	$1,200.00	$1,054.80
	2010	$1,528.60	$1,200.00	$1,190.45
	2015	$1,541.07	$1,200.00	$1,324.39
		$1,510.53	$1,200.00	$1,203.17

Output 6.2.2: Home Value Summary Statistics

	Home Value Statistics					
	2005		**2010**		**2015**	
State	**Mean**	**Median**	**Mean**	**Median**	**Mean**	**Median**
Connecticut	$353,601	$275,000	$404,008	$280,000	$396,875	$270,000
Massachusetts	$401,007	$350,000	$392,878	$319,500	$434,209	$350,000
New Jersey	$378,541	$350,000	$388,920	$320,000	$395,547	$300,000
New York	$304,502	$225,000	$360,828	$250,000	$387,242	$250,000

Output 6.2.3: Electricity Cost Projections Through 2010, 2% Annual Growth

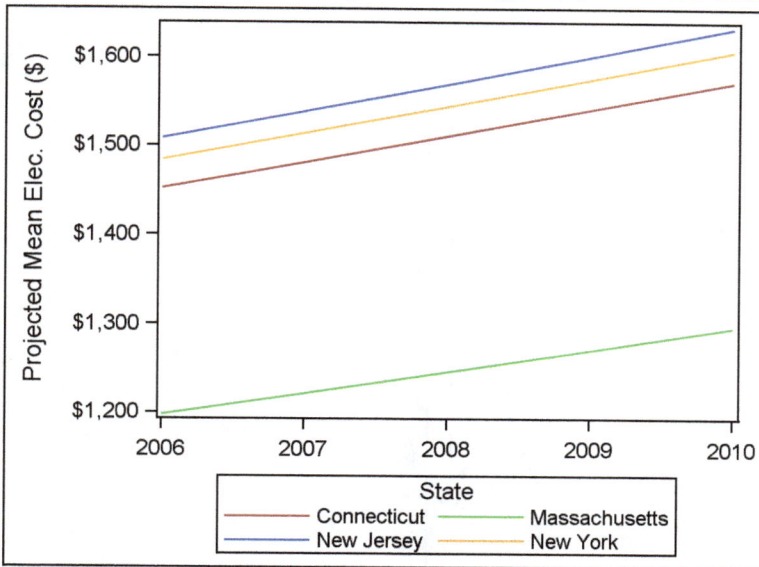

Output 6.2.4: Electricity Cost Projections Versus Actual Value

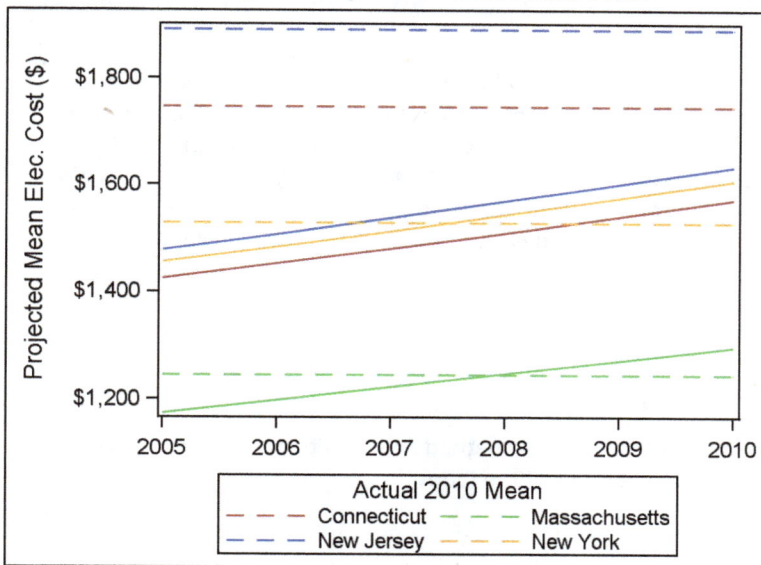

Output 6.2.5: Electricity Cost Projections at 2% Annual Growth, Connecticut

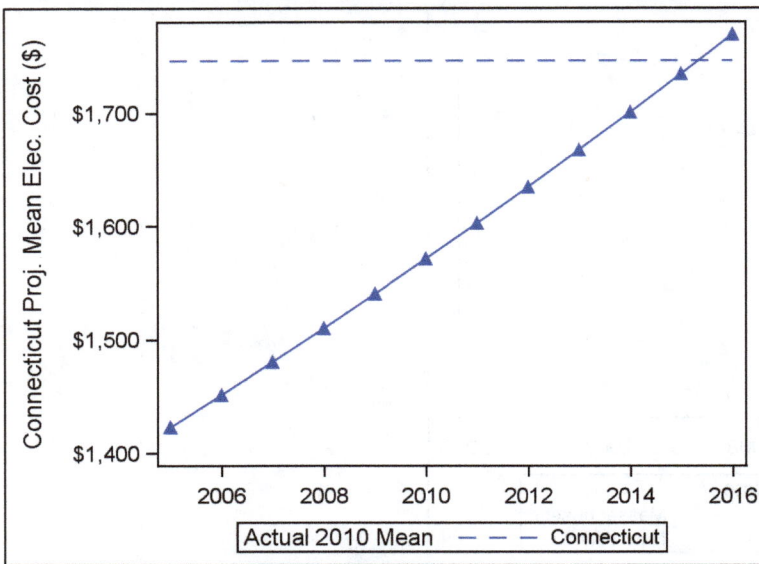

6.3 DO Loops

Section 4.5.3 introduced the DO statement and the DO group, which allows SAS to execute a group of statements once per record, typically as the result of a conditional logic statement. However, often a certain task requires the execution of a group of statements more than once per record, which is typically accomplished by placing the group of statements inside a loop that executes as many times as needed. In other programming languages, these types of loops are often referred to as DO, FOR, or FOREACH loops. In the SAS DATA step, three types of DO loops are available and each of them begins with the keyword DO:

- Iterative
- Conditional
- Iterative-conditional

The primary difference between the three types of loops is the method SAS uses to determine when to stop executing the statements inside the loop, that is, when to exit the loop.

6.3.1 Iterative DO Loops

Iterative DO loops are characterized by the fact that they execute a fixed number of times. One way to define the number of iterations is shown in Program 6.3.1, which uses the data from Program 5.3.5 to project mean electricity costs for North and South Carolina over the ten years following 2005 at a growth rate of 1.5%.

Program 6.3.1: Using a DO Loop for Repeating a Calculation

```
ods listing close;
proc means data=work.ipums2005Cost mean;
  class state;
  where state in ('North Carolina','South Carolina');
  var electric;
  ods output summary= work.means;
run; ❶

data work.projections;
  set work.means;

  year=2005;
  cost=electric_mean;
  output; ❷
```

```
  do j=1 to 10; ❸
    year+1;
    cost=1.015*cost;
    output; ❹
  end; ❸
  keep state year cost; ❺
run;

proc sgplot data= work.projections;
  series❻ x=year y=cost / group=state lineattrs=(pattern=solid);
  yaxis label='Projected Cost';
  xaxis display=(nolabel) integer;
  keylegend / position=topleft location=inside title='' across=1;
run;
```

❶ The MEANS procedure is used to create an initial data set with two records, one containing the mean for Electric in North Carolina, the other for South Carolina.

❷ Before entering the DO loop, new variables for Year and Cost are set to their initial values—the starting year of 2005 and the Electric_Mean value from the first record—and output to the Projections data set.

❸ Every DO loop must begin with the keyword DO and close with the keyword END, just as with the DO group. An iterative DO loop must specify an index variable and its set of values; a common form of this specification is *IndexVariable* = *StartValue* TO *StopValue*. In this example, the loop is indexed by a variable named J for which the values start at 1 and stop at 10. By default, the iterative DO loop increments the indexing by 1, so the values of J are 1, 2, 3, ..., 9, 10.

❹ Inside the DO loop, the value for Year is incremented by 1 each time using a sum statement, and the projected value of Cost is set via a calculation. The OUTPUT statement ensures that both values are output to the Projections data set for each iteration of the loop.

❺ The *IndexVariable*, J, is not needed in the final data set. A KEEP or DROP statement (or option) is useful for subsetting down to the necessary variables.

❻ The resulting data can be viewed with PROC PRINT or PROC REPORT; however, it is also possible to graph the resulting data. The SERIES statement in PROC SGPLOT operates similarly to the SCATTER statement. The resulting plot does not show markers, instead drawing lines through the points—see Output 6.3.1 (the MARKERS option places the scatterplot markers on the graph as well). Other PROC SGPLOT statements and options are similar to those used in previous chapters.

Output 6.3.1: Series Plot of Projections from DO Loop Calculations

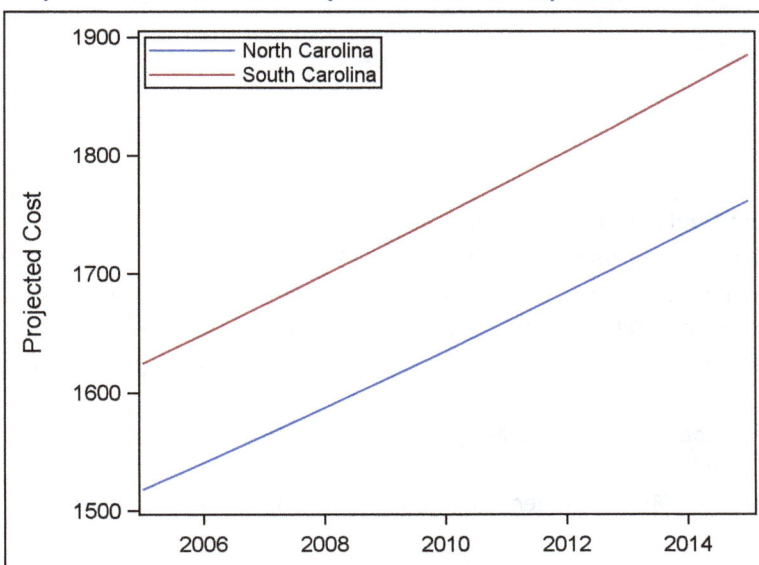

The DO loop in Program 6.3.1 iterates ten times based on the values provided in the DO statement. However, there are multiple ways to specify the values in an iterative DO loop. In addition to the technique shown in Program 6.3.1, the following DO statements show the various techniques for specifying an iterative loop. Each of the following statements iterate over the same set of values, the even numbers from 2 through 10 (inclusive).

1. `do i = 2 to 10 by 2;`
2. `do i = 10 to 2 by -2;`
3. `do i = 2, 4, 6, 8, 10;`

The first and second examples specify starting and stopping values like those used in Program 6.3.1. They introduce the BY keyword for specifying an amount to increment or decrement (using a negative BY value) in the first and second examples, respectively. In each of these cases, incrementing with a BY value results in a value for the *IndexVariable* that is eventually equivalent to the *StopValue*; however, this is not required.

If the BY value is positive; in other words, the loop is incrementing as in the first example, then the loop exits whenever the value of the *IndexVariable* exceeds the *StopValue*. In the second example, the loop is decrementing since the BY value is negative; in this case, the loop exists whenever the value of the *IndexVariable* subceeds the *StopValue*. In the third example, a comma-delimited list is used to specify a collection of values – here the loop exits once it exhausts all values in the list. Any iterative DO loop can only use one of these styles for setting the index variable list. When nesting iterative DO loops, each loop can use a different style, and generally each loop uses a different index variable. Program 6.3.2 demonstrates nesting two DO loops to make projections across several years for two different growth rates.

Program 6.3.2: Using Nested, Iterative DO Loops

```
data work.projections1;
  set work.means;

  do year=2005 to 2015 by 2; ❶
    do rate=0.015,0.025; ❷
      cost=electric_mean*(1+rate)**(year-2005); ❸
      output; ❹
    end;
  end;

  keep state year rate cost;
  format rate percent7.1;
  label rate='Growth Rate'; ❺
run;

proc sgpanel data= work.projections1;
  panelby rate;
  series x=year y=cost / group=state lineattrs=(pattern=solid);
  rowaxis label='Projected Cost';
  colaxis display=(nolabel) integer;
  keylegend / title='';
run; ❻
```

❶ The first DO loop sets the value for year and iterates it from 2005 to 2015 every 2 years.

❷ The second DO loop iterates through two values for the rate of increase, given as a list.

❸ Unlike Program 6.3.1, the calculation of Cost directly refers to variable values for year and rate of increase—the index variables for the nested DO loops. The ** operator is for exponentiation.

❹ The OUTPUT statement occurs inside the inner DO loop, ensuring projections are output for all years and rates of increase.

❺ Any invocation of a DATA step provides the opportunity to assign labels and formats, which are useful for the subsequent display of the Rate variable.

❻ The SGPANEL procedure also includes the SERIES statement. Here the projection plots are separated into panels for the two growth rates.

Output 6.3.2: Panel of Series Plot of Projections from Nested DO Loop Calculations

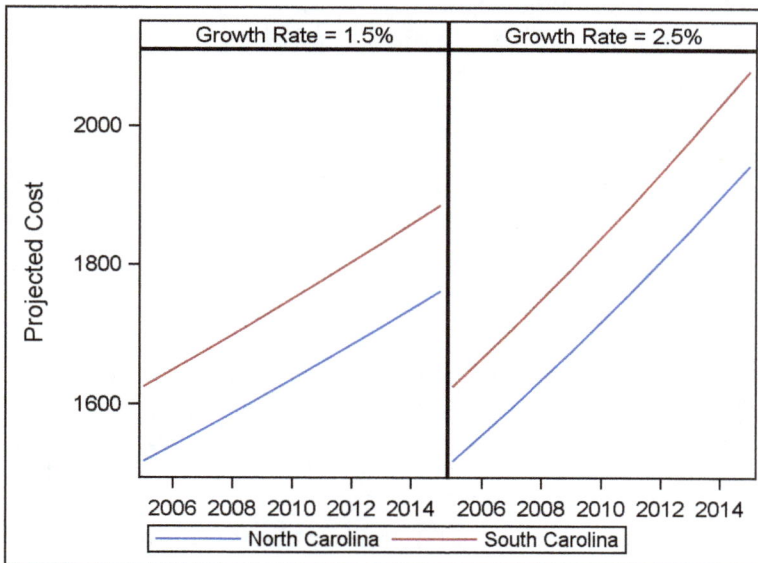

6.3.2 Conditional DO Loops

The DO loops in Programs 6.3.1 and 6.3.2 are iterative since they execute a fixed number of times designated as part of the DO statement. However, it is often the case that the number of iterations needed is not known in advance. Consider projecting an increase in Electric cost in North Carolina at 2% and in South Carolina at 1%. If the goal is to determine when the cost in North Carolina exceeds that of South Carolina, the number of years required is not known until calculations are completed. If the calculations are done in the DATA step, being able to determine when the loop exits based on the value of the projections is necessary.

In the DATA step, there are two ways to implement conditional logic to define the stopping rule for a loop: the DO UNTIL and DO WHILE loops. As with iterative loops, both the DO UNTIL and DO WHILE loops must begin with the keyword DO and conclude with the END statement. They are also similar to iterative DO loops in that they can be nested. In fact, all three types of DO groups can be nested. The syntax for the two conditional DO loops differs from the iterative DO loop in that no index variable is supplied. Instead, they both must include an expression that defines the stopping rule for the loop. However, they differ somewhat in their execution: the DO UNTIL loop iterates until the specified condition is true, checking its condition at the bottom of the loop; the DO WHILE loop iterates while the specified condition is true and checks its condition at the top of the loop.

The DO UNTIL Loop

Program 6.3.3 uses the DO UNTIL loop to make the projections for North Carolina and South Carolina Electric_Mean values as described in the preceding section. It also uses PROC TRANSPOSE, which is introduced in Chapter 5.

Program 6.3.3: Using a DO UNTIL Loop to Condition on a Value Calculated Within the Loop

```
proc transpose data= work.means out= work.means2;
  var electric_mean;
  id state;
run; ❶

data work.projections2;
  set work.means2;
  year=2005;
  output;

  do until(North_Carolina gt South_Carolina); ❷
    year+1;
    North_Carolina=1.02*North_Carolina;
```

```
      South_Carolina=1.01*South_Carolina;
      output; ❸
    end; ❹

  keep year North_Carolina South_Carolina;
run;

proc sgplot data= work.projections2;
  series x=year y=North_Carolina / lineattrs=(pattern=solid);
  series x=year y=South_Carolina / lineattrs=(pattern=solid); ❺
  yaxis label='Projected Cost';
  xaxis display=(nolabel) integer;
  keylegend / position=topleft location=inside title='' across=1;
run;
```

❶ In Programs 6.3.1 and 6.3.2, the projections for the two states can be done independently. In this case, since the scenario requires a comparison for each projection in each year, the mean Electric costs are first transposed into two columns on a single record. State is the ID variable; however, the values of State are not legal variable names as they contain spaces, so the underscore character is inserted as introduced in Program 3.8.1 with PROC IMPORT and Section 5.7.1 with PROC TRANSPOSE.

❷ The condition for the DO UNTIL loop must be placed in parentheses. This loop iterates until the value of the projected Electric cost is greater for North Carolina than for South Carolina.

❸ The projections are completed and output to the new data set. Since the DO UNTIL loop checks its condition at the bottom of the loop, the statements in the loop execute at least once.

❹ The condition specified in the DO UNTIL statement is checked here—so the data set contains exactly one record (the last one) where the value for the North Carolina projection exceeds the one for South Carolina.

❺ In Output 6.3.3, the SGPLOT procedure call creates an overlay of two SERIES plots since the projections are now stored in different variables.

Output 6.3.3: Projecting Costs Conditionally Using a DO UNTIL Loop

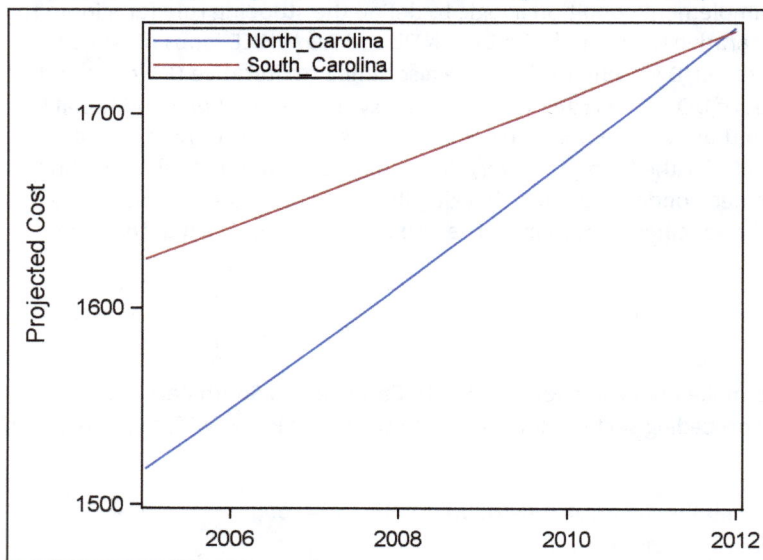

The DO WHILE Loop

Program 6.3.4 demonstrates how to achieve the same result as Program 6.3.3 with a DO WHILE loop.

Program 6.3.4: Using a DO WHILE Loop to Condition on a Value Calculated Within the Loop

```
data work.projections2B;
  set work.means2;
  year=2005;
  output;
```

```
  do while(North_Carolina le South_Carolina);
    year+1;
    North_Carolina=1.02*North_Carolina;
    South_Carolina=1.01*South_Carolina;
    output;
  end;

  keep year North_Carolina South_Carolina;
run;
```

The only changes from Program 6.3.3 to Program 6.3.4 are the change in keyword from UNTIL to WHILE and the reversal of the condition on the values—the resulting data set is the same. The change in the condition reflects the behavior of the DO WHILE loop. Iterating while the condition is true—projecting each year until the value for North Carolina exceeds that of South Carolina—is logically equivalent to projecting each year while the value of North Carolina does not exceed that of South Carolina.

DO UNTIL Versus DO WHILE

It is common for novice SAS programmers to question why both the DO UNTIL and DO WHILE loops exist if the only difference is that the DO UNTIL loop stops when a condition becomes true, compared to the DO WHILE loop stopping when the condition becomes false. As Programs 6.3.3 and 6.3.4 demonstrate, switching between the two conditional loops appears to be relatively simple, merely substituting the complementary condition. The major difference between the two styles of conditional loop is not in their syntax, it is in their execution.

Any time SAS executes a DO WHILE loop, it begins by checking the condition in the DO WHILE statement. If the condition is false, then SAS exits the loop without executing any of the statements in the DO WHILE group. If the condition is true, SAS enters the loop and executes the statements for the first time. SAS then returns to the top of the loop to begin the second iteration where it once again checks the condition to decide whether to continue or exit. Because SAS always checks the condition before executing the statements in a DO WHILE loop, a DO WHILE loop may never execute its statements.

By comparison, any time SAS executes a DO UNTIL loop, because the given condition is not checked until the bottom of the loop, it first must execute the statements inside the loop. Once the statements have been executed, SAS checks the condition in the DO UNTIL statement and, if the condition is true, SAS exits the loop. If the condition is false, SAS re-enters the loop and executes the statements again. Because SAS checks the condition in a DO UNTIL loop at the bottom of the loop, a DO UNTIL loop always executes its statements at least once.

This difference in the minimum number of executions is often the deciding factor when selecting a conditional loop. If the loop must execute at least once, then a DO UNTIL loop is required. To allow for the possibility that the loop never executes, a DO WHILE loop must be used. In all other cases, there is little practical difference between them.

6.3.3 Iterative-Conditional Loops

Beware that both the DO UNTIL and DO WHILE syntax can result in infinite loops. If the expression in a DO UNTIL loop is always false, or the expression in a DO WHILE loop is always true, then the result is an infinite loop. Infinite loops result in a DATA step that does not terminate and can result in a drain on computational resources. One possible way to avoid infinite loops is using iterative-conditional loops as shown in Program 6.3.5. This is a modification of the scenario in Program 6.3.3, projecting (with new growth rates) when the North Carolina projection exceeds that for South Carolina, unless that occurs more than 15 years into the future. This approach uses one DO loop to combine elements of an iterative loop and a conditional loop.

Program 6.3.5: Using an Iterative-Conditional Loop for Multiple Stopping Rules

```
data work.projections3;
  set work.means2;
  year=2005;
  output;
```

```
  do j=1 to 15❶ until(North_Carolina gt South_Carolina) ❷;
    year+1;
    North_Carolina=1.016*North_Carolina;
    South_Carolina=1.012*South_Carolina;
    output;
  end;

  keep year North_Carolina South_Carolina;
run;

proc sgplot data= work.projections3; ❸
  series x=year y=North_Carolina / lineattrs=(pattern=solid);
  series x=year y=South_Carolina / lineattrs=(pattern=solid);
  yaxis label='Projected Cost';
  xaxis display=(nolabel) integer;
  keylegend / position=topleft location=inside title='' across=1;
run;
```

❶ The iterative syntax must come first when combining iterative and conditional syntax. Here the loop is limited to no more than 15 iterations (years).

❷ The conditional logic is unchanged from Program 6.3.3. The DO loop now has two stopping rules—stop after 15 iterations or when the North Carolina projection exceeds that for South Carolina, whichever comes first.

❸ From Output 6.3.5, it is clear the loop stopped after 15 years.

Output 6.3.5: Using an Iterative-Conditional Loop for Multiple Stopping Rules

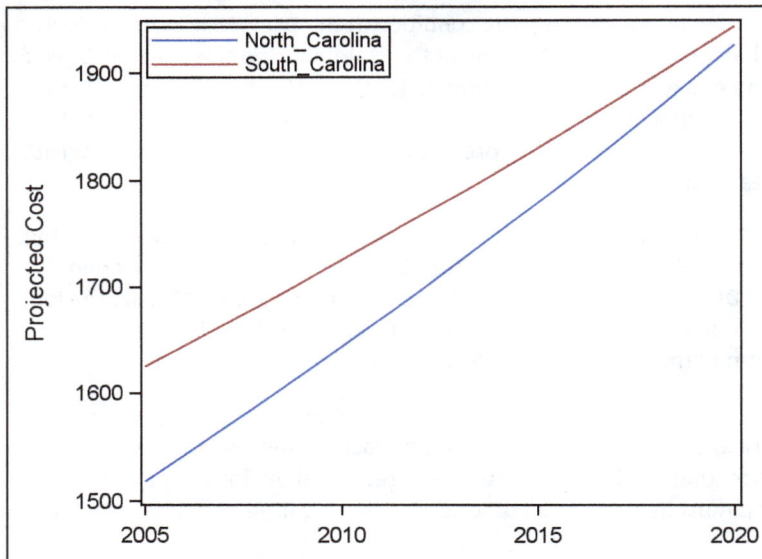

While Program 6.3.5 demonstrates the iterative-conditional loop in a circumstance where an infinite loop was unlikely, these types of loops can be quite beneficial when it is uncertain if a given condition ever changes from true to false (or vice versa). It is also a useful strategy for testing programs on large data sets where an infinite loop is not likely (or even possible) but where the computational time to run the full program is significant.

6.4 Arrays

Section 6.3 introduced DO loops and demonstrated their ability to repeatedly execute a group of statements; however, it is often necessary to execute the same statement for a group of variables. This is not possible with only a DO loop unless the group of variables can be accessed through index values. SAS arrays create numeric indexes for groups of variables in a DATA step with variables indexed in the array having two valid references: by variable name or by array name and position in the array.

6.4.1 One-Dimensional Arrays

Arrays allow for the use of DO loops to simplify a repetitive calculation done on different data set variables, such as deriving the deviations from and ratios to the mean as in Program 5.3.6. Program 6.4.1 revisits a slight variation on this scenario, using an array to simplify the repetitive calculations done in the DATA step.

Program 6.4.1: Calculate Deviations from the Mean Using Arrays

```
proc means data=bookdata.ipums2005basic;
  class state mortgageStatus;
  where state in ('North Carolina','South Carolina');
  var homevalue hhincome mortgagepayment;
  output out= work.means mean=meanHV meanHHI meanMP;
run;

proc sort data=bookdata.ipums2005basic out= work.basic;
  by state mortgagestatus;
  where state in ('North Carolina','South Carolina')
        and scan(mortgagestatus,1) eq 'Yes'; ❶
run;

proc sort data= work.means;
  by state mortgagestatus;
run;

data work.compare;
  merge work.basic(in=inBasic) work.means(in=inMeans);
  by state mortgagestatus;

  array ratios[3]  ❷  HVratio HHIratio MPratio; ❸
  array diffs[3]; ❹
  array means[*] mean:; ❺
  array vals[3] homevalue hhincome mortgagepayment; ❻

  if inMeans eq 1 and inBasic eq 1;

  do j=1 to dim(means); ❼
    diffs[j]=vals[j]-means[j]; ❽
    if means[j] ne 0 ❾ then ratios[j]=vals[j]/means[j];
      else ratios[j]=.;
  end;

  drop j;
run;
```

❶ An additional condition is applied here to remove the cases that produce unusual values for the mean.

❷ After the keyword ARRAY, provide a name for the array—this array is named Ratios. The array name must follow SAS variable naming conventions and cannot be the same as the name of any variable in the PDV. Next is the specification for the dimension of the array, which may be enclosed in brackets, braces, or parentheses. By this specification, Ratios is a one-dimensional array with three elements.

❸ This array statement concludes with the names of the variables to include in the array (known as the elements of the array). So, the reference Ratios[1] is equivalent to referencing HVratio, Ratios[2] is the same as HHIratio, and Ratios[3] is the same as MPratio. When variable names are specified, the number must exactly match the dimension, with a mismatch resulting in a semantic error. As these elements are not members of either data source, they are created during the compilation of the ARRAY statement. By default, created variables are defined as numeric, use the $ immediately after the dimension specification to define an array as character. For any array, all elements must be of the same type.

❹ Without a list of variables declared in the array statement, SAS assigns the array to a set of variables that use the array name as the prefix with a whole number suffix starting at 1—Diffs1, Diffs2, and Diffs3 in this case. These variables are also created at compilation, much like the members of the Ratios array.

❺ When using one-dimensional arrays, using * in place of a specific dimension forces SAS to determine the number of elements by counting the variables provided in the ARRAY statement. Therefore, when using the * to specify the dimension, the variable set must be named in the ARRAY statement. Looking at the

OUTPUT statement in PROC MEANS, the statistics have been strategically named to take advantage of the colon wildcard—corresponding to the name prefix list discussed in Chapter Note 3 in Section 1.7.

❻ The array Vals references values from the original data. As the colon operator takes variables in their order in the PDV for Means, the ordering of the list of variables in this ARRAY statement is chosen to match that ordering.

❼ The DIM function returns the first dimension of an array, which can be applied to any of the arrays in use here as they are all one-dimensional. This statement is equivalent to `DO i = 1 TO 3`, but with added flexibility as changes in array dimensions do not require the DO statement to be updated. This also applies to cases where use of dynamic dimensioning for all arrays may not allow for hardcoding the stopping value.

❽ Referencing an array uses brackets, braces, or parentheses around the subscript value to specify the desired array element. This single computation inside the DO loop replaces the three differencing calculations done in Program 5.3.6. A mismatch in array dimension can cause an error if SAS attempts to access a nonexistent array element. For example, Means[4] is an invalid reference and results in an error message in the log.

❾ Though the WHERE condition in ❶ generally eliminates division by 0 problems, it is good programming practice to take control of scenarios where such errors can occur and eliminate invalid data notes in the log.

The results of Program 6.4.1 are the same as those of Program 5.3.6 on matching records; however, the arrays have replaced the need for repeated assignment statements. As the number of variables changes from three, Program 5.3.6 requires assignment statements to be added/removed as appropriate. In Program 6.4.1 the only necessary updates are in the ARRAY statements in either dimensions and/or variable lists.

It is common for novice programmers to attempt to use the DIM function when declaring an array. This is natural in Program 6.4.1 where the dimensions of all arrays are the same. However, when declaring a one-dimensional array, the dimension must be either an asterisk, a sequence of integers (for example, -6:5), or a positive integer constant (for example, a number token such as 3, which is equivalent to the range 1:3). This means that not only is the DIM function, or any other function or expression, invalid syntax for setting the dimension, so also is any declaration that attempts to use a data set variable name to provide the dimension. Chapter Note 1 in Section 6.8 introduces one way to alleviate this restriction.

Program 6.4.2 revisits a scenario like the projections done in Section 6.3 and again using the data set from Program 5.3.5. In this case, a temporary array is used to store the growth rates, and a second array is used to store the projections for each different rate.

Program 6.4.2: Using a Temporary Array with Initial Values

```
proc means data= work.ipums2005Cost mean;
  class state;
  where state in ('North Carolina','South Carolina');
  var electric;
  ods output summary= work.means;
run;

data work.projections4;
  set work.means;

  array rates[3] _temporary_ ❶ (0.015 0.020 0.025) ❷;
  array proj[3]; ❸

  do year=2006 to 2010;
    do j=1 to dim(rates);
      proj(j)=electric_mean*(1+rates(j))**(year-2005); ❹
    end;
    output; ❺
  end;

  keep state year proj:;
run;
```

```
proc sgpanel data= work.projections4;
  panelby state / novarname;
  series x=year y=proj1 / lineattrs=(pattern=solid) legendlabel='1.5%'; ❻
  series x=year y=proj2 / lineattrs=(pattern=solid) legendlabel='2.0%';
  series x=year y=proj3 / lineattrs=(pattern=solid) legendlabel='2.5%';
  rowaxis label='Projected Cost';
  colaxis display=(nolabel) integer;
  keylegend / title='Growth Rate' down=1;
run;
```

❶ The keyword _TEMPORARY_ is used to indicate to SAS that the array elements are dropped when creating the final data set. Since using the * for dimensioning requires an explicit variable list, the _TEMPORARY_ keyword cannot be used simultaneously with the * for declaring the size of the array.

❷ An array can be initialized with values by placing them in parentheses. The ARRAY statement associates these values with the elements of the array, and they are retained throughout the DATA step.

❸ This array is defined to store the projections under the three different interest rates, and thus also has dimension three.

❹ The inner DO loop is indexed for three iterations, each time using a different growth rate from the temporary array Rates and storing the calculation in the corresponding element of the array Proj.

❺ Here the output occurs after the inner loop is finished—after all three elements (columns) corresponding to the Proj array are computed. Then the outer loop continues the process for the next year.

❻ Note that even if the KEEP statement is removed from Program 6.4.2, no columns relating to the growth rates remain in the output data set, so LEGENDLABEL= is used to provide information to the legend in Output 6.4.2.

Output 6.4.2: Using a Temporary Array with Initial Values

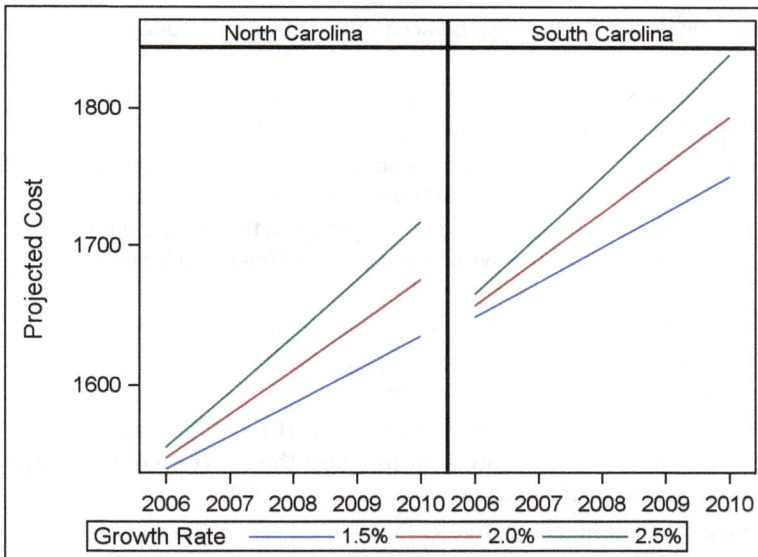

6.4.2 Multidimensional Arrays

In the examples in Section 6.4.1, all of the arrays are one-dimensional—they only use a single dimension and therefore only require a single indexing variable. However, some cases require, or at least are simpler with, a more complex structure. A multidimensional array allows for use of two or more dimensions when structuring the elements. Program 6.4.3 uses data from Program 5.3.5 and shows how to use a multidimensional array to calculate not only deviations from the mean on multiple variables (as was done with one-dimensional arrays in Programs 6.4.1), but it also calculates deviations from the median.

Program 6.4.3: Deviations from the Mean and Median with a Multidimensional Array

```
proc means data= work.ipums2005cost;
  class state mortgageStatus;
  where state in ('North Carolina','South Carolina');
  var electric gas water;
```

```
      output out= work.stats mean=meanE meanG meanW median= medianE medianG medianW;
run;

proc sort data= work.ipums2005cost out= work.sorted2005cost;
  by state mortgagestatus;
  where state in ('North Carolina','South Carolina')
        and scan(mortgagestatus,1) eq 'Yes';
run;

proc sort data= work.stats;
  by state mortgagestatus;
run;

data work.compare2(drop= i j) ❶;
  merge work.sorted2005cost(in=inCost) work.stats(in=inStats);
  by state mortgagestatus;

  array original[3] electric gas water; ❷
  array stats[2,3] ❸ mean: median: ❹;
  array diffs[2,3] EmeanDiff GmeanDiff WmeanDiff
                   EmedianDiff GmedianDiff WmedianDiff; ❺

  if inCost eq 1 and inStats eq 1;

  do i=1 to 3;
    do j=1 to 2;
      diffs[j,i]=original[i]-stats[j,i]; ❻
    end;
  end;
run;
```

❶ Removing the DO loop indexes from the final data set can also be accomplished with the DROP= option on the output data set.

❷ The array for the utility costs is still one-dimensional with three elements indexing the three utilities of interest coming from each household record in Sorted2005Cost.

❸ The Stats array is specified as two dimensional, with the first dimension of size two and the second dimension is of size three, giving a total of six elements. An array dimension of [3,2] can also be used for six elements, so the choice depends largely on the practical associations between the array elements. With two-dimensional arrays, the first dimension indexes rows and the second dimension indexes columns.

❹ Using the colon wildcard, the first three elements of the Stats array are the mean variables, and the second three are the median variables—that is, the first row is means and the second row is medians. A [3,2] dimension does not provide a useful division of these six variables.

❺ The Diffs array is built with the same fundamental structure of the stats array. The variable names are defined here explicitly, and they are written in the ARRAY statement to reflect the structure of the array. This is not required, but it does enhance readability of the code.

❻ The differences are handled with a single calculation. The outer DO loop moves across the columns (or utility costs) in the multi-dimensional arrays, while the inner do loop moves across the rows (or statistics). So, the index on the single-dimensional array for the original utility costs matches the column dimension on the multi-dimensional arrays. This should follow as a one-dimensional array can be considered a multi-dimensional array with a single row.

The concepts of temporary arrays and initializing with values discussed in Program 6.4.2 are valid for multidimensional arrays as well. In addition to introducing the flexible nature of arrays, this section also mentions several limitations; namely, arrays can only use numeric indices, must be of fixed size, cannot mix numeric and character data values, and must be based on variables in the data set. SAS hash objects provide an alternative to arrays that removes these limitations, but hash objects are beyond the scope of this text. Chapter Note 2 in Section 6.8 gives a brief explanation of the SAS hash object. Nonetheless, the flexibility of SAS arrays makes them an invaluable asset for working with data, and the next section looks at the use of arrays to reshape data sets.

6.5 Interchanging Row and Column Information with Arrays

Section 5.7 introduced the concept of exchanging row and column information via PROC TRANSPOSE; however, it is not the only method. In fact, in some cases a single TRANSPOSE procedure is insufficient to achieve the desired data structure. Using the power and flexibility of the DATA step, arrays provide an alternative technique when restructuring a data set. Program 6.5.1 demonstrates the use of arrays to reproduce the results of Program 5.7.3.

Program 6.5.1: Reshaping a Data Set with Arrays

```
data work.UtilityVertical;
  set work.ipums2005utility; ❶

  array util[*] Electric --❷ Fuel; ❸
  array names[4] $ _temporary_ ('electric' 'gas' 'water' 'fuel'); ❹

  do i = 1 to dim(util); ❺
    utility = names[i]; ❻
    cost = util[i]; ❻
    output; ❼
  end;

  format cost dollar.; ❽
  drop electric -- fuel i; ❽
run;

proc print data= work.UtilityVertical(obs=8) noobs;
run;
```

❶   The SET statement reads in the IPUMS2005Utility data set created in Program 5.6.3.

❷   Recall from Chapter Note 3 in Section 1.7, the double dash accesses variables with a name range list. The first variable in the list is Electric. SAS adds variables to the list in the order they appear in the PDV until it encounters the last variable in the list, Fuel.

❸   The ARRAY statement defines the array Util without explicitly declaring the number of elements. The dimension is the number of variables that appear in the name range list Electric -- Fuel. Examining the IPUMS2005Utility data set reveals there are four variables in the list: Electric, Gas, Water, and Fuel.

❹   A second array, Names, contains character constants. The _TEMPORARY_ keyword is used to retain the values and drop the associated variables.

❺   Recall, the DIM function returns the number of elements in the array.

❻   To match the results of Program 5.7.3, the assignment statements create the Utility and Cost variables by retrieving the values from the relevant arrays: Names for the utility names and Util for the utility costs. Remember to ensure the elements of the arrays are in the correct order so that it is appropriate to link the first element in the Names array with the first element in the Util array.

❼   Because there are four utilities per value of Serial, each of which must be its own record, the OUTPUT statement writes the current PDV to the data set each time the information for a new utility is copied from the array variables into the Utility and Cost variables. Failure to include the OUTPUT statement results in a data set that only contains the last utility, rather than each of them.

❽   The TRANSPOSE procedure inherited the format for Cost from the columns included in the VAR statement. (Since they each had the same format, if formats are different on the transposed columns no format is assigned to the result.) When reshaping records in the DATA step, the FORMAT statement allows for the application of the appropriate formatting. Similarly, the TRANSPOSE procedure automatically removes the transposed columns. In the DATA step, a KEEP or DROP allows for the production of similar results.

Output 6.5.1 is identical to Output 5.7.3, demonstrating that PROC TRANSPOSE and array approaches produce identical results.

Output 6.5.1: Reshaping IPUMS2005Utility with Arrays

serial	utility	cost
2	fuel	$1,300
3	fuel	$9,993
4	fuel	$9,993
5	fuel	$9,993
6	fuel	$9,993
7	fuel	$9,993
8	fuel	$9,993
9	fuel	$9,993

Some programmers might find the array approach of Program 6.5.1 more intuitive, while other programmers might prefer the PROC TRANSPOSE method of Program 5.7.3. However, there are cases where a single TRANSPOSE step is insufficient for producing the desired results and frequently requires follow-up code that includes DATA step processing. In those cases, the use of arrays in the DATA step is more direct and typically the preferred method. This is particularly true for large data sets where the computational needs of PROC TRANSPOSE can become a limiting factor.

To demonstrate such a scenario, consider the following data comparing the national average utility costs with the average utility costs for North and South Carolina. Input Data 6.5.2 shows the structure of this data set—each state has both state-level and national-level statistics for all utilities in a single record. However, it may be preferable to see the data restructured so that each state has a separate record for each utility, with the state and national averages in two columns, and the deviation between the two computed as another column, as shown in Table 6.5.1. Programs 6.5.2 through 6.5.4 demonstrate various approaches—with varying degrees of success—to compute the deviations from the means.

Input Data 6.5.2: State and National Average Utility Costs in a Horizontal Structure

State	Electric Mean	Water mean	Gas Mean	Fuel Mean	National Electric Mean	National Water Mean	National Gas Mean	National Fuel Mean
North Carolina	$1,518	$372	$1,220	$686	$1,360	$446	$1,094	$938
South Carolina	$1,624	$398	$1,079	$588	$1,360	$446	$1,094	$938

Table 6.5.1: State and National Average Utility Costs in a Vertical Structure

State	Utility	State Average	National Average	Difference (State - National)
North Carolina	Electric	$1,518	$1,360	158
North Carolina	Water	$372	$446	-74
North Carolina	Gas	$1,220	$1,094	126
North Carolina	Fuel	$686	$938	-252
South Carolina	Electric	$1,624	$1,360	265
South Carolina	Water	$398	$446	-48
South Carolina	Gas	$1,079	$1,094	-16
South Carolina	Fuel	$588	$938	-349

Program 6.5.2 attempts to produce a vertical structure useful for comparisons via PROC TRANSPOSE. Because PROC TRANSPOSE must group the records within each value of State, the records must either be sorted or indexed according to the value of State. Since only two states are present, North Carolina and South Carolina, it is easy to confirm the data set is already sorted.

Program 6.5.2: Attempting to Produce Table 6.5.1 via PROC TRANSPOSE

```
proc transpose data = BookData.horizontal2005utilitystats out = work.UtilityVertical;
  by State;
run;

proc report data = work.UtilityVertical(obs = 5);
  columns State _Name_ Col1;
  define state / display;
  define _name_ / display 'Original Column Name';
  define Col1 / display 'Original Value';
run;
```

Output 6.5.2 shows the results of the TRANSPOSE procedure in Program 6.5.2. It is clear that the resulting data set is structured vertically; however, all utility means are stored in a single column, COL1, rather than in separate columns. While the ID statement and options such as RENAME= and NAME= allow for further customization of the _NAME_ and COL1 variables, there is no option or statement in PROC TRANSPOSE that can restructure the data to match Table 6.5.1—a vertical data set with statewide and national means in separate columns.

Output 6.5.2: Using BY-Group Processing on Input Data 6.5.2

State	Original Column Name	Original Value
North Carolina	Emean	1517.7328
North Carolina	Wmean	372.3612
North Carolina	Gmean	1220.2278
North Carolina	Fmean	686.23776
North Carolina	NationalE	1359.958

While Table 6.5.1 can result from the application of multiple TRANSPOSE procedures and an additional step for data management, it can become a computationally intensive solution for large data sets due to the need for multiple steps. Instead, arrays allow for a more efficient solution since each record only needs to undergo a single transposition. Program 6.5.3 demonstrates the use of arrays to exchange the row and column information in this data set.

Program 6.5.3: Reshaping a Data Set with Arrays

```
data work.Utility2005Summary;
  set BookData.Horizontal2005UtilityStats;

  array StateMeans[*] Emean--Fmean; ❶
  array national[*] NationalE -- NationalF; ❶

  do Utility = 1 to dim(StateMeans); ❷
    StateMean = StateMeans[Utility]; ❸
    NationalMean = National[Utility];
    Deviation = StateMean - NationalMean; ❹
    output; ❺
  end;

  keep State Utility StateMean NationalMean Deviation; ❻
run;
```

```
proc report data = work.Utility2005Summary split='*'; ❼
  columns State Utility StateMean NationalMean Deviation;
  define State / display;
  define Utility / display 'Utility';
  define StateMean / display 'State Average' format = dollar7.;
  define NationalMean / display 'National Average' format = dollar7.;
  define Deviation / display 'Difference*❼(State - National)' format = comma6.;
run;
```

❶ Due to the flexibility of the DATA step, there are many ways to carry out the transposition. Here an array with four elements holds each state's average utility cost, and a second array holds the national means. Generally, there is a tradeoff between the simplicity of the array structure and of the associated DO loops.

❷ Since both arrays have the same dimension, the DIM function can be used on either to set the DO loop ending value to iterate through the arrays.

❸ An assignment statement copies the information from the current array element (for example, from Emean when i = 1) and places it into the variable StateMean.

❹ To calculate the deviations, a third assignment statement is included. Note that while the calculation could use array references, it uses the newly created variables StateMean and NationalMean. There is no computational difference in the two approaches. No matter which technique is used, this assignment must occur inside the DO loop, and prior to the OUTPUT statement, for the Deviation values to appear on the transposed records.

❺ The DO loop iterations work across the four utilities on each record, assigning values for StateMean, NationalMean, Deviation, and Utility (the index variable). To ensure that each of these results is established as a record in the output data set, the OUTPUT statement is the last statement in the DO group, writing the current PDV to the data set each time the DO loop completes an iteration.

❻ The KEEP statement ensures only the necessary variables remain in the UtilityVertical data set. The original state and national summary statistic variables are not needed after the transposition.

❼ The SPLIT= option provides a way to control when PROC REPORT wraps the text in a long label. Its behavior is like the DLM= option when reading raw data—with the exception that SPLIT= only accepts a single character. Note that the * does not appear in the printed label; it is only used to delimit the label into separate lines.

Output 6.5.3 shows that while the structure of the records is correct—vertically arranged with state-level means adjacent to the national-level means—the results are still not ideal. The column from the DO loop index variable Utility in Output 6.5.3 does not provide any way to assign meaning to the records unless there is a legend somewhere that indicates what the Utility numbers correspond to. It is clear that Program 6.5.3 handled the transposition efficiently, but it did not provide the desired detail to make the results useful.

Output 6.5.3: Reshaping a Data Set with Arrays

State	Utility	State Average	National Average	Difference (State - National)
North Carolina	1	$1,518	$1,360	158
North Carolina	2	$372	$446	-74
North Carolina	3	$1,220	$1,094	126
North Carolina	4	$686	$938	-252
South Carolina	1	$1,624	$1,360	265
South Carolina	2	$398	$446	-48
South Carolina	3	$1,079	$1,094	-16
South Carolina	4	$588	$938	-349

This lack of clarity is less likely to be an issue when using PROC TRANSPOSE since it automatically includes a variable, _NAME_, that distinguishes the rows in the transposed data set by identifying the source column form the original data set. The contents of the _NAME_ column may not be superior to the data set produced using the array-based method; however, when using array-based methods, additional programming statements in the DATA step are available to achieve the desired clarity. Improvements over what PROC TRANSPOSE can provide require invocation of another step—typically a DATA step. As with the transposition itself, several approaches are available for including the relevant information. The most common are conditional logic or formats—both of which have been covered in previous sections of this text. Program 6.5.4 constructs and uses a format to improve the Utility variable stored in the data set.

Program 6.5.4: Using PROC FORMAT to Include Utility Names

```
proc format;
  value util 1 = 'Electric' ❶
             2 = 'Water' ❶
             3 = 'Gas' ❶
             4 = 'Fuel' ❶
  ;
run;

data work.Utility2005SummaryNamed;
  set BookData.Horizontal2005UtilityStats;

  ❷
  array StateMeans[*] Emean--Fmean;
  array national[*] NationalE -- NationalF;

  do i = 1 to dim(StateMeans);
    Utility = put(i, util.); ❸
    StateMean = StateMeans[i];
    NationalMean = National[i];
    Deviation = StateMean - NationalMean;
    output;
  end;

  keep State Utility❹ StateMean NationalMean Deviation;
run;
```

❶ The DO loops in Program 6.5.3 already took advantage of the indexing of the utilities in the arrays. The format Util associates index values with the utility names.

❷ Any statements associated with the original transposition (except for the creation of the index variable) remain the same—this includes the ARRAY statements, DO loops, original assignment statements, and OUTPUT statement that follow.

❸ The PUT function maps the index values {1, 2, 3, 4} to the Utility names using the user-defined formats. It is also possible to achieve a result that is similar in appearance by leaving the Utility variable as the index variable for the DO loop and formatting it in a format statement; however, the results have important differences. See Chapter Note 3 in Section 6.8 for a discussion of these differences.

❹ Utility remains in the KEEP list because it now contains the names rather than numeric codes.

Using the same PROC REPORT from Program 6.5.3 with the data set created in Program 6.5.4 generates Output 6.5.4, demonstrating that the array-based approach produces the requisite data structure along with the customized identification information missing from Output 6.5.3.

Output 6.5.4: Using PROC FORMAT to Include Utility Names

State	Utility	State Average	National Average	Difference (State - National)
North Carolina	Electric	$1,518	$1,360	158
North Carolina	Water	$372	$446	-74
North Carolina	Gas	$1,220	$1,094	126

State	Utility	State Average	National Average	Difference (State - National)
North Carolina	Fuel	$686	$938	-252
South Carolina	Electric	$1,624	$1,360	265
South Carolina	Water	$398	$446	-48
South Carolina	Gas	$1,079	$1,094	-16
South Carolina	Fuel	$588	$938	-349

Because the desired results often vary widely from scenario to scenario, no single approach to exchanging row and column information is best. The TRANSPOSE procedure offers many benefits, including the automatic inclusion of important information to distinguish the newly created rows and the automatic dropping of the transposed columns, but is computationally inefficient for large data sets and sacrifices programming flexibility. The extensive list of DATA step programming statements, including arrays, provides a flexibility that allows for a wider range of applications. For more complicated scenarios, it is common to sketch out what should happen during the transposition, then make it work using one or more OUPTUT statements with a single observation. Then populate arrays with the necessary variables and replace variable names with array references in a DO loop. The exercises at the end of this chapter provide additional practice on array usage, as does Chapter 7 where arrays are used as a technique to read in more complicated raw data structures.

6.6 Generating Tables with PROC REPORT

Thus far, the REPORT procedure has been used as a means of displaying a data table in forms similar to the PRINT procedure. However, PROC REPORT is much more powerful and can be used to build summary tables with a variety of structures, mimicking those that can be produced in other procedures, but with more flexibility.

6.6.1 Basic Report Structures and Summaries Using a Single Group Variable

As seen earlier in Program 4.3.3, when only numeric variables are included in the COLUMN statement, PROC REPORT defaults to producing a summary (the sum) on those columns. These summaries can be separated across categories via use of the GROUP usage option as part of the DEFINE statement for the variable(s) used to define the categories. Program 6.6.1 uses the GROUP option in a DEFINE statement to produce a summary report on household incomes and utility costs, split across levels of the MortgageStatus variable, as shown in Output 6.6.1.

Program 6.6.1: Generating a Summary Report with the GROUP Option

```
proc report data= work.ipums2005cost; ❶
   column mortgageStatus❷ hhincome electric gas water fuel❸;
   define mortgageStatus / group❹;
run;
```

❶   This data set is originally created in Program 5.3.5.

❷   MortgageStatus is a character variable, which is defined with the default usage of DISPLAY. That default usage can be overridden by the GROUP option, which is legal for both character and numeric variables.

❸   The other five variables in the column statement are all numeric and have the default usage of ANALYSIS and, due to the use of GROUP with MortgageStatus, are summarized using the default statistic SUM.

❹   The GROUP option does the following: treats each distinct, formatted value of the variable as a category, orders the categories, and consolidates all records for the category into a single row, if possible. Consolidation is possible when all other columns defined in the report have an ANALYSIS usage or are designated with a grouping usage—either with GROUP or ACROSS (covered later in this section). For a more detailed discussion of the usage options available, see Chapter Note 4 in Section 6.8.

Output 6.6.1: Generating a Summary Report with the GROUP Option

mortgageStatus	hhIncome	electric	gas	water	fuel
N/A	1.1278E10	2.86E8	1.12E8	4.5E7	1.5E7
No, owned free and clear	1.6089E10	3.77E8	2.14E8	9.16E7	6.03E7
Yes, contract to purchase	498224239	1.47E7	6.75E6	3.7E6	1.14E6
Yes, mortgaged/ deed of trust or similar debt	4.5943E10	8.37E8	4.2E8	2.12E8	8.8E7

Application of the GROUP option to the MortgageStatus variable operates much like including it in a CLASS statement in PROC MEANS or a TABLE statement in PROC FREQ. However, this behavior is contingent on the structure and uses for the other columns in the table. If, for example, the variable State is added to the column set in Program 6.6.1, the use of State would have to be modified—since it is a character variable its default use is DISPLAY. Therefore, no summary statistic is available for it, so no consolidation occurs for it by default.

Statistics can be selected for variables that are summarized by PROC REPORT, and the list of keywords available to select them is effectively the same as those available in PROC MEANS. Program 6.6.2 includes DEFINE statements on all variables to set the statistic for each, along with establishing labels and formats.

Program 6.6.2: Setting Statistics in a Summary Report
```
proc report data= work.ipums2005cost;
   column mortgageStatus hhincome electric gas water fuel;
   define mortgageStatus / group 'Mortgage Status';
   define hhincome / mean❶ 'Avg. Household Income'❷ format=dollar10.❸;
   define electric / mean 'Avg. Electricity Cost' format=dollar10.;
   define gas / mean 'Avg. Gas Cost' format=dollar10.;
   define water / mean 'Avg. Water Cost' format=dollar10.;
   define fuel / mean 'Avg. Fuel Cost' format=dollar10.;
run;
```

❶ Statistic keywords can be used in a DEFINE statement for any variable with the usage ANALYSIS.

❷ For some uses of statistic keywords, column headers corresponding to the statistic are provided automatically, though in this case they are not. (The default label associated with the variable is used when no alternative is provided.) For either situation, custom labels can be set in the DEFINE statement.

❸ Formats can be assigned in the define statement with the FORMAT= option. For this example, assignment of the labels and formats can be done with LABEL and FORMAT statements as well. In general, labels and formats set in DEFINE statements override those set in LABEL and FORMAT statements in PROC REPORT.

Output 6.6.2: Setting Statistics in a Summary Report

Mortgage Status	Avg. Household Income	Avg. Electricity Cost	Avg. Gas Cost	Avg. Water Cost	Avg. Fuel Cost
N/A	$37,181	$1,082	$884	$387	$770
No, owned free and clear	$53,569	$1,270	$1,114	$409	$953
Yes, contract to purchase	$51,068	$1,518	$1,096	$488	$786
Yes, mortgaged/ deed of trust or similar debt	$84,204	$1,542	$1,158	$480	$966

Often several statistics on a single variable are desired, and two methods for constructing such summaries are considered. The first method involves use of aliases for variable names, allowing a variable to be displayed in multiple columns with different DEFINE statements associated with each. Program 6.6.3 uses this method to produce a set of four statistics on the electricity variable.

Program 6.6.3: Producing Multiple Statistics on a Single Variable Using Aliases

```
proc report data= work.ipums2005cost;
   column mortgageStatus electric=num❶ electric=middle electric=mean❷ electric=sd;
   define mortgageStatus / group 'Mortgage Status';
   define num❸ / n 'Number of Observations' format=comma8.;
   define middle / median 'Median Electricity Cost' format=dollar10.;
   define mean❷ / mean❷ 'Mean Electricity Cost' format=dollar10.;
   define sd / std 'Standard Deviation' format=dollar10.;
run;
```

❶ Any variable listed in the column statement can be assigned an alias in the form: *variable-name=alias*. Aliases must form legal SAS names that are not otherwise in use in the report, either from the data set, defined as new variables, or assigned as other aliases.

❷ The statistic keywords are permissible as alias names, their separate roles are correctly identified by the compiler. In certain instances, as shown in the next example, the statistic keyword can be made to fill both roles.

❸ For any column that is aliased, the corresponding DEFINE statement must refer to the alias. Since each use of the variable Electric is aliased, a DEFINE statement referring to it is ignored for this case.

Output 6.6.3: Producing Multiple Statistics on a Single Variable Using Aliases

Mortgage Status	Number of Observations	Median Electricity Cost	Mean Electricity Cost	Standard Deviation
N/A	264,782	$840	$1,082	$821
No, owned free and clear	297,008	$1,080	$1,270	$898
Yes, contract to purchase	9,650	$1,200	$1,518	$980
Yes, mortgaged/ deed of trust or similar debt	542,950	$1,200	$1,542	$996

In Program 6.6.3 the Electric variable was aliased for all four of its uses; however, to avoid any ambiguous references, it is sufficient to alias three of the references and leave the other as a direct reference to Electric. Whenever aliasing is used on a variable, it is considered a good programming practice to alias all instances of that variable to avoid confusion and enhance readability.

It is also possible to produce the same result as Output 6.6.3 without using any aliasing at all. To produce several statistics on a single variable, the corresponding statistic keywords can be nested with the variable in the COLUMN statement as demonstrated in Program 6.6.4.

Program 6.6.4: Producing Multiple Statistics on a Single Variable by Nesting

```
proc report data= work.ipums2005cost;
   column mortgageStatus electric,(n median mean std)❶;
   define mortgageStatus / group 'Mortgage Status';
   define electric / 'Electricity Cost'❷;
   define n❸ / 'N. Obs.' format=comma8.;
   define median / 'Median' format=dollar10.;
   define mean / 'Mean' format=dollar10.;
   define std / 'Std. Dev.' format=dollar10.;
run;

proc report data= work.ipums2005cost;
   column mortgageStatus (n median mean std),electric❶;
   define mortgageStatus / group 'Mortgage Status';
   define electric / 'Electricity'❷;
   define n / 'N. Obs.' format=comma8.❹;
   define median / 'Median Cost' format=dollar10.;
   define mean / 'Mean Cost' format=dollar10.;
   define std / 'Std. Dev.' format=dollar10.;
run;
```

❶ The comma operator is used to nest one set of definitions with another. Here the set of statistic keywords, given as a parenthetical list, is nested with the electricity variable. The second item(s) listed are nested beneath distinct levels of the first item(s).

❷ The Electric variable is at the top level in the first report and at the secondary level in the second report. While the summary statistics are the same in each report, the header structure is different and good choices for labels likely differ for the two structures.

❸ For either structure, DEFINE statements for the statistics refer to the statistic keywords in the position where variable names or aliases are placed. The statistic keyword is not given as an option, but other options used previously are available to set attributes for these columns.

❹ Depending on nesting order, some options may not apply as expected. Output 6.6.4B shows the format applied to the column for N is not comma8. (nor is it dollar10.). Since the columns are defined by the Electric variable nested within the statistic, the format applied is the default format for the Electric variable.

Output 6.6.4A: Producing Multiple Statistics on a Single Variable by Nesting Keywords in a Variable

	Electricity Cost			
Mortgage Status	N. Obs.	Median	Mean	Std. Dev.
N/A	264,782	$840	$1,082	$821
No, owned free and clear	297,008	$1,080	$1,270	$898
Yes, contract to purchase	9,650	$1,200	$1,518	$980
Yes, mortgaged/ deed of trust or similar debt	542,950	$1,200	$1,542	$996

Output 6.6.4B: Producing Multiple Statistics on a Single Variable by Nesting a Variable in Keywords

	N. Obs.	Median Cost	Mean Cost	Std. Dev.
Mortgage Status	Electricity	Electricity	Electricity	Electricity
N/A	264782	$840	$1,082	$821
No, owned free and clear	297008	$1,080	$1,270	$898
Yes, contract to purchase	$9,650	$1,200	$1,518	$980
Yes, mortgaged/ deed of trust or similar debt	542950	$1,200	$1,542	$996

For the two reports generated by Program 6.6.4, there is an obvious preference for the first structure. The version generated by Program 6.6.5 likely makes this preference stronger with a simple modification to the DEFINE statement for the Electric variable.

Program 6.6.5: Removing Header Labels in Nested Column Structures

```
proc report data= work.ipums2005cost;
  column mortgageStatus electric,(n median mean std);
  define mortgageStatus / group 'Mortgage Status';
  define electric / ' ';
  define n / 'Number of Observations' format=comma8.;
  define median / 'Median Electricity Cost' format=dollar10.;
  define mean / 'Mean Electricity Cost' format=dollar10.;
  define std / 'Standard Deviation' format=dollar10.;
run;
```

Output 6.6.5: Removing Header Labels in Nested Column Structures

Mortgage Status	Number of Observations	Median Electricity Cost	Mean Electricity Cost	Standard Deviation
N/A	264,782	$840	$1,082	$821
No, owned free and clear	297,008	$1,080	$1,270	$898
Yes, contract to purchase	9,650	$1,200	$1,518	$980
Yes, mortgaged/ deed of trust or similar debt	542,950	$1,200	$1,542	$996

By providing a null label for the Electric variable, the top row containing that header is eliminated in Output 6.6.5 as compared to Output 6.6.4A. Here the information lost with the removal of that label is now included in the other column labels; however, it could also be added via titles, footnotes, or other PROC REPORT techniques covered in Chapter 7.

Understanding how nesting order changes the structure of the table is important when using the ACROSS option in the next section, and it is also of value when constructing multiple statistics on multiple variables, as shown in Program 6.6.6.

Program 6.6.6: Producing Multiple Statistics on Multiple Variables by Nesting

```
proc report data= work.ipums2005cost;
  column mortgageStatus (electric gas),(mean median);
  define mortgageStatus / group 'Mortgage Status';
  define electric / 'Elec. Cost';
  define gas / 'Gas Cost';
  define mean / 'Mean';
  define median / 'Median';
run;

proc report data= work.ipums2005cost;
  column mortgageStatus (mean median),(electric gas);
  define mortgageStatus / group 'Mortgage Status';
  define electric / 'Elec.';
  define gas / 'Gas';
  define mean / 'Mean Cost';
  define median / 'Median Cost';
run;
```

Output 6.6.6A: Producing Multiple Statistics on Multiple Variables by Nesting Keywords in Variables

Mortgage Status	Elec. Cost		Gas Cost	
	Mean	Median	Mean	Median
N/A	$1,082	$840	$884	$600
No, owned free and clear	$1,270	$1,080	$1,114	$840
Yes, contract to purchase	$1,518	$1,200	$1,096	$720
Yes, mortgaged/ deed of trust or similar debt	$1,542	$1,200	$1,158	$840

Output 6.6.6B: Producing Multiple Statistics on Multiple Variables by Nesting Variables in Keywords

Mortgage Status	Mean Cost		Median Cost	
	Elec.	Gas	Elec.	Gas
N/A	$1,082	$884	$840	$600
No, owned free and clear	$1,270	$1,114	$1,080	$840
Yes, contract to purchase	$1,518	$1,096	$1,200	$720
Yes, mortgaged/ deed of trust or similar debt	$1,542	$1,158	$1,200	$840

Depending on whether the preference is to have the same statistic in adjacent columns or the same variable in adjacent columns, different nesting structures are chosen. Also note the differences in choices for column heading labels. In certain instances, the hierarchical structure of the column labels may be undesirable and may be removed (depending on the structure of the nesting).

For a simplified header structure when displaying common set of statistics on several variables, aliases can be used instead of nesting, as shown in Program 6.6.7.

Program 6.6.7: Using Aliases to Allow for Individual Column Headings

```
proc report data= work.ipums2005cost;
   column mortgageStatus electric=midEl electric=meanEl gas=midGas gas=meanGas;
   define mortgageStatus / group 'Mortgage Status';
   define midEl / median 'Median Elec. Cost' format=dollar10.;
   define meanEl / mean 'Mean Elec. Cost' format=dollar10.;
   define midGas / median 'Median Gas Cost' format=dollar10.;
   define meanGas / mean 'Mean Gas Cost' format=dollar10.;
run;
```

Output 6.6.7: Using Aliases to Allow for Individual Column Headings

Mortgage Status	Median Elec. Cost	Mean Elec. Cost	Median Gas Cost	Mean Gas Cost
N/A	$840	$1,082	$600	$884
No, owned free and clear	$1,080	$1,270	$840	$1,114
Yes, contract to purchase	$1,200	$1,518	$720	$1,096
Yes, mortgaged/ deed of trust or similar debt	$1,200	$1,542	$840	$1,158

6.6.2 Additional Summaries Using BREAK and RBREAK Statements

Additional summary rows can be added to a table generated with PROC REPORT in a variety of locations via several methods. The most basic of these methods is a whole-report summary generated by the RBREAK (shorthand for report break) statement, which is demonstrated in Program 6.6.8, which adds a summary line to the end of the table shown in Output 6.6.8.

Program 6.6.8: Adding a Whole-Report Summary Using the RBREAK Statement

```
proc report data= work.ipums2005cost;
   column mortgageStatus electric,(n median mean std);
   rbreak after❶ / summarize❷;
   define mortgageStatus / group 'Mortgage Status';
   define electric / ' ';
   define n / 'Number of Observations' format=comma8.❸;
   define median / 'Median Electricity Cost' format=dollar10.;
   define mean / 'Mean Electricity Cost' format=dollar10.;
   define std / 'Standard Deviation' format=dollar10.;
run;
```

❶ The RBREAK statement requires a location, which can be either BEFORE or AFTER. BEFORE places the break line between the column headers and the first row of the report, AFTER places it at the end of the report. (If BY-group processing is in place, these conventions apply to each BY group.)

❷ The SUMMARIZE option provides a summary on any analysis variable across all observations in the data set, corresponding to the statistic keyword assigned to that column.

❸ By default, values in the break line are formatted corresponding to the assigned format for that column. Checking Output 6.6.8, this is a case where a format that is adequate for the data rows is inadequate for the summary row. Recall, Chapter Note 3 in Section 2.12 discusses how SAS handles values that exceed the specified format width.

Output 6.6.8: Adding a Whole-Report Summary Using the RBREAK Statement

Mortgage Status	Number of Observations	Median Electricity Cost	Mean Electricity Cost	Standard Deviation
N/A	264,782	$840	$1,082	$821
No, owned free and clear	297,008	$1,080	$1,270	$898
Yes, contract to purchase	9,650	$1,200	$1,518	$980
Yes, mortgaged/ deed of trust or similar debt	542,950	$1,200	$1,542	$996
	1114390	$1,080	$1,360	$951

If more than one grouping variable is in use, table rows provide summaries for each combination of levels of the variables. It is possible to produce higher order summaries on individual group variables with the BREAK statement. Program 6.6.9 produces summary statistics across State and MortgageStatus combinations (each reduced in scope to create a smaller table for display), while also including summaries across each value of State using a BREAK statement, and a whole-report summary using the RBREAK statement.

Program 6.6.9: Adding Summaries Using the BREAK Statement

```
proc format;
  value $Mort_Status
    'No'-'Nz'='No'
    'Yes'-'Yz'='Yes'
  ;
run;

proc report data= work.ipums2005cost;
  where state in ('North Carolina','South Carolina');
  column state mortgageStatus electric,(n median mean std);
  define state / group 'State';
  define mortgageStatus / group 'Mortgage Status' format=$Mort_Status.;
  define electric / '';
  define n / 'Number of Observations' format=comma8.;
  define median / 'Median Electricity Cost' format=dollar10.;
  define mean / 'Mean Electricity Cost' format=dollar10.;
  define std / 'Standard Deviation' format=dollar10.;
  break after❶ state❷ / summarize❸ suppress❹;
  rbreak after / summarize;
run;
```

❶ Like the RBREAK statement, the BREAK statement requires a location: BEFORE or AFTER.

❷ The BREAK statement also requires a variable is defined with either the GROUP or ORDER usage in the report. The AFTER and BEFORE locations are relative to the blocks of rows for each distinct level of this variable

❸ The SUMMARIZE option is also available in the BREAK statement to provide summaries as in the RBREAK statement.

❹ By default, the value of the BREAK variable is repeated on the break rows in the output table. The SUPPRESS option leaves this column blank on the break rows—Output 6.6.9A shows the table without the SUPPRESS option, 6.6.9B is the result of using the SUPPRESS option. Styling options discussed in Chapter 7 allow for more ways to differentiate summary rows from other rows in the table.

Output 6.6.9A: Adding Summaries Using the BREAK Statement—Without the SUPPRESS Option

State	Mortgage Status	Number of Observations	Median Electricity Cost	Mean Electricity Cost	Standard Deviation
North Carolina	N/A	8,548	$1,080	$1,327	$834
	No	9,382	$1,200	$1,416	$803
	Yes	17,303	$1,440	$1,667	$892
North Carolina		35,233	$1,320	$1,518	$868
South Carolina	N/A	3,807	$1,200	$1,426	$862
	No	5,074	$1,320	$1,543	$880
	Yes	8,283	$1,560	$1,766	$898
South Carolina		17,164	$1,440	$1,624	$896
		52,397	$1,320	$1,553	$879

Output 6.6.9B: Adding Summaries Using the BREAK Statement—With the SUPPRESS Option

State	Mortgage Status	Number of Observations	Median Electricity Cost	Mean Electricity Cost	Standard Deviation
North Carolina	N/A	8,548	$1,080	$1,327	$834
	No	9,382	$1,200	$1,416	$803
	Yes	17,303	$1,440	$1,667	$892
		35,233	$1,320	$1,518	$868
South Carolina	N/A	3,807	$1,200	$1,426	$862
	No	5,074	$1,320	$1,543	$880
	Yes	8,283	$1,560	$1,766	$898
		17,164	$1,440	$1,624	$896
		52,397	$1,320	$1,553	$879

6.6.3 Grouping in Columns with the ACROSS Usage Option

In the REPORT procedure, group categories can also be cast into columns using the ACROSS usage option. Program 6.6.10 provides two versions for restructuring the table created in Program 6.6.9 by re-designating the usage of the State variable as ACROSS (along with other modifications).

Program 6.6.10: Using an ACROSS Variable

```
proc report data= work.ipums2005cost;
  where state in ('North Carolina','South Carolina');
  column mortgageStatus state❶,(electric=num electric=mid electric=mean
    electric=sd) ❷;
  define state / across❶ 'State Electricity Costs';
  define mortgageStatus / group 'Mortgage Status' format=$Mort_Status.;
  define num / n 'N' format=comma8.;
  define mid / median 'Median' format=dollar10.;
  define mean / mean 'Mean' format=dollar10.;
```

```
   define sd / std 'Std. Dev.' format=dollar10.;
   rbreak after / summarize;
run;

proc report data= work.ipums2005cost;
   where state in ('North Carolina','South Carolina');
   column mortgageStatus state,electric,(n median mean std) ❸;
   define state / across 'State Electricity Costs';
   define mortgageStatus / group 'Mortgage Status' format=$Mort_Status.;
   define electric / '' ❹;
   define n / 'N' format=comma8.;
   define median / 'Median' format=dollar10.;
   define mean / 'Mean' format=dollar10.;
   define std / 'Std. Dev.' format=dollar10.;
   rbreak after / summarize; ❺
run;
```

❶ The State variable is designated as an ACROSS variable. If the COLUMN statement included only the MortgageStatus and State variables in this structure, the summary would consist of a cross-tabulation of frequencies in each category.

❷ The ACROSS usage allows for nesting. Here a set of statistics on the Electric variable is defined via aliasing.

❸ In this structure, the statistic keywords are nested under the Electric variable, which is nested under the State variable. By default, this hierarchy is present in the structure of the column headers.

❹ A null column label helps simplify the header structure and make the structures match for these reports.

❺ Here the RBREAK statement still plays the role of providing a full-report summary on all of the columns, which are now created by levels of the State variable in combination with the chosen statistics on the Electric variable.

Output 6.6.10: Using an ACROSS Variable

| Mortgage Status | State Electricity Costs | | | | | | | |
| | North Carolina | | | | South Carolina | | | |
	N	Median	Mean	Std. Dev.	N	Median	Mean	Std. Dev.
N/A	8,548	$1,080	$1,327	$834	3,807	$1,200	$1,426	$862
No	9,382	$1,200	$1,416	$803	5,074	$1,320	$1,543	$880
Yes	17,303	$1,440	$1,667	$892	8,283	$1,560	$1,766	$898
	35,233	$1,320	$1,518	$868	17,164	$1,440	$1,624	$896

With the ability to nest multiple entities together in the COLUMN specification, more than one variable can be cast with the ACROSS usage, as in Program 6.6.11.

Program 6.6.11: Using Multiple ACROSS Variables

```
proc format;
   value MetroStatus
      0 = "Unknown"
      1 = "Non-Metro"
      2-4 = "Metro"
   ;
run;

proc report data= work.ipums2005cost;
   where state in ('North Carolina','South Carolina');
   column mortgageStatus state❶,metro❷,electric❸;
   define state / across 'State'; ❶
   define metro / across 'Metro Status' format=MetroStatus.; ❷
   define mortgageStatus / group 'Mortgage Status' format=$Mort_Status.;
   define electric / mean 'Avg. Elec. Cost' format=dollar10.; ❸
run;
```

❶ The State variable is designated as an ACROSS variable at the highest level in the nesting hierarchy. The label assigned to this variable is at the top, spanning the set of levels for that variable.

❷ The Metro variable is the next variable in the nesting hierarchy and is also assigned the usage of ACROSS. As with most other options or statements that assign a categorical role to a variable, the levels of an ACROSS variable are defined with respect to its assigned format. The label assigned to this variable forms a spanning header across all of its levels within each level of the previous variable (for each State in this example).

❸ Electric is the last variable in the nesting hierarchy and is assigned the summary statistic of MEAN. The designated label appears under each occurrence of levels of the previous two variables, creating the rather complex header structure shown in Output 6.6.11.

Output 6.6.11: Using Multiple ACROSS Variables

	State					
	North Carolina			South Carolina		
	Metro Status			Metro Status		
	Metro	Non-Metro	Unknown	Metro	Non-Metro	Unknown
Mortgage Status	Avg. Elec. Cost	Avg. Elec. Cost	Avg. Elec. Cost	Avg. Elec. Cost	Avg. Elec. Cost	Avg. Elec. Cost
N/A	$1,275	$1,427	$1,750	$1,398	$1,529	$1,408
No	$1,389	$1,440	$1,652	$1,506	$1,633	$1,524
Yes	$1,617	$1,765	$2,091	$1,713	$1,898	$1,817

As in previous examples, nesting of elements in the COLUMN statement can produce excessively complicated header structures. Careful assignment of labels in column definitions, including the use of null labels, leads to much more compact, readable structures. Program 6.6.12 produces a simplified version of Output 6.6.11 by removing two labels and improving others, while also changing the hierarchy of the nesting. Also note the use of the ORDER= option to revert to the internal ordering of the encodings of the Metro variable (as of the writing of this text, the default ordering for GROUP and ACROSS variables is FORMATTED).

Program 6.6.12: Improving Header Structure when Using Multiple ACROSS Variables

```
proc report data= work.ipums2005cost;
  where state in ('North Carolina','South Carolina');
  column mortgageStatus electric,state,metro;
  define state / across 'Mean Electricity Costs';
  define metro / across '' format=MetroStatus. order=internal;
  define mortgageStatus / group 'Mortgage Status' format=$Mort_Status.;
  define electric / mean '' format=dollar10.;
run;
```

Output 6.6.12: Improving Header Structure when Using Multiple ACROSS Variables

	Mean Electricity Costs					
	North Carolina			South Carolina		
Mortgage Status	Unknown	Non-Metro	Metro	Unknown	Non-Metro	Metro
N/A	$1,750	$1,427	$1,275	$1,408	$1,529	$1,398
No	$1,652	$1,440	$1,389	$1,524	$1,633	$1,506
Yes	$2,091	$1,765	$1,617	$1,817	$1,898	$1,713

Programs 5.7.5 and 5.7.6 use data built on a transposed version of the utility cost data. Program 6.6.13 uses a similar data structure and produces a report on average utility costs similar to Output 6.6.2.

Program 6.6.13: Working with the Transposed Utility Data

```
proc transpose data= work.ipums2005utility
               out= work.Utility(rename=(col1=Cost _name_=Utility));
  by serial;
  var electric--fuel;
run;

data work.ipums2005cost2;
  merge BookData.ipums2005basic work.Utility;
  by serial;
  utility=propcase(utility);
run; ❶

proc report data= work.ipums2005cost2;
  column mortgageStatus cost,Utility❷;
  define mortgageStatus / group 'Mortgage Status';
  define Utility / across 'Average Utility Cost';
  define cost / mean ''; ❸
  where cost lt 9000; ❹
run;
```

❶ See Section 5.7 to review the details of this combination of PROC TRANSPOSE and the DATA step. The Utility variable is created as part of the transposing process—it is a re-naming of the default _NAME_ variable and its values have been refined via use of the PROPCASE function.

❷ With the Utility variable cast in the ACROSS usage, it may be expected to have it placed at the top level of a nesting hierarchy with the Cost variable (which stores the utility cost values) nested underneath. However, to get simplified headers using a null label on cost, the order must be reversed.

❸ The Cost variable is assigned the MEAN summary statistic and, to simplify the headers, a null label. It is also possible to place the MEAN keyword at the bottom of the nesting in the Column statement and assign it a null label in a DEFINE statement for the keyword. The summary statistics are the same; however, the simplification of headers does not work.

❹ Missing utility values have not been properly assigned in the data set as they were in Program 6.6.1. Here, a WHERE statement suffices for removing them.

Output 6.6.13: Working with the Transposed Utility Data

	Average Utility Cost			
Mortgage Status	**Electric**	**Fuel**	**Gas**	**Water**
N/A	$1,082	$770	$884	$387
No, owned free and clear	$1,270	$953	$1,114	$409
Yes, contract to purchase	$1,518	$786	$1,096	$488
Yes, mortgaged/ deed of trust or similar debt	$1,542	$966	$1,158	$480

6.6.4 Generating Data Sets from PROC REPORT Summaries

Any tables produced by PROC REPORT can be converted to data sets using the OUT= option in the PROC REPORT statement. The data set specified mimics the table generated by PROC REPORT to a certain degree; however, in certain cases, it may contain rows or columns not displayed in the report table. Further, some of the values in each column may be stored in the data set differently than they are displayed in the report table. Finally, some of the variable names in the data set may be unexpected, particularly for columns generated by special operations in the COLUMN statement. Program 6.6.14 illustrates this, replicating the report produced by Program 6.6.9, with the additional steps of generating an OUT= data set and using PROC PRINT to display it.

Program 6.6.14: Generating an Output Data Set with PROC REPORT

```
proc report data= work.ipums2005cost out= work.reportData;
  where state in ('North Carolina','South Carolina');
  column state mortgageStatus electric,(n median mean std);
  define state / group 'State';
  define mortgageStatus / group 'Mortgage Status' format=$Mort_Status.;
  define electric / '';
  define n / 'Number of Observations' format=comma8.;
  define median / 'Median Electricity Cost' format=dollar10.;
  define mean / 'Mean Electricity Cost' format=dollar10.;
  define std / 'Standard Deviation' format=dollar10.;
  break after state / summarize;
  rbreak after / summarize;
run;

proc print data= work.reportData noobs label;
run;
```

Output 6.6.14: Generating an Output Data Set with PROC REPORT

State	Mortgage Status	Number of Observations	Median Electricity Cost	Mean Electricity Cost	Standard Deviation
North Carolina	N/A	8,548	$1,080	$1,327	$834
	No	9,382	$1,200	$1,416	$803
	Yes	17,303	$1,440	$1,667	$892
North Carolina		35,233	$1,320	$1,518	$868
South Carolina	N/A	3,807	$1,200	$1,426	$862
	No	5,074	$1,320	$1,543	$880
	Yes	8,283	$1,560	$1,766	$898
South Carolina		17,164	$1,440	$1,624	$896
		52,397	$1,320	$1,553	$879

State	Mortgage Status	_C3_	_C4_	_C5_	_C6_	_BREAK_
North Carolina	N/A	$8,548	$1,080	$1,327	$834	
North Carolina	No, owned free and clear	$9,382	$1,200	$1,416	$803	
North Carolina	Yes, contract to purchase	$17303	$1,440	$1,667	$892	
North Carolina		$35233	$1,320	$1,518	$868	state
South Carolina	N/A	$3,807	$1,200	$1,426	$862	
South Carolina	No, owned free and clear	$5,074	$1,320	$1,543	$880	
South Carolina	Yes, contract to purchase	$8,283	$1,560	$1,766	$898	
South Carolina		$17164	$1,440	$1,624	$896	state
		$52397	$1,320	$1,553	$879	_RBREAK_

Contrasting the two tables in Output 6.6.14, the first being the output table generated by PROC REPORT and the second the result of printing the OUT= data set generated by the PROC REPORT, several differences are noteworthy. Given that the LABEL option is active in PROC PRINT, it is clear that the column labels defined in the report are not preserved in the data set—labels default to the original labels from the input data set used in PROC REPORT. Further, all of the columns corresponding to summary statistics on the Electric variable use neither the analysis variable name nor the statistic keyword. The columns are referenced via generic names of the form _C#_, where # is the position number for the generated column (a naming convention available for all columns that is used in Chapter 7). If the statistic columns are generated using aliases, the aliases form the variable names.

The additional variable, _BREAK_, is included to track which records correspond to break lines in the report, with _RBREAK_ being a special value given to the line generated by the RBREAK statement, and lines generated by the BREAK statement identified by the variable used. Note that the formats applied to the values of the variables correspond to the formats from the original data, much like the behavior of the labels. So, the three levels of MortgageStatus are not the formatted versions in the report, but the unformatted versions. For multiple values belonging to the same format category, each value is represented by the first internal value. For all statistics on the Electric variable, the format used is the format used in the first DATA step of Program 6.6.1, so even the N statistic receives a dollar format.

Being able to view the report as a data set is helpful when doing computations inside PROC REPORT, a concept covered in Chapter 7. With some care, it is also useful to generate summary data sets from PROC REPORT for use in other procedures, much like examples with the MEANS and FREQ procedures in previous chapters, starting in Section 3.5. Program 6.6.15 uses the OUT= data set generated in Program 6.6.14 to make a panel graph.

Program 6.6.15: Using the OUT= Data Set from PROC REPORT

```
proc sgpanel data= work.reportData;
  panelby state / novarname;
  format mortgageStatus $Mort_Status.❶;
  vbar mortgageStatus / response=_c4_ legendlabel='Mean'❷
                        discreteoffset=-0.2 barwidth=0.4;
  vbar mortgageStatus / response=_c5_ legendlabel='Median'
                        discreteoffset=0.2 barwidth=0.4;
  rowaxis label='Electricity Cost';❸
  colaxis label='Mortgage Status' labelpos=left;
run;
```

❶ As noted above, formats used in the report are not preserved. To make the chart categories match the report categories, the $Mort_Status format must be reassigned to the MortgageStatus variable.

❷ Column names for each statistic use a generic column numbering form and have no labels. Renaming with RENAME= or labeling with a LABEL statement is valid, but the LEGENDLABEL= option is also useful.

❸ Again, given that the labels created in the report are not preserved, labeling of the axes may be necessary.

Output 6.6.15: Using the OUT= Data Set from PROC REPORT

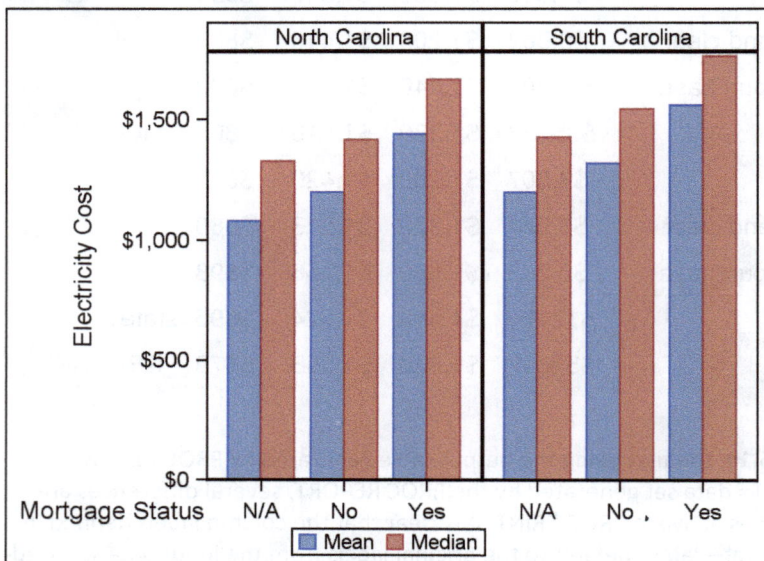

6.7 Wrap-Up Activity

Use the lessons and examples contained in this and previous chapters to create the results shown in Section 6.2.

Data

This Wrap-Up Activity uses the data from the Wrap-Up Activity for Chapter 5.

Scenario

Using the skills mastered so far, including those from previous chapters, generate the reports shown in Output 6.2.1 and 6.2.2. The graphs shown in Output 6.2.3, 6.2.4, and 6.2.5 are built from data generated assuming a 2% annual growth rate in electricity costs.

6.8 Chapter Notes

1. *Macro Variables in Array Definitions.* When declaring an array, SAS syntax requires a numeric constant or, in the case of a one-dimensional array, the * character to define the dimension. DATA step variables are not valid. However, it is possible to include a variable when defining the dimension, but it must be a macro variable. Macro variables are interpreted by the compiler in the SAS macro facility. This is a different compilation than occurs during the compilation phase introduced in Section 2.9. While a complete treatment of the SAS macro facility is well beyond the scope of this text, SAS provides multiple ways to define macro variables and the simplest of those is demonstrated here. Program 6.8.1 below revisits program 6.5.1 to produce the same results.

Program 6.8.1: Using a Macro Variable in an Array Definition

```
%let❶ MyDim = 4;❷
data work.UtilityVertical;
  set work.ipums2005utility;

  array util[&MyDim]❸ Electric -- Fuel;
  array names[&MyDim]❹ $ _temporary_ ('electric' 'gas' 'water' 'fuel');

  do i = 1 to &MyDim;❹
    utility = names[i];
    cost = util[i];
    output;
  end;

  format cost dollar.;
  drop electric -- fuel i;
run;
```

❶ The percent sign is used to distinguish macro language elements, like this %LET statement, from other SAS language elements such as an OPTIONS or TITLE statement. Here it triggers the SAS macro facility to compile this statement

❷ This %LET statement defines a SAS macro variable. The assignment statement works similarly to the DATA step assignment statement with one notable exception—rather than storing the variable in a data set, SAS stores the variable in a way that makes it accessible across all DATA and PROC steps.

❸ Just as the percent sign differentiates macro language from non-macro language, the ampersand distinguishes a macro variable from a DATA step variable. The ampersand also triggers the SAS macro facility just as the percent sign does. This instructs SAS to look up the value of MyDim from this set of variables and return its value—4—before the ARRAY statement is compiled as described in Section 2.9.

❹ As with DATA step variables, macro variables can be used repeatedly to avoid hardcoding a value. In this case, if new utility categories are added in the future then the %LET statement is the only line that requires an update to ensure the rest of the program uses the updated value.

Macro variables, and the SAS macro facility in general, are an extensive topic. However, small applications such as the %LET statement can have a large impact on the flexibility in a SAS program. See the SAS Documentation or other resources to learn more.

2. *Hash Objects.* Unlike an array, which is simply an additional method for referencing a specific group of variables—the array elements—a hash object is a data structure that contains variable values. As a result, hash objects can require substantially more memory since they must store the data values as well as the variable names. However, it is reasonable to consider hash objects nothing more than uber-arrays in that the hash object is indexed with key variables and communicates readily with the PDV. Since hash objects are indexed with PDV variables, there is no restriction to only use numeric indices. This greatly increases the areas of application since variables such as a numeric patient ID, name of a state, or both, are available for looking up data values. Hash objects, and the associated hash object iterators used to move from one hash object entry to the next, are examples of DATA step component objects. For more information about hash objects, hash iterators, and DATA step component objects in general, see the SAS Documentation.

3. *Applying Formats Versus Creating Character Values Using Formats and the PUT Function.* In Programs 6.5.3 and 6.5.4, the variable distinguishing among utilities was established in two different ways, but other methods combining aspects of these two ideas are possible. Consider the following alternative:

Program 6.8.2: Formats in FORMAT Statements Versus PUT Functions

```
proc format;
  value util 1 = 'Electric'
             2 = 'Water'
             3 = 'Gas'
             4 = 'Fuel'
  ;
run;

data Work.Utility2005SummaryNamed;
  set BookData.Horizontal2005UtilityStats;

  array StateMeans[*] Emean -- Fmean;
  array national[*] NationalE -- NationalF;

  do Utility = 1 to dim(StateMeans);
    StateMean = StateMeans[Utility];
    NationalMean = National[Utility];
    Deviation = StateMean - NationalMean;
    output;
  end;
  format Utility Util.;
  keep State Utility StateMean NationalMean Deviation;
run;
```

Applying the Util format to the Utility variable makes the result appear to be the same as that of Program 6.5.4, but it is not. Since Utility remains as the DO loop index, taking on the values of 1, 2, 3, or 4, it is still a numeric variable internally exactly as in Program 6.5.3 with a different display format. The advantage of this approach is that the values of Utility are simple, which is helpful in conditional logic or re-formatting values. And it is also helpful in cases where an internal order that is different from the alphabetical order of the character values is desired. A major drawback is that the user-defined format must be provided to all users of the data set to make it function as intended. If storage space is not an issue, an effective compromise is to create two variables—for example, leave Utility as defined above without formatting it and assign a different variable, say UtilityChar, with the PUT function as in Program 6.5.4. Now no format needs to be transmitted, but the advantages of having a full character description and a simple numeric encoding are both present in the data.

4. *Usage Options in PROC REPORT.* The DEFINE statement in PROC REPORT has six options for the usage of the column it applies to.

   a. *DISPLAY.* The DISPLAY usage corresponds to an output table that displays a row for each record read from the input data set (and possibly more with insertions on break lines). All character variables are assigned the default usage of DISPLAY.

   b. *ORDER.* The ORDER usage is similar to the DISPLAY usage in that it displays a row for each record read from the input data set; however, it orders the rows subject to ordering options in use and the structure of the other elements of the COLUMN statement. If any variables with the DISPLAY usage appear in the COLUMN statement before a variable with the ORDER usage, no ordering of the rows is done. When multiple variables are assigned the ORDER usage, the nesting of the ordering corresponds to the left-to-right ordering in the COLUMN statement. Values for these variables are displayed once per group, much like the output of PROC PRINT when using a common BY and ID statement as in Program 2.3.7.

   c. *GROUP.* The GROUP usage is similar to the ORDER usage in that it orders the distinct values of the variables, and if more than one variable has the GROUP usage, it is a nested ordering corresponding to the left-to-right ordering in the COLUMN statement. The primary difference between ORDER and GROUP is that GROUP condenses multiple observations corresponding to each level of the GROUP variable (or combinations of levels of multiple GROUP variables) into a single summary row when possible. Observations cannot be consolidated into groups if any other column is assigned the DISPLAY or ORDER usage and, in these cases, the GROUP usage mimics ORDER.

   d. *ANALYSIS.* For numeric variables, the default usage is ANALYSIS and the default summary statistic is SUM. PROC REPORT condenses multiple observations into a single row when all variables in the column statement have the ANALYSIS usage, or when other report items that do not have the ANALYSIS usage have either the GROUP, ACROSS, or COMPUTED usage assigned. Even when PROC REPORT cannot condense multiple rows, such as when a report item is assigned the DISPLAY or ORDER usage, it calculates the summary statistic on the single value in the row. The DEFINE statement for any ANALYSIS variable can include a statistic keyword to change the default summary. Any column assigned a statistic keyword as an option in its DEFINE statement is given the ANALYSIS usage (if appropriate), and any column generated by nesting of a statistic keyword also has the ANALYSIS usage (if appropriate).

   e. *ACROSS.* The ACROSS usage constructs a column for each distinct value of the variable, summarizing the frequency in each category if no other variables are associated with it in the COLUMN statement. Multiple ACROSS variables can be nested in the COLUMN statement, producing columns for each distinct combination of variable values. If an ANALYSIS variable is nested with an ACROSS variable, it is summarized in the column according to the assigned statistic.

   f. *COMPUTED.* The COMPUTED usage corresponds to variables not in the input data set that are created during PROC REPORT execution. These variables are subject to the structure of the table induced by the variables assigned any of the other five usage options and therefore do not affect the consolidation of multiple observations into summary rows.

If more than one usage is assigned in a DEFINE statement, the last usage listed is applied (if valid). If a statistic keyword and the ANALYSIS usage appear in the same define statement, their order is irrelevant—the statistic keyword is used. For additional information about usage concepts, see the SAS Documentation.

6.9 Exercises

Concepts: Multiple Choice

1. How many records are in the output data set after submission of the following DATA step?

```
data work.loopA;
  do i=1 to 4;
    do i=1 to 3;
      output;
    end;
  end;
run;
```

   a. 0
   b. 3
   c. 4
   d. It is an infinite loop

2. How many records are in the output data set after submission of the following DATA step?

```
data work.loopB;
  i=3;
  do until(i eq 3);
    do i=1 to 3;
      output;
    end;
  end;
run;
```

   a. 0
   b. 3
   c. 4
   d. It is an infinite loop

3. How many records are in the output data set after submission of the following DATA step?

```
data work.loopC;
  do i=1 to 5 while(i eq 3);
    output;
  end;
run;
```

   a. 0
   b. 3
   c. 4
   d. It is an infinite loop

4. If the following program is submitted, which one of the answer choices contains a correct reference to the values contained in the variable Profit2010?

```
data work.test;
  array profits[11] profit2007 - profit2017;
run;
```

   a. Profits[3]
   b. Profits[profit2010]
   c. Profits[2010]
   d. Profits[4]

5. What is the result of submitting the following program?

```
data work.Annual;
  month = 12;
  array year[month];
run;
```

   a. The ARRAY statement adds the variables Year1, Year2, …, Year12 to the PDV and associates them with an array named Year.
   b. The ARRAY statement adds the variables Month1, Month2, …, Month12 to the PDV and associates them with an array named Year.
   c. The array Year is not created due to a syntax error—DATA step variables cannot be used to set the number of elements in an array.
   d. The array Year is not created due to a syntax error—the elements of the array must be explicitly named in the ARRAY statement.

6. Which one of the following is true about a SAS array?
   a. It is saved with the data set.
   b. It can be used in a KEEP or DROP list.
   c. It exists only for the duration of the DATA step.
   d. It can be used in both DATA and PROC steps.

7. Which of the answer choices provides the correct DO statement to complete the provided program so that it processes the elements of the Weekly array?

```
data work.stats;
  set revenue(keep = mon tue wed thu fri);
  array weekly[5] mon tue wed thu fri;
  <insert DO statement here>
    total = weekly[i] * 0.25;
    output;
  end;
run;
```

   a. `DO i = 1 to 5;`
   b. `DO Weekly[i] = 1 to 5;`
   c. `DO i = Mon, Tue, Wed, Thu, Fri;`
   d. `DO i = 'Mon', 'Tue', 'Wed', 'Thu', 'Fri';`

8. The COLUMN statement in PROC REPORT contains the patient's age (Age), primary care physician's name (PhysName), and unpaid balance (Unpaid). If the following answer choices show a complete set of DEFINE statements for a REPORT procedure, which answer choice shows a set of DEFINE statements that create a report with the mean Age and median Unpaid for each value of PCP?

   a.
```
define PhysName / order;
define age / mean;
define unpaid / median;
```
   b.
```
define PhysName / display;
define age / mean;
define unpaid / median;
```
   c.
```
define age / mean;
define unpaid / median;
```
   d.
```
define PhysName / group;
define age / mean;
define unpaid / median;
```

9. Given the ARRAY statement below, which of the following answer choices correctly identifies the variable referenced by Char[2,3]?

```
array Char[3,4] $ Date1 - Date12;
```

   a. Date5
   b. Date6
   c. Date7
   d. Date8

10. Which of the following usages on any column prevents PROC REPORT from compressing multiple observations with the same value for a GROUP variable into a single row?

   a. ACROSS
   b. ANALYSIS
   c. DISPLAY
   d. None of the above

Concepts: Short-Answer

1. One of the benefits of OUT= options, OUTPUT statements, and ODS OUTPUT statements is the ability to validate the results of a procedure using the data sets they create.

   a. What aspects of independent reports generated with PROC REPORT can be validated using PROC COMPARE on data sets created with the OUT= option?
   b. What are some aspects of reports that cannot be validated with PROC COMPARE? How should each one be validated?

2. For each of the following scenarios, which loop type (iterative, conditional, or iterative-conditional) is best-suited?

   a. Compute the time required for an investment of $50/month to generate $1,000 in accrued interest.
   b. Determine the number of times necessary to flip a fair coin until it lands on Heads 100 times in a row.
   c. Count the number of visits each patient had at a given doctor's office last year.
   d. Carry out an operation on every element in a SAS array.

3. For which of the following is the use of a DATA step array helpful? If it is not, why not?

   a. Apply labels to a set of variables.
   b. Convert a set of variables containing temperature data from Fahrenheit to Celsius.
   c. Copy values from a sequence of variables (Day1, Day2, ...) into multiple records for a single variable (Day).
   d. Set the formats for a set of variables.

4. Explain why each of the following statements is true or false about the BREAK and RBREAK statements in the REPORT procedure.

   a. The RBREAK and BREAK statements are interchangeable.
   b. The RBREAK statement must be associated with a column with either the GROUP or ORDER usage.
   c. The BREAK statement must be associated with a column with either the GROUP or ORDER usage.
   d. Using the SUMMARIZE option in the RBREAK statement places summary statistics at the end of the report.
   e. The SUMMARIZE option in the BREAK statement is only valid when the BREAK statement is applied to a numeric variable.

5. Complete each of the following DATA steps to create the values in Diff, which are differences in the corresponding Value and Mean variables in the SummaryData set. The only ARRAY statement(s) permitted are those given. All references in any assignment statement to any of the data set variables (input or output) must refer to these arrays.

SummaryData

Value1	Value2	Value3	Value4	Mean1	Mean2	Mean3	Mean4
10	14	20	31	12	12	25	30
12	15	22	33	12	15	24	31

Diff

Diff1	Diff2	Diff3	Diff4
-2	2	-5	1
0	0	-2	2

a.
```
data work.Diff;
  set SummaryData;
  array values(4) value1-value4
  array means(4) mean1-mean4;
  array diff(4);
  <insert statements here>
run;
```

b.
```
data work.Diff;
  set SummaryData;
  array vars(8) value1-value4 mean1-mean4;
  array diff(4);
  <insert statements here>
run;
```

c.
```
data work.Diff;
  set SummaryData;
  array vars(8) value1 mean1 value2 mean2
                value3 mean3 value4 mean4;
  array diff(4);
  <insert statements here>
run;
```

d.
```
data work.Diff;
  set SummaryData;
  array vars(2,4) value1-value4
                  mean1-mean4;
  array diff(4);
  <insert statements here>
run;
```

e.
```
data work.Diff;
  set SummaryData;
  array vars(3,4) value1-value4
                  mean1-mean4
                  diff1-diff4;
  <insert statements here>
run;
```

Programming Basics

1. Use the Sashelp.Baseball data set to produce the following report. Only the records for League=American are shown.

League at the End of 1986	Division at the End of 1986	Team at the End of 1986	Times at Bat in 1986	Hits in 1986	Home Runs in 1986	1987 Salary in $ Thousands
			390.1	103.4	11.1	425.0
American	East	Baltimore	338.5	89.1	10.9	415.0
		Boston	499.8	137.8	13.8	855.0
		Cleveland	456.5	130.3	12.8	550.0
		Detroit	398.8	105.7	15.3	420.0
		Milwaukee	351.4	91.8	8.4	232.5
		New York	436.5	124.4	16.6	875.0
		Toronto	468.6	130.4	15.2	787.5
American	East		414.3	113.3	13.0	532.5
	West	California	398.1	101.8	12.5	462.5
		Chicago	387.6	96.7	8.5	185.0
		Kansas City	367.9	93.9	9.5	325.0
		Minneapolis	405.8	107.4	14.6	300.0
		Oakland	413.8	105.8	12.7	512.5
		Seattle	414.6	106.6	12.8	300.0
		Texas	391.3	105.5	13.6	235.0
American	West		396.3	102.4	12.0	325.0

2. Use the Sashelp.Cars data set to produce the following report.

	MPG Means					
	Asia		Europe		USA	
Type	City	Highway	City	Highway	City	Highway
Sedan/Wagon	22.8	29.8	19.5	27.0	20.7	28.6
Sports	20.2	26.6	17.7	25.1	16.9	24.2
Truck/SUV	17.5	21.8	14.5	18.7	15.6	20.2

3. The DATA step below uses the binomial distribution to generate a single random outcome that simulates one toss of a fair coin. The variable Outcome has a value of either 0 or 1, which can be assigned to heads and tails, respectively. (This assignment is arbitrary.)

```
data work.CoinFlip;
  Outcome = rand('binomial',0.50,1);
run;
```

a. Modify the program above so that it keeps track of the number of tails that occur and the proportion of tails that occur in 50 simulated coin tosses.

b. In the provided program, the value of 0.50 controls the expected proportion of times that Outcome=1 occurs. Further modify the program from (a) to track the number and proportion of tails that occur using values of 0.10, 0.25, and 0.40 for the expected proportion.

c. Produce a graph similar to the one shown below to demonstrate the long-run behavior of this coin-flipping simulation. (Note: due to the use of RAND, graphs may vary.)

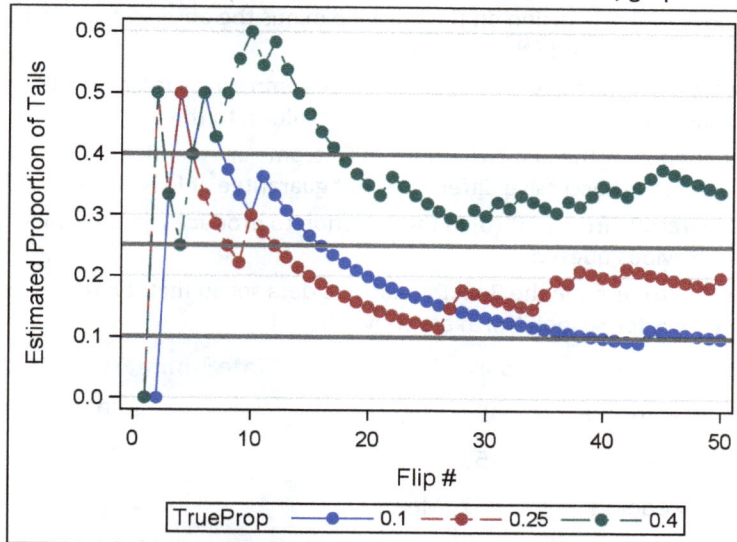

4. The DATA step below uses the binomial distribution to generate a single random outcome that simulates one toss of a fair coin. The variable Outcome has a value of either 0 or 1, which can be assigned to heads and tails, respectively. (This assignment is arbitrary.) In the provided program, the value of 0.50 controls the expected proportion of times that Outcome=1 occurs.

```
data work.CoinFlip;
   Outcome = rand('binomial',0.50,1);
run;
```

a. The data set BookData.CoinLimits provides records that include the TrueProp and Limit variables. Modify the above program using either iterative or conditional loops so that it uses the value of TrueLimit in place of the 0.50 and then simulates coin flips until the absolute difference between the observed proportion of tails and the expected proportion of tails is at least as small as the value of the Limit variable. Design the program so that it arrives at a solution for any valid value of TrueProp and Limit.

b. Repeat part (a) using a different iterative or conditional approach.

c. Produce a graph like the one shown below to demonstrate the differences between the scenarios in this coin-flipping simulation.

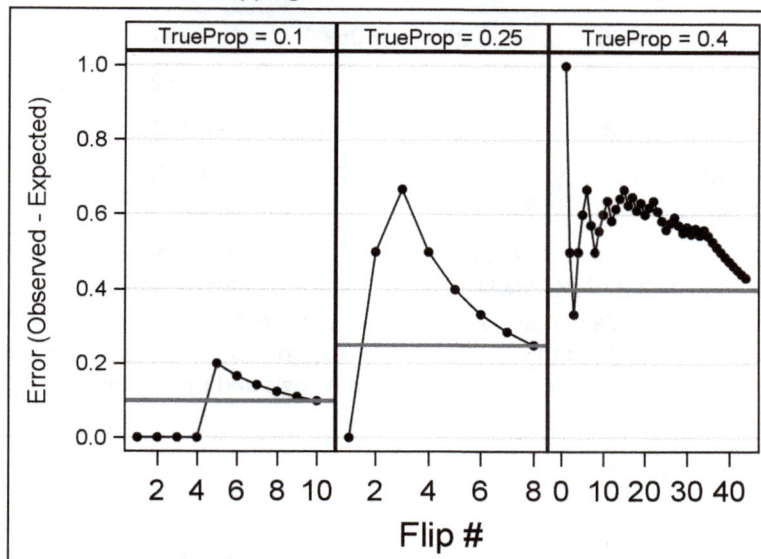

5. The previous question asks for a program that uses the difference between the observed proportion and true proportion as the basis for when to stop simulating coin flips. In practice, the true value from a population is unknown. Instead, it would be more practical to use successive differences, for example, the observed proportion on the third flip minus the observed proportion on the second flip, to determine when to conclude the simulation.

   a. Repeat part (a) of the previous question by comparing successive differences, instead of the difference between the true and observed values, to the value of Limit.

   b. What happens if the first two values of Outcome are identical? Modify the program to continue using successive differences but guarantee at least 10 tosses of the coin.

   c. Plot the results from part (b) in this question to produce a graph similar to the one in part (c) of the previous question.

6. Use the DATA step to reshape the BookData.Scores data set so that it has the following structure and contents. (Only rows for School=North are shown here.)

School	Subject	Score	Stat	StateSummary	Deviation
North	Math	453	Mean	450	3
North	Math	453	Median	455	-2
North	Reading	457	Mean	460	-3
North	Reading	457	Median	460	-3
North	Science	260	Mean	240	20
North	Science	260	Median	246	14

7. The Sashelp.LeuTest data set contains information about genes for cases of acute leukemia. Records with Y = 1 have acute lymphoblastic leukemia and records with Y = -1 have acute myeloid leukemia. The variables X1 through X7129 represent measurements on 7,129 genes.

   a. Based on this information, calculate the mean values for genes 1 through 7,128 separately for each leukemia type, then use only one-dimensional arrays to calculate the differences between each variable and its average. Produce the following report. (Note: only the first 5 records and a selected set of columns are shown here.)

Acute Leukemia Type	Gene 1	Gene 1 Deviation	Gene 100	Gene 100 Deviation	Gene 7128	Gene 7128 Deviation
Myeloid	-0.238	-0.513	0.115	-0.174	-0.220	-0.161
Myeloid	0.476	0.201	0.067	-0.221	0.236	0.295
Myeloid	1.132	0.857	0.321	0.032	-0.467	-0.409
Myeloid	-1.266	-1.540	-0.833	-1.121	-0.500	-0.442
Myeloid	-0.036	-0.310	-0.027	-0.315	-0.431	-0.373

   b. Reproduce the report from part (a) of this question by using only multidimensional arrays with dimensions of 9x24x33 instead of using one-dimensional arrays.

   c. Compare the code necessary to complete the report using the two array-based approaches in parts (a) and (b) in this question. Which approach used simpler DO group logic and why is this relationship between simpler DO group logic and the complexity of the arrays expected?

Case Studies

For additional practice, multiple case studies are available in addition to the IPUMS CPS case study used in the chapters. See Section 8.6 to apply the skills from this chapter to the Clinical Trials Case Study. For additional case studies, including extensions to the IPUMS CPS case study, see the author pages.

Chapter 7: Advanced DATA Step Concepts

7.1 Learning Objectives

At the conclusion of this chapter, mastery of the concepts covered in the narrative includes the ability to:

- Employ line hold specifiers to create a single observation from multiple rows or to create multiple observations from a single row

- Apply the RETAIN statement to prevent a PDV value from being reset across iterations of the DATA step

- Apply the DATA step sum statement to add an accumulator variable to the PDV

- Use FIRST. and LAST. variables, created when BY groups are used in the DATA step, to condition on location within BY groups

- Use the STYLE= option to set styles at various locations in tables generated by the REPORT procedure

- Implement COMPUTE blocks to create new columns or add other content to tables generated in PROC REPORT

- Use the CALL DEFINE function to set styles and other attributes in COMPUTE blocks in PROC REPORT

- Apply the LIBNAME statement to retrieve data from Excel and Access sources

Use the concepts of this chapter to solve the problems in the wrap-up activity. Additional exercises and case-studies are also available to test these concepts.

7.2 Case-Study Activity

Continuing the case study covered in the previous chapters, the tables and graphs shown below are based on the data assembled for the Wrap-Up Activities in Chapters 5 and 6. The objective, as described in the Wrap-Up Activity in Section 7.7, is to create the reports shown in Output 7.2.1 and 7.2.2, which are modifications of

Output 6.2.1 and 6.2.2. The tables shown in Output 7.2.3 and 7.2.4 are based on similar projections used to make the graphs shown in Output 6.2.3, 6.2.4, and 6.2.5.

Output 7.2.1: Electricity Cost Summary Statistics

State	Year	Mean	Median	Diff: Mean-Median	Std. Dev.
Connecticut	2005	$1,423.00	$1,200.00	$223.00	$979.81
	2010	$1,746.45	$1,560.00	$186.45	$1,215.27
	2015	$1,717.85	$1,440.00	$277.85	$1,384.61
		$1,638.04	$1,440.00	$198.04	$1,223.86
Massachusetts	2005	$1,173.31	$960.00	$213.31	$862.60
	2010	$1,242.80	$1,080.00	$162.80	$999.99
	2015	$1,369.06	$1,080.00	$289.06	$1,275.12
		$1,269.12	$1,080.00	$189.12	$1,078.57
New Jersey	2005	$1,477.86	$1,200.00	$277.86	$1,137.00
	2010	$1,888.13	$1,560.00	$328.13	$1,424.46
	2015	$1,749.70	$1,440.00	$309.70	$1,428.80
		$1,711.41	$1,320.00	$391.41	$1,353.26
New York	2005	$1,454.35	$1,200.00	$254.35	$1,054.80
	2010	$1,528.60	$1,200.00	$328.60	$1,190.45
	2015	$1,541.07	$1,200.00	$341.07	$1,324.39
		$1,510.53	$1,200.00	$310.53	$1,203.17
				Mean Exceeds Median Value by More Than 25%	

Output 7.2.2: Home Value Summary Statistics

State	2005 Mean	2005 Median	2010 Mean	2010 Median	2015 Mean	2015 Median
Connecticut	$353,601	$275,000	$404,008	$280,000	$396,875	$270,000
Massachusetts	$401,007	$350,000	$392,878	$319,500	$434,209	$350,000
New Jersey	$378,541	$350,000	$388,920	$320,000	$395,547	$300,000
New York	$304,502	$225,000	$360,828	$250,000	$387,242	$250,000
				Mean Exceeds Median by More Than 25%		
				Mean Exceeds Median by More Than 50%		

Output 7.2.3: Electricity Cost Projections, 2% Annual Increase

State	2005	2006	2007	2008	2009	2010
Connecticut	$1,423.00	$1,451.46	$1,480.49	$1,510.10	$1,540.30	$1,571.11
Massachusetts	$1,173.31	$1,196.78	$1,220.71	$1,245.13	$1,270.03	$1,295.43

State	Mean Cost					
	2005	**2006**	**2007**	**2008**	**2009**	**2010**
New Jersey	$1,477.86	$1,507.42	$1,537.57	$1,568.32	$1,599.69	$1,631.68
New York	$1,454.35	$1,483.44	$1,513.11	$1,543.37	$1,574.24	$1,605.72

Output 7.2.4: Electricity Cost Projections Versus Actual Value

State	Mean Cost					
	2005	**2006**	**2007**	**2008**	**2009**	**2010**
Connecticut	$1,593.55	$1,451.46	$1,480.49	$1,510.10	$1,540.30	$1,571.11
Massachusetts	$1,210.13	$1,196.78	$1,220.71	$1,245.13	$1,270.03	$1,295.43
New Jersey	$1,690.77	$1,507.42	$1,537.57	$1,568.32	$1,599.69	$1,631.68
New York	$1,493.28	$1,483.44	$1,513.11	$1,543.37	$1,574.24	$1,605.72
				Projected Values Exceeding the Actual 2010 Mean		

7.3 Reading Complex Raw Data Structures

The previous chapters use raw data files with a variety of characteristics. Some files include no delimiter, while other files use one or more delimiters. In both of those cases, some variables require an informat while other variables do not. And in those cases, records may contain missing values for some of the variables. One characteristic common to most of these data sets is the intent for each raw record to correspond to exactly one record in the resulting SAS data set.

To alter these structures after reading the data, Chapters 5 and 6 present techniques for restructuring SAS data sets by exchanging row and column information—either via PROC TRANSPOSE or with a DATA step array. However, it is often much more computationally efficient to create a SAS data set with the desired structure directly from the raw file, alleviating the need to read in the data set only to then operate on it in a separate step. Data files that benefit from this data reading approach are broadly classified as either having

- a single row of text that should appear as multiple observations in a SAS data set, or
- multiple rows of text that should appear as a single observation in a SAS data set.

The methods to read raw data files in either of these cases depend on the same concept: additional methods to explicitly control loading of records into the input buffer.

7.3.1 Creating A Single Observation from Multiple Rows

Programs 5.7.3 and 6.5.1 each used the Ipums2005Utility data set when demonstrating how to use PROC TRANSPOSE and arrays to exchange row and column information in a SAS data set. Table 7.3.1 displays the first two records of this data set as a reminder of its structure and as a reminder that earlier in the text the utility values such as 9,993 and 9,997 were replaced with missing values.

Table 7.3.1: Review of Utility Cost Data Set Structure

Serial	Electric	Gas	Water	Fuel
2	1800	.	700	1300
3	1320	2040	120	.

Input Data 7.3.1 shows this utility cost data from 2005 in a raw file with a different form than is available in Utility Cost 2005.txt read in Program 5.3.1 to create Ipums2005Utility. Unlike the previous examples, the 12 records displayed in Input Data 7.3.1 reveal this file has a vertical structure where the value for Serial, Utility, and Cost are on separate records. Note that the blank fifth line indicates a missing value; in this case the value of Cost for Serial=2 and Utility=Gas.

Input Data 7.3.1: Utility2005ComplexA Text File (First 12 Records)

```
2
$1,800
Electricity
2

Gas
2
$700
Water
2
$1,300
Fuel
```

To successfully read the Utility2005ComplexA.txt file shown in Input Data 7.3.1 and store it as shown in Table 7.3.1, the DATA step must access information from multiple rows in the raw file in order to create a single observation in the resulting SAS data set. Program 7.3.1 uses line pointer controls as a first step to achieving this.

Program 7.3.1: Single SAS Observations from Three Raw Records Using Relative Line Pointer Controls

```
data work.UtilityA2005v1;
   infile rawdata('Utility2005ComplexA.txt') missover❶;
   input Serial
         /❷ cost : comma.
         /❷ Utility: $11.;
   label Cost='Utility Cost' Utility='Utility Type';
   format cost dollar8.;
run; ❸

proc print data = work.UtilityA2005v1(obs = 4) label noobs;
run;
```

❶ Since blank lines denote missing values, it is necessary to use an alternative option to prevent SAS from using the default action of FLOWOVER when encountering such raw records. Either TRUNCOVER or MISSOVER is appropriate here since list input is used for all variables in the INPUT statement.

❷ In the INPUT statement, the forward slash acts as a relative line pointer control. It explicitly moves the line pointer SAS uses to track the current observation in the input buffer. The forward slash advances the line pointer by one line in the raw file and loads that record. While syntax does not require line pointer controls to appear at the beginning of a new line, debugging is often easier when the layout of the INPUT statement logically aligns with the raw file(s) in use.

❸ With no OUTPUT statement in the DATA step, the implicit OUTPUT at the bottom of each iteration writes the current PDV to the data set UtilityA2005v1. Since the INPUT statement contains two relative line pointer controls, it uses three lines of raw text to build the PDV from which the DATA step creates a single record.

Output 7.3.1 shows the first four records of UtilityA2005v1 created from the first 12 rows of UtiltyComplex2005A.txt. This approach is much more efficient than reading Utility2005ComplexA.txt in as a single column then restructuring it with arrays in the DATA step or with PROC TRANSPOSE. Note that unlike the relative column pointer controls introduced in Chapter 3, relative line pointer controls can only advance the line pointer. It is not possible to return to an earlier line with relative line pointer controls. For a discussion of how this is equivalent to using multiple INPUT statements, see Chapter Note 1 in Section 7.8.

Output 7.3.1: Creating a Single SAS Observation from Three Raw Records

Serial	Utility Cost	Utility Type
2	$1,800	Electricity
2	.	Gas
2	$700	Water
2	$1,300	Fuel

As Output 7.3.1 shows, the data set Program 7.3.1 creates is not identical to versions used in earlier chapters—the Cost and Utility columns are interchanged. Of course, interchanging the Cost and Utility variable names in a VAR statement in the PROC PRINT of Program 7.3.1 would create the desired output, but would not update the UtilityA2005v1 data set itself. To control the order of columns in the UtilityA2005v1 data set, the relative line pointer is not enough. While including additional statements such as FORMAT or LENGTH prior to the INPUT statement would set the order of variables in the PDV, line pointer controls provide an alternative approach.

Program 7.3.2 updates Program 7.3.1 by replacing the relative line pointer controls with absolute line pointer controls. Like the absolute column pointer controls introduced in Chapter 3, absolute line pointer controls allow the INPUT statement to read a designated line. It is possible to use multiple INPUT statements in a DATA step, each of which can use absolute line pointer controls, relative line pointer controls, or a combination of the two. Here, examples are limited to using one type of line pointer control in a single input statement, more advanced combinations are beyond the scope of this book.

Program 7.3.2: Single SAS Observations from Three Raw Records with Absolute Line Pointer Controls

```
data work.UtilityA2005v2;
   infile rawdata('Utility2005ComplexA.txt') missover;
   input Serial
         #3 Utility: $11.
         #2 cost : comma.;
   label Cost='Utility Cost' Utility='Utility Type';
   format cost dollar8.;
run;

proc print data = work.UtilityA2005v2(obs = 4) label noobs;
run;
```

Output 7.3.2: Single SAS Observations from Three Raw Records with Absolute Line Pointer Controls

Serial	Utility Type	Utility Cost
2	Electricity	$1,800
2	Gas	.
2	Water	$700
2	Fuel	$1,300

The presence of absolute line pointer controls in Program 7.3.2 triggers modifications to the raw data reading process. Whenever the DATA step encounters an INPUT statement containing absolute line pointer controls, it modifies the input buffer to contain multiple lines and uses the largest line pointer value to define the number of lines. The absolute line pointers are absolute row positions in the current input buffer, not rows in the raw data file. While it is also true that absolute column pointers are column positions in the current input buffer, this usually corresponds to column position in the raw file as well. Therefore, within the input buffer absolute has the same meaning for rows or columns, in the raw data file it does not.

On the first iteration of the DATA step in Program 7.3.2, the INPUT statement causes the first three records of the raw data file to be loaded into the input buffer. First, it reads Serial from the first row of the input buffer—default row position is 1 without any row pointer controls. Next, Utility is read from the third row of the input buffer, followed by Cost coming from the second row. When the second iteration of the DATA step begins, the input buffer is reset. Upon encountering the INPUT statement, SAS loads the next three records into the input buffer. Now the input buffer contains a new block of three lines—records 4 through 6 from the original raw file—and continues reading data as it did for the first iteration. Given that the block of rows loaded is the same size on all iterations, it is imperative that every intended observation consists of the same number of raw records.

While, Programs 7.3.1 and 7.3.2 successfully use line pointer controls to create a single SAS observation from multiple raw records, both leave the final data set in a vertical structure with four observations for each Serial since each value of Utility is represented in separate observations. Recall from Input Data 7.3.1 that the twelve lines shown are all associated with the same value of Serial, and each subsequent set of twelve records includes all necessary information for each subsequent value of Serial. Rather than relying on PROC TRANSPOSE to restructure these data sets, Program 7.3.3 uses absolute line pointer controls to selectively read the utility cost values directly into one record per value of Serial with variables for each utility: Electric, Gas, Water, and Fuel.

Program 7.3.3: Single SAS Observations from 12 Raw Records Using Absolute Line Pointer Controls

```
data work.UtilityA2005v3;
  infile rawdata('Utility2005ComplexA.txt') missover;
  input ❶serial
        ❷#2 Electric : comma.
        ❸#5 Gas : comma.
        ❸#8 Water : comma.
        ❸#11 Fuel : comma.
        ❹#12 ;
  label Serial = 'Serial'
        Electric = 'Electric Cost'
        Gas = 'Gas Cost'
        Water = 'Water Cost'
        Fuel = 'Fuel Cost'
        ; ❺
  format Electric Gas Water Fuel dollar7.; ❺
run;

proc print data = work.UtilityA2005v3(obs = 4) label noobs;
  var serial Electric Gas Water Fuel;
run;
```

❶ Serial appears on lines #1, #4, #7, and #10, but the INPUT statement only needs to read it once. Since the default location is the first row, no line pointer control is necessary.

❷ The second line always refers to the cost of electricity, so it is explicitly assigned to a variable named Electric.

❸ Because the intent is to store utility costs in specifically named variables, the third line of raw text—which contains the literal value 'Electricity' is redundant. Similarly, the value of Serial is already in the PDV and does not need to be reread. As a result, the INPUT statement can move directly to the fifth line to read the utility cost associated with the Gas variable. The remaining utility costs are read similarly by skipping the repetitive values of Serial and the unnecessary utility names. Again, knowing that all blocks of 12 rows have the same structure is important to ensuring the correct observations are produced.

❹ Since the last record contains only the unnecessary literal value 'Fuel', it is common for novice programmers to attempt to skip that line by not including it in the INPUT statement. However, recall the largest absolute line pointer value defines the number of rows in the input buffer. Pointing the INPUT statement to line 12 without reading any values successfully communicates to SAS that records appear in groups of 12, not 11, without using the 12th raw record in each block to produce any SAS data.

❺ Remember that assigning labels and formats in the DATA step ensures these default values are available for future procedures.

Output 7.3.3 shows the results of the PROC REPORT step in Program 7.3.3. The absolute line pointer controls allow the INPUT statement to access a subset of each set of twelve raw records and use them to build a single record in the PDV.

Output 7.3.3: Creating a Single SAS Observation from 12 Raw Records

Serial	Electric Cost	Gas Cost	Water Cost	Fuel Cost
2	$1,800	.	$700	$1,300
3	$1,320	$2,040	$120	.
4	$1,440	$120	.	.
5	.	.	.	.

Programs 7.3.1 through 7.3.3 all use list input, either simple or modified, in the INPUT statements. However, just as column pointer controls are available in the INPUT statement regardless of input style, so too are line pointer controls. Input Data 7.3.4 shows the first eight records from Utility2005ComplexB.txt, which contains the same information as the Utility2005ComplexA.txt file in a modified form. (The ruler is not part of the raw file.) Program 7.3.4 demonstrates the use of formatted input, column pointer controls, and line pointer controls in the same INPUT statement to read this file.

Input Data 7.3.4: First Eight Records of Utility2005ComplexB.txt

```
----+----1----+----2
2
 $1,800   Electricity
2

         Gas
2
   $700   Water
2
 $1,300   Fuel
```

Program 7.3.4: Using Absolute Line Pointer Controls and Column Pointer Controls in Formatted Input

```
data work.UtilityB2005v1;
   infile rawdata('Utility2005ComplexB.txt') truncover❶;
   input serial
           ❷#2 ❸@10 Utility $11.  ❹ @1 cost comma7.;
run;
```

❶ Since formatted input uses column positions, and some column sets may contain only a partial value, the TRUNCOVER option replaces the MISSOVER option.

❷ The second line contains both the value of Cost and the value of Utility. Note that either relative or absolute line pointer controls are acceptable here.

❸ An absolute column pointer control moves the pointer to column 10 where the field for Utility begins.

❹ Another absolute column pointer returns to the first column before the field for Cost is read. In both ❸ and ❹ relative column pointer controls are acceptable.

Program 7.3.4 produces a data set identical to the Output 7.3.2 since the column pointer controls allow the INPUT statement to read Utility before Cost. Swapping the order of Cost and Utility in the INPUT statement produces a result identical to that of Program 7.3.1. Thus, all three data sets are the same except for column order.

7.3.2 Creating Multiple Observations from A Single Row

Section 7.3.1 focuses on cases where the information for a single SAS data set observation is spread across multiple records in the raw source file. However, some raw files contain information corresponding to multiple observations in a single record. Input Data 7.3.5 shows the first 70 columns of the first four lines from the Utility2005ComplexC.txt file, which is one possible example of such a structure. (The ruler is not part of the raw file.) Each record in this file contains the complete utility information for three different values of the variable Serial.

Input Data 7.3.5: Partial Records from Utility2005ComplexC.txt

```
----+----1----+----2----+----3----+----4----+----5----+----6----+----7
2,Electricity,"$1,800",2,Gas,,2,Water,$700,2,Fuel,"$1,300",3,Electrici
5,Electricity,,5,Gas,,5,Water,,5,Fuel,,6,Electricity,"$1,320",6,Gas,$6
8,Electricity,"$2,160",8,Gas,,8,Water,$480,8,Fuel,,9,Electricity,"$1,2
11,Electricity,"$1,440",11,Gas,,11,Water,$250,11,Fuel,,12,Electricity,
```

Unlike the raw data structures in Section 7.3.1, using line pointer controls to allow the INPUT statement to access multiple lines is no longer helpful for the structure of Input Data 7.3.5. However, recall from Section 3.7.2 after the INPUT statement populates its variables, all remaining information in the input buffer is released. Program 7.3.5 introduces line hold specifiers as a possible approach to this scenario. The UtilityC2005 data set created by Program 7.3.5 is identical to the data sets created in Programs 7.3.2 and 7.3.4.

Program 7.3.5: Using Double Trailing @ to Read Multiple SAS Observations from a Single Raw Record

```
data work.UtilityC2005;
   infile rawdata('Utility2005ComplexC.txt') dsd❶;
   input serial Utility: $11. cost : comma. @@❷;
run;
```

❶ The DSD option interprets the sequential commas as a missing value and ignores the commas appearing as part of the Cost values inside the quoted strings.

❷ The double trailing @ is a line hold specifier. It instructs the INPUT statement to keep the current input buffer until it encounters either (a) an end-of-line marker or (b) an INPUT statement. If additional INPUT statements are used in the DATA step, whether the line is retained in the input buffer depends on if and how line-hold specifiers appear in the additional INPUT statements. For additional details, see Chapter Note 2 in Section 7.8.

Note that since the double trailing @ is used to hold records in the input buffer across iterations of the DATA step, the INPUT statement continues processing records past the end of each line. As a result, the note shown in Log 7.3.5 appears whenever the double trailing @ is used.

Log 7.3.5: Partial Log of Program 7.3.5 Indicating Double Trailing @ was Used

```
NOTE: SAS went to a new line when INPUT statement reached past the end of a line.
```

Of course, it is the default FLOWOVER option that allows SAS to read past the current line and read the following record into the input buffer. In previous examples the MISSOVER and TRUNCOVER options prevent SAS from loading new records to fill in missing data at the end of the line. Therefore, be aware that the MISSOVER option cannot be used with the double trailing @—it generates the following error in the SAS log since the combination of MISSOVER and the double trailing @ guarantee an infinite loop.

```
ERROR: The INFILE statement MISSOVER option and the INPUT statement double trailing @
option, are being used in an inconsistent manner.  The execution of the DATA STEP is
being terminated to prevent an infinite loop condition.
```

While SAS allows the use of the TRUNCOVER option simultaneously with the double trailing @ in some scenarios, by default it also results in an infinite loop. However, TRUNCOVER does not generate an error (or even a warning) when used in conjunction with the double trailing @.

Input Data 7.3.6 contains yet another common raw data structure in which information for several data set observations occurs on a single line of raw text. Serial is only printed once at the beginning of the line followed by the four pairs of Utility and Cost values. The following examples demonstrate several approaches to handle data of this form.

Input Data 7.3.6: First Four Records in Utility2005ComplexD.txt.txt.

```
2,Electricity,"$1,800",Gas,,Water,$700,Fuel,"$1,300"
3,Electricity,"$1,320",Gas,"$2,040",Water,$120,Fuel,
4,Electricity,"$1,440",Gas,$120,Water,,Fuel,
5,Electricity,,Gas,,Water,,Fuel,
```

The two common approaches to data sets like the one shown in Input Data 7.3.6 are to either recover the vertical structure and produce results like Output 7.3.1, or to maintain the horizontal structure and produce results like those of Output 7.3.3. Program 7.3.6 uses line hold specifiers to recover the vertical structure.

Program 7.3.6: Using @ to Reshape a Single Raw Record into Multiple SAS Observations

```
data work.UtilityD2005v1;
  infile rawdata('Utility2005ComplexD.txt') dsd;
  input serial @; ❶
  input Utility:$11. cost : comma. @; ❷
  output; ❸
  input Utility:$11. cost : comma. @; ❹
  output; ❹
  input Utility:$11. cost : comma. @; ❹
  output; ❹
  input Utility:$11. cost : comma. @; ❹
  output; ❹
run;
```

❶ This INPUT statement reads the value of Serial from the input buffer. The single trailing @ is a line hold specifier that prevents SAS from releasing the input buffer after the INPUT statement reads Serial. This value is common to all records created from the information in the current input buffer.

❷ The second INPUT statement reads the first pair of Utility and Cost values from the input buffer. The single trailing @ again instructs SAS to hold the input buffer. Note that while this INPUT statement can be removed if the variables here are placed in ❶, for clarity in this example they appear separately to demonstrate the four pairs of INPUT and OUTPUT statements.

❸ Now that the values of Serial, Utility, and Cost are in the PDV, the OUTPUT statement writes them to the UtilityD2005v1 data set.

❹ Each pair of INPUT and OUTPUT statements reads one of the remaining pairs of Utility and Cost values. This is necessary since, unlike the double trailing @, the single trailing @ does not hold a record across iterations of the DATA step. Once the DATA step iteration is complete, SAS releases the input buffer and any remaining information is lost.

Variables like Serial that apply to each data set observation built from a single raw record are often called common variables in this context. In contrast, variables such as Utility and Cost are referred to as unique variables since each set of values from a raw record is used in exactly one data set observation. There is no limit on the number of either common or unique variables.

While the results of Program 7.3.6 produce a data set identical to the one shown in Output 7.3.1, the program itself can be written in a better form. Specifically, the repetitive blocks of INPUT/OUTPUT statements are difficult to maintain as the number of sets of unique variables increases. Program 7.3.7 uses a DO loop to provide a more efficient and elegant approach than Program 7.3.6, while still producing the desired data set.

Program 7.3.7: Using A DO Loop to Improve Efficiency of Program 7.3.6

```
data work.UtilityD2005v1(drop = i); ❶
  infile rawdata('Utility2005ComplexD.txt') dsd;
  input serial @;
  do i = 1 to 4; ❶
    input Utility : $11. cost : comma. @; ❷
    output;
  end; ❶
run;
```

❶ The DO loop replaces the need for sequential blocks of INPUT and OUTPUT statements. Since the index variable is unimportant, the DROP= option prevents it from appearing in the final data set.

❷ Keeping the common and unique variables in separate INPUT statements in Program 7.3.6 supports a more natural transition to the DO loop structure.

Programs 7.3.6 and 7.3.7 reproduce the vertical structure, which is commonly required by many statistical analyses in SAS. However, other analyses may require the horizontal layout of Output 7.3.3. Program 7.3.8 demonstrates the use of the single trailing @ and arrays to read data directly into the horizontal layout.

Program 7.3.8: Using Arrays with Single Trailing @

```
data work.UtilityD2005v2(drop = i Utility);
  infile rawdata('Utility2005ComplexD.txt') dsd;
  input serial @;
  array cost[4] Electric Gas Water Fuel; ❶

  do i = 1 to 4;
    input Utility : $11. cost[i]❷ : comma. @;❸
  end; ❹
run;
```

❶ The Cost array declares four new numeric variables: Electric, Gas, Water, and Fuel. The ARRAY statement occurs after the INPUT statement for Serial only to ensure the order of variables in the data set matches previous examples.

❷ The array reference Cost[i] replaces the name of a specific variable in the INPUT statement. This allows the INPUT statement to dynamically assign each new cost value to the appropriate type of utility.

❸ The values of Utility are not placed into an array, which causes the values to be overwritten when the INPUT statement reads each successive value of Utility. This is intentional since the goal is only to read the utility cost values, but the nature of the data precludes the use of column pointer controls to skip this field. Utility appears in the DROP= option as it is extraneous in the final result.

❹ Unlike Program 7.3.7, the DO loop does not contain an OUTPUT statement. Each iteration of the DO loop sets the Cost value for one utility, so the record is only fully populated after all iterations of the DO loop. The implicit OUTPUT at the end of each DATA step iteration sends the observations to the final data set.

Thus far in this section, programs that use the single trailing @ explicitly assume the number of unique variable sets is constant for each value of the group variable. For example, in Programs 7.3.7 and 7.3.8 the DO loops explicitly use the fact that there are four pairs of Utility and Cost values. Similarly, in Program 7.3.6 there are four blocks of INPUT/OUTPUT statements that correspond to the four pairs of values. However, when using the double trailing @ no consideration is given to the number of observations that appear in a single line of raw text.

This raises the question of how to adapt Programs like 7.3.6 and 7.3.7 in cases where the number of sets of unique variables is not constant across all observations—Input Data 7.3.9 shows such a scenario.

Input Data 7.3.9: First Four Records of Utility2005ComplexE.txt

```
2,Electricity,"$1,800",Water,$700,Fuel,"$1,300"
3,Electricity,"$1,320",Gas,"$2,040",Water,$120
4,Electricity,"$1,440",Gas,$120
5
```

Note the similarities to Input Data 7.3.6—the only difference is the inclusion of missing values in Input Data 7.3.6, whereas Input Data 7.3.9 has missing values excluded. However, note that even in the case of Serial=5 where all data was missing, it is important that the value of Serial appears in the raw file. Program 7.3.9 demonstrates one approach for reading the data in Input Data 7.3.9 and creating a vertically structured data set.

Program 7.3.9: Reading Raw Records When the Number of Variables is not Constant

```
data work.UtilityE2005v1;
   infile rawdata('Utility2005ComplexE.txt') dsd missover; ❶
   input serial Utility : $11. cost : comma. @; ❷
   do while(not missing(cost)); ❸
     output; ❹
     input Utility : $11. cost : comma. @; ❺
   end; ❻
run;

proc report data = work.UtilityE2005v1(obs = 4);
   columns serial Utility cost;
   define serial / display 'Serial';
   define Utility / display 'Utility Type';
   define cost / display format= dollar7. 'Utility Cost';
run;
```

❶ Like Input Data 7.3.6, Input Data 7.3.9 has quoted strings containing the delimiter, so DSD is still necessary. Some records now contain fewer fields than necessary to complete an observation, so the MISSOVER (or TRUNCOVER) option is needed to prevent SAS from reading the next line when it encounters a short record.

❷ The INPUT statement attempts to read in a full set of values. If no Utility or Cost data is available, as in the case of Serial=5, MISSOVER sets these values to missing.

❸ Since the number of Utility and Cost pairs is no longer fixed, an iterative loop is not appropriate. Recall the DO WHILE loop tests its condition before entering the loop. As such, this DO WHILE statement checks to see whether Cost is missing before executing any statements inside the DO group. Only when the INPUT statement successfully reads a nonmissing value for Cost does SAS execute the statements that appear in the loop.

❹ It may appear strange to begin with the OUTPUT statement here; however, reaching this point means the first pair of Utility and Cost values are already in the PDV. The OUTPUT statement writes them out to the data set before attempting to read in a new set of values.

❺ This INPUT statement reads in the next pair of Utility and Cost variables, with MISSOVER setting them to missing if no more pairs are available.

❻ The DO WHILE loop continues reading and outputting records while the last Cost value read is nonmissing. Once a missing value is read, the DO WHILE loop exits without executing its statements, so the missing values are not written to the data set.

The Program 7.3.9 produces the table shown in Output 7.3.9; note the difference between it and Output 7.3.1— – the missing values no longer appear. Because they are not included in the raw data set, they are not automatically created in the SAS data set.

Output 7.3.9: Reading Raw Records When the Number of Variables is not Constant

Serial	Utility Type	Utility Cost
2	Electric	$1,800
2	Water	$700
2	Fuel	$1,300
3	Electric	$1,320

It is worth noting that even though the missing values do not appear in Output 7.3.9, applying PROC TRANSPOSE to this data set can produce them in certain scenarios. Program 7.3.10 demonstrates this using the methods first introduced in Program 5.7.5.

Program 7.3.10: Applying PROC TRANSPOSE to the Results of Program 7.3.9

```
proc transpose data = work.UtilityE2005v1
                out = work.UtilityE2005Wide(drop = _name_);
  by serial;
  id Utility;
  var cost;
run;

proc report data = work.UtilityE2005Wide(obs = 4);
  columns serial Electric Gas Water Fuel;
  define serial   / display 'Serial';
  define Electric / display format= dollar7. 'Electric Cost';
  define Gas      / display format= dollar7. 'Gas Cost';
  define Water    / display format= dollar7. 'Water Cost';
  define Fuel     / display format= dollar7. 'Fuel Cost';
run;
```

Output 7.3.10 shows the results of the transposition on the first four records. Note that the missing values are present in the transposed data set because the ID statement creates a column for each distinct value of Utility, with all of those columns populated on each record. Be aware that this produced the desired results only because at least one value of Electric, Gas, Water, and Fuel appeared in the source data. If this PROC TRANSPOSE is applied to a subset of the data where no records contain a cost for Fuel, then the Fuel column would not exist in the output data set. This is important to keep in mind for dynamic data sets—the output may change over time as new records are added to the database.

Output 7.3.10: Applying PROC TRANSPOSE to the Results of Program 7.3.9

Serial	Electric Cost	Gas Cost	Water Cost	Fuel Cost
2	$1,800	.	$700	$1,300
3	$1,320	$2,040	$120	.
4	$1,440	$120	.	.
6	$1,320	$600	$240	.

Reshaping the results of Program 7.3.9 with PROC TRANSPOSE in Program 7.3.10 is not actually a necessary step. Program 7.3.11 extends Program 7.3.9 using conditional logic in the DO WHILE loop to reshape the raw data as it is read in using a single DATA step.

Program 7.3.11: Revising Program 7.3.9—Including Conditional Logic to Obtain a Horizontal Structure

```
data work.UtilityE2005v2;
  infile rawdata('Utility2005ComplexE.txt') dsd missover;
  input serial Utility : $11. cost : comma. @; ❶
  do while(not missing(cost)); ❶
    select(Utility); ❸
      when('Electricity') electric = cost; ❹
      when('Gas') gas = cost; ❹
      when('Water') water = cost; ❹
      when('Fuel') fuel = cost; ❹
      otherwise putlog 'QCNOTE: Unknown value for Utility' Utility=; ❹
    end; ❸
    input Utility: $11. cost : comma. @; ❺
  end; ❷
  output; ❻
  drop Utility cost; ❼
run;
```

❶ The values of Serial, Utility, and Cost are read from the input buffer and the single trailing @ holds the current record in the input buffer.

❷ The DATA step enters the DO WHILE loop and immediately tests the condition. If the value of Cost is not missing, then SAS enters the loop.

❸ Conditional logic determines which Utility is associated with the current value of Cost. Recall the presence of a select expression in the SELECT statement forces SAS to compare the value of the expression to the expressions included in the following WHEN statements. The SELECT group concludes with the END statement.

❹ Based on the current value of Utility, the current value of Cost is assigned to the complementary utility-specific variable. This case uses an OTHERWISE statement since certain records may have missing values or values outside the set provided in the WHEN statements for Utility. The OTHERWISE statement provides a diagnostic note to the SAS log in this case.

❺ The next value of Utility and Cost are read from the current input buffer, and the single trailing @ continues to hold the current line in the input buffer.

❻ After the DO WHILE loop completes the final iteration, the current record is OUTPUT. Note that this appears outside the loop here instead of inside the loop as it did in Program 7.3.9. Because the intent of this program is to store all values of Cost across four variables in a single record, it is imperative that the output does not occur inside the loop to ensure each value of Serial is associated with only a single record. In this case, using the implicit output at the bottom of each DATA step iteration is also appropriate.

❼ Due to the now-horizontal structure, the Utility and Cost variables are extraneous and potentially confusing and should be dropped.

The results of Program 7.3.11 are identical to Output 7.3.10. Note that while the single trailing @ is used to hold a line in the input buffer, it releases the line at the end of an iteration of the DATA step. Similarly, the PDV is reset between DATA step iterations, which ensures that no information is carried forward and used when creating a future record.

7.4 Working Across Records in the DATA Step

Often when working with data sets, computations require access to more than a single record at a time. Common examples include comparing a current observation to a past observation, calculating running totals, or carrying out separate analyses based on grouping variables. This section provides several DATA step techniques for carrying information from an iteration of the DATA step forward into subsequent iterations. While the tasks given in the examples can be achieved with various procedures, the goal is to build an understanding of how to expand use the DATA step as a computational tool.

7.4.1 RETAIN Statement

Recall that, by default, during DATA step processing most variables in the PDV are reset—or reinitialized—to missing values at the top of each DATA step iteration. This applies to variables read in (from a raw file or SAS data set) or new variables created during the current DATA step. One of the effects of this automatic reset is that these values are only available for computations within the current record, thus computations using information from multiple records require special statements, functions, or options. Program 7.4.1 demonstrates the results of a typical naïve attempt at calculating a running total and compares it to a correct approach using the RETAIN statement.

Program 7.4.1: Comparing Attempts for Calculating Running Totals

```
data work.ipums2005Utility;
  infile RawData('Utility Cost 2005.txt') dlm='09'x dsd firstobs=4;
  input serial electric:comma. gas:comma. water:comma. fuel:comma.;
  format electric gas water fuel dollar.;
  if electric ge 9000 then electric=.;
  if gas ge 9000 then gas=.;
  if water ge 9000 then water=.;
  if fuel ge 9000 then fuel=.;
run;
```

```
proc transpose data= work.ipums2005Utility
                out= work.Utility2005Vert(rename=(col1=Cost _name_=Utility));
  by serial;
  var electric--fuel;
run;

data work.Totals;
  set work.Utility2005Vert;
  Tot1 = Tot1 + Cost; ❶
  Tot2 = sum(Tot2, Cost); ❷
  retain❸ Tot3 0❹;
  Tot3 = sum(Tot3, Cost); ❺
run;

proc print data = work.Totals(obs = 5);
run;
```

❶ As shown in Output 7.4.1 below, this naïve approach results in a value of Tot1 that is always missing. Tot1 is created as the target variable in an assignment statement, so the DATA step initializes it to missing at each iteration. As a result, this statement is always equivalent to adding a missing value to the current value of Cost. When using the addition operator, or any arithmetic operator, the default is for operations including a missing value to result in missing values. So regardless of the value of Cost, Tot1 is always assigned a missing value.

❷ The SUM function is shown here only to contrast with the use of the addition operator. The default behavior of the SUM function is that operations including missing values only result in missing values if all input values are missing. Like Tot1, Tot2 is reinitialized to missing on every iteration, so its value is the same as adding the current value of Cost to missing with SUM, resulting in the current value of Cost on each record.

❸ Variables that appear in the RETAIN statement are not reset between iterations of the DATA step—SAS preserves the value for use during the next iteration. The RETAIN statement cannot affect variables read from files such as SAS data sets read via SET, MERGE, MODIFY, or UPDATE statements or variables read using the INPUT statement, as the reading process overwrites any potentially retained value. While RETAIN is a compile-time statement—its position in the DATA step does not impact execution—it is positioned here to ensure Tot3 appears after Tot1 and Tot2 in the data set.

❹ This RETAIN statement also provides an initial value of 0 for the variable Tot3. When retaining a variable, it is expected that its value is explicitly set at least once, otherwise unexpected results can occur, including notes indicating the variable is uninitialized. (A semantic error.)

❺ Now that the value of Tot3 is retained across iterations of the DATA step. The SUM function is more appropriate than the addition operator for calculating the running total to prevent Tot3 from being missing if Cost is missing.

Output 7.4.1 compares the values of Tot1, Tot2, and Tot3 and, from the results, it is clear that the naïve approaches used to create Tot1 and Tot2 are ineffective as they did not use the RETAIN statement. However, the combination of the RETAIN statement to preserve the value across iterations and the SUM function to accommodate potentially missing values of Cost are successful.

Output 7.4.1: Comparing Running Total Variables from Program 7.4.1

Obs	serial	Utility	Cost	Tot1	Tot2	Tot3
1	2	electric	$1,800	.	1800	1800
2	2	gas	.	.	.	1800
3	2	water	$700	.	700	2500
4	2	fuel	$1,300	.	1300	3800
5	3	electric	$1,320	.	1320	5120

Program 7.4.1 generates a note in the log, shown in Log 7.4.1 below. As with other notes informing the programmer about internal SAS decisions, such as when implicit data conversion occurs, SAS only prints the note once per DATA step. To determine the source of the note and the number of records that contributed, SAS provides the number of incidences along with the associated line and column numbers. Note that the line numbers SAS references in Log 7.4.1 do not correspond to the line numbers in the submitted program, but instead to the line numbers in the SAS log itself.

In the case of Program 7.4.1, missing values resulted from operations on missing values 4,636,248 times due to 15th column in the assignment statement in Line 149, corresponding to the addition operator. Similarly, Line 150 contributed 1,869,695 times due to its 10th column, where the SUM function is positioned. While no change occurs in the resulting data set in Program 7.4.1, the explicit handling of missing values, data conversion, or other note-generating statements via conditional logic is often considered a good programming practice since it provides clear communication to other programmers updating the code and prevents SAS from generating such notes.

Log 7.4.1: Partial Log from Program 7.4.1—Note Regarding Operations on Missing Values

```
147   data work.Totals;
148     set Utility2005Vert;
149     Tot1 = Tot1 + Cost;
150     Tot2 = sum(Tot2, Cost);
151     retain Tot3 0;
152     Tot3 = sum(Tot3, Cost);
153   run;

NOTE: Missing values were generated as a result of performing an operation on
      missing values.
      Each place is given by: (Number of times) at (Line):(Column).
      4636248 at 149:15   1869695 at 150:10
```

Another common use of the RETAIN statement is to compare values across records, such as determining the maximum or minimum value observed up to and including the current record. In clinical trials, where subjects are often measured at multiple time points, techniques such as best observation carried forward (BOCF), last observation carried forward (LOCF), and worst observation carried forward (WOCF) are used regularly and require the retention and comparison of values within a subject. Program 7.4.2 demonstrates a comparison across records in a data set using the RETAIN statement.

Program 7.4.2: Using a RETAIN Statement to Compare Values Across Records

```
data work.WOCF;
  set work.Utility2005Vert;
  retain MaxCost 0 MaxUtility; ❶
  if Cost ge MaxCost ❷ then do; ❸
     MaxUtility = Utility; ❹
     MaxCost = Cost; ❹
   end;
run;

proc print data = work.WOCF(obs = 6);
  var serial Utility Cost MaxUtility MaxCost;
run;
```

❶ Multiple variables are valid in a single RETAIN statement. As with LABEL, FORMAT, and other similar statements, it is a good programming practice to only have one RETAIN statement per DATA step.

❷ The IF statement compares the current values of Cost and MaxCost. Since MaxCost is initialized to zero and the RETAIN statement holds its previous value in the PDV, the current value of MaxCost is the largest of all the previous values of Cost, so the condition is only met when the current value of Cost is a new maximum.

❸ When the IF condition is met and a new maximum is found, multiple updates are made. Recall the DO group is necessary to execute multiple statements when the IF condition is met.

❹ To set the new maximum, the current values of Utility and Cost are copied into the variables MaxUtility and MaxCost, respectively. Since both variables appear in the RETAIN statement, these values are available for future iterations. Note that even though no initial value is present for MaxUtility in the RETAIN statement, the assignment statement in this DO group ensures an initial value is provided.

Output 7.4.2 shows the results of tracking the highest observed cost and its associated utility category. When the sixth observation occurs, the assignment statements update the retained value of MaxUtility and MaxCost to reflect the new maximum utility cost and category. Note that while the DATA step does provide access to the MAX function, it operates across columns and not across rows. As such, the MAX function cannot produce an equivalent result with the data structured vertically.

Output 7.4.2: Carrying Forward the Highest Observed Cost

Obs	serial	Utility	Cost	MaxCost	MaxUtility
1	2	electric	$1,800	1800	electric
2	2	water	$700	1800	electric
3	2	fuel	$1,300	1800	electric
4	3	electric	$1,320	1800	electric
5	3	gas	$2,040	2040	gas
6	3	water	$120	2040	gas

All records must be checked to ensure the maximum is found. However, the last record contains all the required information and, for the sake of efficiency, is the only one that should be output. SAS provides multiple methods to identify the final record in the data set, and Program 7.4.3 demonstrates outputting the final record with the commonly used END= option.

Program 7.4.3: Selecting the Final Observation with END=
```
data work.WOCFLast;
  set work.Utility2005Vert end = Last; ❶
  retain MaxCost 0 MaxUtility;
  if Cost ge MaxCost then do;
     MaxUtility = Utility;
     MaxCost = Cost;
   end;
  if Last = 1; ❷
run;

proc report data = work.WOCFLast;
  columns MaxCost MaxUtility; ❸
  define MaxCost    / display 'Highest Cost';
  define MaxUtility / display 'Utility';
run;
```

❶ The END= option adds a user-named temporary variable to the PDV, which takes the value 1 only for the last record in the data set and the value 0 for all other records. Because the variable is temporary, the DATA step does not write it to the output data set. The variable name must follow naming conventions and should be a name that is otherwise not in use during the DATA step.

❷ The END= option from ❶ names the temporary variable Last, so this subsetting IF statement only allows the final record in the data set to be output. The OBS= data set option is no longer needed in PROC REPORT to limit the output produced since the WOCFLast data set only contains a single record.

❸ Because only the final record appears in the data set, the Serial, Utility, and Cost values of the final record are not of interest. It is a good programming practice to drop them as they are spurious values held over from the last record.

Output 7.4.3 shows the highest value of Cost—stored in the variable MaxCost—and the associated utility— stored in the variable MaxUtility.

Output 7.4.3: Viewing the Maximum Value and Corresponding Utility for the Full Data Set

Highest Cost	Utility
7080	gas

7.4.2 DATA Step Sum Statement

The first demonstration of the RETAIN statement in Section 7.4.1 is in the creation of a running total by simultaneously using a SUM function with a variable that appears in the RETAIN statement. This particular use of the RETAIN statement—carrying out an iterative calculation across successive records—is so common that SAS provides a streamlined method for its implementation: the DATA step sum statement. Program 7.4.4 demonstrates its use and extends the results of Program 7.4.1.

Program 7.4.4: Using a DATA Step Sum Statement to Create a Running Total

```
data work.DataSum;
  set work.Utility2005Vert;
  Tot1 = Tot1 + Cost;
  Tot2 = sum(Tot2, Cost);
  retain Tot3 0;
  Tot3 = sum(Tot3, Cost);
  ❶ Tot4 +❷ Cost❸;
run;

proc print data = work.DataSum(obs = 5);
run;
```

❶ The DATA step sum statement does not begin with a keyword. Instead, it begins with the name of a variable referred to as an accumulator variable. To ensure the statement operates as intended, the accumulator variable should not appear in a data set listed in the SET, MERGE, UPDATE, or MODIFY statements or be modified by any other programming statements.

❷ The accumulator variable must be followed by the addition operator.

❸ After the addition operator, include any valid expression that SAS can resolve to a numeric value.

As Output 7.4.4 shows, the accumulator variable Tot4 contains the same values as Tot3. SAS issues an implicit RETAIN statement for all accumulator variables and initializes them to zero. Furthermore, operations on missing values no longer result in missing values, so the SAS log after the submission of Program 7.4.4 contains no missing value notes related to Tot4.

Output 7.4.4: DATA Step Sum Statement Versus Other Running Total Calculation Methods

Obs	serial	Utility	Cost	Tot1	Tot2	Tot3	Tot4
1	2	electric	$1,800	.	1800	1800	1800
2	2	water	$700	.	700	2500	2500
3	2	fuel	$1,300	.	1300	3800	3800
4	3	electric	$1,320	.	1320	5120	5120
5	3	gas	$2,040	.	2040	7160	7160

One of the most common uses of the DATA step sum statement is to include a counter when operating on a data set. When using conditional loops such as in Program 7.3.11, there is no built-in index to track the number of times the loop iterates. Program 7.4.5 revisits Program 7.3.11 and demonstrates one possible method to track the number of nonmissing utility costs associated with each value of Serial.

Program 7.4.5: Using and Resetting an Accumulator Variable

```
data work.UtilityE2005v2;
   infile rawdata('Utility2005ComplexE.txt') dsd missover;
   input serial Utility : $11. cost : dollar7. @;
   do while(not missing(cost));
      select(Utility);
         when('Electricity') electric = cost;
         when('Gas') gas = cost;
         when('Water') water = cost;
         when('Fuel') fuel = cost;
         otherwise;
      end;
      UtilityCount + 1; ❶
      input Utility: $11. cost : dollar7. @;
   end;

   output;
   call missing(UtilityCount); ❷
   drop Utility cost;
run;

proc report data = work.UtilityE2005v2(obs = 4);
   columns serial Electric Gas Water Fuel UtilityCount;
   define serial       / display 'Serial';
   define Electric      / display format= dollar7. 'Electric Cost';
   define Gas           / display format= dollar7. 'Gas Cost';
   define Water         / display format= dollar7. 'Water Cost';
   define Fuel          / display format= dollar7. 'Fuel Cost';
   define UtilityCount / display;
run;
```

❶ The DATA step sum statement inside the DO WHILE loop creates the accumulator variable UtilityCount. Each time the loop iterates—while Cost is nonmissing—this statement increments the value of UtilityCount by one.

❷ To prevent the value from incrementing across all records in the data set, the CALL MISSING statement invokes the MISSING routine which, as expected, sets the value of UtilityCount to missing. This statement appears after the OUTPUT statement to ensure the value of UtilityCount is only set to missing after the DATA step writes the current PDV to the data set and before the next iteration begins.

Output 7.4.5 is similar to Output 7.3.10 but it includes the new UtilityCount column. Note that once the record is transposed, it is possible to calculate the same value by simply counting the number of nonmissing utility-specific variables within a record. However, the approach using the DATA step sum statement is common due to its simplicity.

Output 7.4.5: Using and Resetting an Accumulator Variable

Serial	Electric Cost	Gas Cost	Water Cost	Fuel Cost	UtilityCount
2	$1,800	.	$700	$1,300	3
3	$1,320	$2,040	$120	.	3
4	$1,440	$120	.	.	2
5	.	.	.	.	.

7.4.3 BY-Group Processing

BY-group processing is the method SAS uses to take advantage of SAS data sets that are already sorted—or at least grouped—by one or more variables. BY-group processing is available in the DATA step, such as to carry out an interleave or a match-merge, or in PROC steps, such as with PROC TRANSPOSE to exchange row and column information separately for each BY group. When a BY statement appears in a DATA step, it adds temporary variables to, and changes the way SAS resets, the PDV. These changes provide another way to work across records in a DATA step.

When SAS encounters a BY statement in the DATA step, it automatically tracks whether a record is the first or last observation in its BY group using two automatic variables—FIRST.*variable* and LAST.*variable*—for each variable in the BY statement. The values of FIRST.*variable* and LAST.*variable* are either 0 or 1, similar to the temporary variable created by the END= option or the IN= data set option. Program 7.4.6 demonstrates BY-group processing to calculate separate running totals for each unique value of Serial.

Program 7.4.6: Demonstrating FIRST. and LAST. Variable Usage

```
data work.Subtotals;
  set work.Utility2005Vert;
  by serial; ❶

  if first.serial eq 1 then call missing(TotalCost); ❷
  TotalCost + Cost; ❸

  first = first.serial; ❹
  last = last.serial; ❹
run;

proc report data = work.subtotals(obs = 6);
  columns serial Utility cost totalcost first last;
  define serial      / display 'Serial';
  define Utility      / display 'Utility';
  define Cost         / display 'Cost';
  define TotalCost    / display 'Cumulative Utility ';
  define first        / display 'FIRST.Serial';
  define last         / display 'LAST.Serial';
run;
```

❶ The BY statement invokes BY-group processing. For each variable in the BY statement, SAS creates a FIRST.*variable* and LAST.*variable* as temporary variables.

❷ This IF-THEN statement resets the value of TotalCost to missing at the beginning of each new BY group—indicated by a value for FIRST.Serial of 1. Here Serial is the only BY variable, if multiple variables are used, setting of FIRST. and LAST. variables is more complex, a concept discussed in the next example.

❸ The DATA step sum statement creates the accumulator variable TotalCost and uses it to track the cumulative total for Cost.

❹ Since FIRST.*variable* and LAST.*variable* are temporary variables, these assignment statements save them as user-defined variables, so SAS writes them to the Subtotals data set. This is done here so PROC REPORT can display the values for this example.

As Output 7.4.6 shows, the value of FIRST.Serial is always 1 for the first record in the BY group and 0 for all other records. Similarly, LAST.Serial is 1 only for the last record in the BY group and 0 for all other records. By extension, any record that is neither first nor last in its BY group has values of 0 for both FIRST.Serial and LAST.Serial. Any BY group consisting of a single observation—such as a value of Serial associated with only a single value of Utility—would have values of 1 for both FIRST.Serial and LAST.Serial. As the IF-THEN statement in Program 7.4.6 demonstrates, one of the most common uses of FIRST.*variable* and LAST.*variable* is to conditionally carry out statements for records based on their position within the BY group.

Output 7.4.6: Displaying the Values of FIRST.Serial and LAST.Serial

Serial	Utility	Cost	Cumulative Expense	FIRST.Serial	LAST.Serial
2	electric	$1,800	1800	1	0
2	water	$700	2500	0	0
2	fuel	$1,300	3800	0	1
3	electric	$1,320	1320	1	0
3	gas	$2,040	3360	0	0
3	water	$120	3480	0	1

The method BY-group processing uses to assign values to FIRST.*variable* and LAST.*variable* is rather intuitive when only a single BY variable is present. To explore how SAS assigns values to these variables when two or more BY variables are present, Program 7.4.7 constructs a BY grouping on four variables in the Sashelp.Cars data set.

Program 7.4.7: Investigating FIRST.*variable* and LAST.*variable* Values for Multiple BY Variables

```
proc sort data=sashelp.cars out= work.cars;
  by make type drivetrain Origin; ❶
run;

data work.carsFirstLast;
  set work.cars;
  by make type drivetrain; ❶

  FMake=first.make; ❷
  LMake=last.make; ❷
  FType=first.type; ❷
  LType=last.type; ❷
  FDrive=first.drivetrain; ❷
  LDrive=last.drivetrain; ❷

  keep make model type drivetrain FMake--LDrive;
  rename drivetrain=drive;
run;

proc print data= work.carsFirstLast(firstobs=24 obs=28);
run;

proc print data= work.carsFirstLast(firstobs=217 obs=221);
run;
```

❶ As with other uses of the BY statement, BY-group processing does not require identical BY statements. (See Program 2.3.6.) However, if the BY groups are not in order, SAS issues an error as soon as it encounters nonsequential BY values. See Chapter Notes 3 and 4 in Section 7.8 for a discussion of how to use the GROUPFORMAT or NOTSORTED options.

❷ As in Program 7.4.6, assignment statements store the temporary FIRST.*variable* and LAST.*variable* values for each of the six variables.

Outputs 7.4.7A and 7.4.7B show portions of the sorted data set CarsFirstLast and each of the FIRST.*variable* and LAST.*variable* values. In general, when the value of any BY variable changes, SAS assigns a value of 1 to its associated FIRST.*variable* value and to all FIRST.*variable* values associated with any variable that follow this updated BY variable in the BY statement. In the case of Program 7.4.7, in addition to the process described above for Make, if the value of Type changes then SAS sets FIRST.Type and FIRST.Drivetrain to 1. If the value of Drivetrain changes, then SAS only updates FIRST.Drivetrain. A similar process sets each of the LAST.*variable* values when a BY variable changes in the next record.

Output 7.4.7A: Investigating FIRST.*variable* and LAST.*variable* Values (Records 24 through 28)

Obs	Make	Model	Type	drive	FMake	LMake	FType	LType	FDrive	LDrive
24	Audi	TT 1.8 convertible 2dr (coupe)	Sports	Front	0	0	0	1	0	1
25	Audi	A6 3.0 Avant Quattro	Wagon	All	0	0	1	0	1	0
26	Audi	S4 Avant Quattro	Wagon	All	0	1	0	1	0	1

Obs	Make	Model	Type	drive	FMake	LMake	FType	LType	FDrive	LDrive
27	BMW	X3 3.0i	SUV	All	1	0	1	0	1	0
28	BMW	X5 4.4i	SUV	All	0	0	0	1	0	1

Output 7.4.7B: Investigating FIRST.*variable* and LAST.*variable* Values (Records 217 through 221)

Obs	Make	Model	Type	drive	FMake	LMake	FType	LType	FDrive	LDrive
217	Land Rover	Discovery SE	SUV	All	0	0	0	0	0	0
218	Land Rover	Freelander SE	SUV	All	0	1	0	1	0	1
219	Lexus	GX 470	SUV	All	1	0	1	0	1	0
220	Lexus	LX 470	SUV	All	0	0	0	0	0	0
221	Lexus	RX 330	SUV	All	0	0	0	1	0	1

When working with complex BY-group processing scenarios, remember that assignment statements can store the values to help determine the correct usage of the BY groups. One application of BY-group processing with multiple BY variables is shown in Program 7.4.8 to simultaneously find the maximum home value and total home value for various combinations of levels of State, Metro, and MortgageStatus in the IPUMS2005Basic data set.

Program 7.4.8: Determining Total and Maximum Home Value within Subgroups

```
proc sort data=bookdata.ipums2005basic out= work.Basic2005;
  by state metro mortgageStatus;
  where HomeValue lt 9999999;
run;

data work.HomeValueStats(drop = homeValue);
  set work.Basic2005;
  by state metro mortgageStatus; ❶
  retain MaxValue .; ❷

  if first.mortgageStatus then call missing(TotalValue, MaxValue); ❸
  TotalValue+HomeValue; ❹
  if HomeValue gt MaxValue then MaxValue = HomeValue; ❺
  if last.mortgageStatus; ❻
run;

proc report data = work.HomeValueStats(obs = 6);
  columns state metro mortgageStatus MaxValue TotalValue;
  define state          / order 'State'; ❼
  define metro          / order 'Metro'; ❼
  define mortgageStatus / order 'Mortgage Status'; ❼
  define MaxValue       / display 'Maximum Home Value' format=dollar10.;
  define TotalValue     / display 'Total Home Value' format=dollar10.;
run;
```

❶ Use a BY statement to invoke BY-group processing on the sorted data set.

❷ As in previous programs, retain and explicitly initialize a variable to hold the maximum home value.

❸ Since the values of TotalValue and MaxValue are to be determined for each distinct combination of State, Metro, and MortgageStatus, their values must be reset at the start of each BY group. The BY group changes each time the variable lowest in the sort hierarchy has its FIRST. variable set to 1, so this IF-THEN statement conditions on FIRST.MortgageStatus.

❹ Use a DATA step sum statement to calculate the cumulative values of HomeValue.

⑤ Use an IF-THEN statement to compare current values of HomeValue and MaxValue and update MaxValue when appropriate.

⑥ The subsetting IF only returns the final value in each BY group.

⑦ The ORDER usage suppresses repeated printing of BY values from the DATA step.

Output 7.4.8 shows the cumulative totals and maximums are reset between groups, as expected. If additional values, such as State- or Metro-level summary values are of interest, then including additional conditional logic statements is a simple modification.

Output 7.4.8: Viewing the Within-Group Maximum and Total HomeValue Amounts

State	Metro	Mortgage Status	Maximum Home Value	Total Home Value
Alabama	0	No, owned free and clear	$1,000,000	$62,742,500
		Yes, contract to purchase	$350,000	$1,850,000
		Yes, mortgaged/ deed of trust or similar debt	$875,000	$70,177,500
	1	No, owned free and clear	$1,000,000	$191,345,000
		Yes, contract to purchase	$350,000	$4,037,500
		Yes, mortgaged/ deed of trust or similar debt	$875,000	$191,047,500

7.5 Customizations in the REPORT Procedure

This section explores tools to expand the overall utility of the REPORT procedure. Methods to set styles at various levels of the report, all the way down to the individual cells, are discussed and include changing of fonts, colors, sizes, and other attributes. Computations within PROC REPORT are also covered, which encompasses construction of new columns, inserting custom text, and setting styles.

7.5.1 Defining Styles in PROC REPORT

A wide variety of style attributes can be modified in the code when using the REPORT procedure, Style definitions can be included in several different statements in PROC REPORT, and the styles can be targeted to distinct locations in the table. Program 7.5.1 modifies the result of Program 6.6.9, adding STYLE options in the PROC REPORT statement.

Program 7.5.1: Adding Styles in the PROC REPORT Statement

```
proc format;
  value $Mort_Status
    'No'-'Nz'='No'
    'Yes'-'Yz'='Yes'
  ;
run;

proc report data= work.ipums2005cost❶
            style(header)=[fontfamily='Arial Black' backgroundcolor=gray55
                          color=white]❷
            style(column)=[fontfamily='Georgia' backgroundcolor=grayDD
                          fontsize=10pt]❸
            style(summary)=[backgroundcolor=grayAA fontweight=bold
                          fontstyle=italic]❹;
  where state in ('North Carolina','South Carolina');
  column state mortgageStatus electric,(n median mean std);
  define state / group 'State';
  define mortgageStatus / group 'Mortgage Status' format=$Mort_Status.;
  define electric / '';
  define n / 'Number of Observations' format=comma8.;
  define median / 'Median Electricity Cost' format=dollar10.;
  define mean / 'Mean Electricity Cost' format=dollar10.;
```

```
    define std / 'Standard Deviation' format=dollar10.;
    break after state / summarize;
    rbreak after / summarize;
run;
```

❶ See Program 5.3.5 for the code that generates this data set.

❷ The general form of the style option is: STYLE(*location*)=[*style-element1=style1 style-element2=style2* ...].
   Here, the HEADER target applies to the table cells that label the columns. Those cells are assigned text in
   the Arial Black font (FONTFAMILY=), with a text color of white (COLOR=), and a relatively dark gray
   background color (BACKGROUNDCOLOR=). Any attribute not chosen in a STYLE= option inherits its default
   value—for tables displayed in the text, this is determined by the CustomSapphire style. (See Chapter Note
   9 in section 1.7.)

❸ This style option uses the COLUMN location where the data/summary values reside. The FONTSIZE=
   option is included here, which allows for standard units for font sizing as seen with graphics. (See Section
   3.4.)

❹ The SUMMARY location is all rows generated by BREAK or RBREAK statements. FONTWEIGHT= and
   FONTSTYLE= are used to set the font to be bold and italic, respectively, and the background color for the
   cells is set to a somewhat darker gray than the rest of the data column. The 10-point Georgia font that
   appears on the break lines is inherited from the attributes set in the options in ❸ targeting the COLUMN
   location—break lines are part of the data columns.

Output 7.5.1: Adding Styles in the PROC REPORT Statement

State	Mortgage Status	Number of Observations	Median Electricity Cost	Mean Electricity Cost	Standard Deviation
North Carolina	N/A	8,548	$1,080	$1,327	$834
	No	9,382	$1,200	$1,416	$803
	Yes	17,303	$1,440	$1,667	$892
North Carolina		*35,233*	*$1,320*	*$1,518*	*$868*
South Carolina	N/A	3,807	$1,200	$1,426	$862
	No	5,074	$1,320	$1,543	$880
	Yes	8,283	$1,560	$1,766	$898
South Carolina		*17,164*	*$1,440*	*$1,624*	*$896*
		52,397	*$1,320*	*$1,553*	*$879*

In the PROC REPORT statement other locations available are REPORT, LINES, and CALLDEF, with REPORT being
the default if the location is omitted. The concepts of LINES and CALLDEF (short for CALL DEFINE) are covered
later in this section. As seen with the relationship between the COLUMN and SUMMARY locations in Output
7.5.1—SUMMARY inherits COLUMN styles that are not set—it is useful to think of the styles set in PROC
REPORT as a styling hierarchy. The style template (again, CustomSapphire for default output in this text) is at
the top of the hierarchy, establishing a full set of styles for the output table. Styles set in the PROC REPORT
line update the selected attributes in the chosen location throughout the report. Styles can be set in DEFINE
statements, updating style attributes for the column corresponding to it, and in BREAK or RBREAK statements,
updating styles in the summary lines generated by those statements. Program 7.5.2 includes specific styling
for the State column in its DEFINE statement and the whole-report summary line in the RBREAK statement to
demonstrate locations available for each and show inheritance of style attributes.

Program 7.5.2: Adding Styles in DEFINE, BREAK, or RBREAK Statements

```
proc report data= work.ipums2005cost
            style(header)=[fontfamily='Arial Black' backgroundcolor=gray55
                           color=white]
            style(column)=[fontfamily='Georgia' backgroundcolor=grayDD fontsize=10pt]
            style(summary)=[backgroundcolor=grayAA fontweight=bold fontstyle=italic];
  where state in ('North Carolina','South Carolina');
  column state mortgageStatus electric,(n median mean std);
  define state / group 'State' style(header)=[fontsize=14pt] ❶
                 style(column)=[fontfamily='Arial' backgroundcolor=white] ❷;
  define mortgageStatus / group 'Mortgage Status' format=$Mort_Status.;
  define electric / '';
  define n / 'Number of Observations' format=comma8.;
  define median / 'Median Electricity Cost' format=dollar10.;
  define mean / 'Mean Electricity Cost' format=dollar10.;
  define std / 'Standard Deviation' format=dollar10.;
  break after state / summarize;
  rbreak after / summarize style=[backgroundcolor=gray88] ❸;
run;
```

❶ The HEADER location in the DEFINE statement still refers to the column label cell, but only for the column referenced by the DEFINE statement. In Output 7.5.2, the font size is set to 14 point only for the label of State in the first column.

❷ The COLUMN location refers to the data in the column, but again only for the column referenced by the DEFINE statement. Note that the styles applied by the STYLE(SUMMARY)= option in the PROC REPORT statement override the corresponding style settings (BACKGROUNDCOLOR=) for this column on the break lines. HEADER and COLUMN are the only locations available for styling in a DEFINE statement, if the location is omitted the default is both locations simultaneously.

❸ The styling locations available for a BREAK or RBREAK statement are LINES and SUMMARY, with both locations being the default if no location is given (the concept of LINES is covered later in this section). Here the background color change only applies to the whole-report summary line, and all other attributes for that line are inherited from the STYLE(SUMMARY)= option in the PROC REPORT statement and the default template styles.

Output 7.5.2: Adding Styles in DEFINE, BREAK, or RBREAK Statements

State	Mortgage Status	Number of Observations	Median Electricity Cost	Mean Electricity Cost	Standard Deviation
North Carolina	N/A	8,548	$1,080	$1,327	$834
	No	9,382	$1,200	$1,416	$803
	Yes	17,303	$1,440	$1,667	$892
North Carolina		*35,233*	*$1,320*	*$1,518*	*$868*
South Carolina	N/A	3,807	$1,200	$1,426	$862
	No	5,074	$1,320	$1,543	$880
	Yes	8,283	$1,560	$1,766	$898
South Carolina		*17,164*	*$1,440*	*$1,624*	*$896*
		52,397	*$1,320*	*$1,553*	*$879*

In addition to sizes for text, physical sizes of table cells can also be set. In Output 7.5.1, the choice of font and style has the state names wrapping on the break lines because of the limited space in the cell. Program 7.5.3 modifies the width of the columns to allow the state names to span across on the break lines, while still allowing the table to fit within the page margins.

Program 7.5.3: Changing Physical Sizes Using Style Options

```
proc report data= work.ipums2005cost
              style(header)=[fontfamily='Arial Black' backgroundcolor=gray55
                         color=white fontsize=10pt]
              style(column)=[fontfamily='Georgia' backgroundcolor=grayDD fontsize=10pt
                         cellwidth=.90in❶]
              style(summary)=[backgroundcolor=grayAA fontweight=bold fontstyle=italic];
  where state in ('North Carolina','South Carolina');
  column state mortgageStatus electric,(n median mean std);
  define state / group 'State' style(column)=[cellwidth=1.2in]❷;
  define mortgageStatus / group 'Mortgage Status' format=$Mort_Status.;
  define electric / '';
  define n / 'Number of Observations' format=comma8.
              style(column)=[cellwidth=1.1in]❸;
  define median / 'Median Electricity Cost' format=dollar10.;
  define mean / 'Mean Electricity Cost' format=dollar10.;
  define std / 'Standard Deviation' format=dollar10.;
  break after state / summarize;
  rbreak after / summarize style=[backgroundcolor=gray88];
run;
```

❶ As expected, the CELLWIDTH= option controls the width of the column. In this case, it is applied to the COLUMN location, but its effect is on the whole column—header and data portions.

❷ Here, a width of 1.2 inches is applied to the COLUMN location for the State column. It is important to match this location to the one used in the PROC REPORT line so it properly overrides for this column. The default location (both COLUMN and HEADER) can also be used.

❸ The width of the column for the frequency statistic is also expanded, this is to accommodate the size of the column label. The table in Output 7.5.3 was designed to fit on a standard 8.5-inch-wide page with left and right margins of 1.25 inches, or 6 inches of available width. With four columns at a width of 0.9 inches, one at 1.2 inches, and another at 1.1 inches, the total width of the cells is 5.9 inches—*not counting the borders*. Many components need to be coordinated when trying to size a table: fonts, font sizes, font weights, borders, and other attributes.

Output 7.5.3: Changing Physical Sizes Using Style Options

State	Mortgage Status	Number of Observations	Median Electricity Cost	Mean Electricity Cost	Standard Deviation
North Carolina	N/A	8,548	$1,080	$1,327	$834
	No	9,382	$1,200	$1,416	$803
	Yes	17,303	$1,440	$1,667	$892
North Carolina		*35,233*	*$1,320*	*$1,518*	*$868*
South Carolina	N/A	3,807	$1,200	$1,426	$862
	No	5,074	$1,320	$1,543	$880
	Yes	8,283	$1,560	$1,766	$898
South Carolina		*17,164*	*$1,440*	*$1,624*	*$896*
		52,397	*$1,320*	*$1,553*	*$879*

A style attribute does not need to be set to a single, explicit value, it can be defined conditionally using a format. Program 7.5.4 uses formats to change font colors and weights based on the data value in the table cell.

Program 7.5.4: Setting Styles Conditionally Using Formats

```
proc format;
  value costR 1500-high=cxFF0000; ❶
  value Rbold 1500-high=bold; ❷
run;

proc report data= work.ipums2005cost
            style(header)=[fontfamily='Arial Black' backgroundcolor=gray55
                           color=white]
            style(column)=[fontfamily='Georgia' backgroundcolor=grayDD fontsize=10pt]
            style(summary)=[backgroundcolor=grayAA fontweight=bold fontstyle=italic];
  where state in ('North Carolina','South Carolina');
  column state mortgageStatus electric,(n median mean std);
  define state / group 'State';
  define mortgageStatus / group 'Mortgage Status' format=$Mort_Status.;
  define electric / '';
  define n / 'Number of Observations' format=comma8.;
  define median / 'Median Electricity Cost' format=dollar10.
            style(column)=[color=costR. fontweight=Rbold.]❸;
  define mean / 'Mean Electricity Cost' format=dollar10.
            style(column)=[color=costR. fontweight=Rbold.]❸;
  define std / 'Standard Deviation' format=dollar10.;
  break after state / summarize;
  rbreak after / summarize;
run;
```

❶ Here, an RGB color code corresponding to a bright red is assigned to the value range of 1,500 and above, with no assignment to any other values. Any legal color name or code can be placed on the right side of the equal sign for any value range to make conditional color assignments. The costR name is chosen in an effort to indicate that this format is used to make certain cost values appear in red.

❷ Values assigned to format values or ranges are not limited to colors—any style attribute value is permissible provided it is later assigned to a style attribute for which it is legal. Here the value range of 1,500 and above is also assigned to a value of bold, which is appropriate for use in the FONTWEIGHT= attribute.

❸ In the styling for the COLUMN location for both the mean and median statistics, the costR format is used to change values of 1500 or over to red by assigning it to the COLOR= attribute. Any number shown in red is also given a bold weight by using the Rbold format in the FONTWEIGHT= attribute. All other values remain in their previously assigned color and weight.

Output 7.5.4: Setting Styles Conditionally Using Formats

State	Mortgage Status	Number of Observations	Median Electricity Cost	Mean Electricity Cost	Standard Deviation
North Carolina	N/A	8,548	$1,080	$1,327	$834
	No	9,382	$1,200	$1,416	$803
	Yes	17,303	$1,440	$1,667	$892
North Carolina		*35,233*	*$1,320*	*$1,518*	*$868*
South Carolina	N/A	3,807	$1,200	$1,426	$862
	No	5,074	$1,320	$1,543	$880
	Yes	8,283	$1,560	$1,766	$898
South Carolina		*17,164*	*$1,440*	*$1,624*	*$896*
		52,397	*$1,320*	*$1,553*	*$879*

The broadest set of styles available is through a style template. Program 7.5.5 uses the Journal template supplied by SAS to set the styles, with Output 7.5.5 showing the result.

Program 7.5.5: Using an Existing Style Template

```
ods rtf file='Using the Journal Template.rtf' style=journal;
proc report data= work.ipums2005cost;
  where state in ('North Carolina','South Carolina');
  column state mortgageStatus electric,(n median mean std);
  define state / group 'State';
  define mortgageStatus / group 'Mortgage Status' format=$Mort_Status.;
  define electric / '';
  define n / 'Number of Observations' format=comma8.;
  define median / 'Median Electricity Cost' format=dollar10.;
  define mean / 'Mean Electricity Cost' format=dollar10.;
  define std / 'Standard Deviation' format=dollar10.;
  break after state / summarize;
  rbreak after / summarize;
run;
ods rtf close;
```

Output 7.5.5: Using an Existing Style Template

State	Mortgage Status	Number of Observations	Median Electricity Cost	Mean Electricity Cost	Standard Deviation
North Carolina	N/A	8,548	$1,080	$1,327	$834
	No	9,382	$1,200	$1,416	$803
	Yes	17,303	$1,440	$1,667	$892
North Carolina		*35,233*	*$1,320*	*$1,518*	*$868*
South Carolina	N/A	3,807	$1,200	$1,426	$862
	No	5,074	$1,320	$1,543	$880
	Yes	8,283	$1,560	$1,766	$898
South Carolina		*17,164*	*$1,440*	*$1,624*	*$896*
		52,397	*$1,320*	*$1,553*	*$879*

If STYLE= options are added to Program 7.5.5, any style attributes not set inherit their values from the Journal style, as opposed to the results of Program 7.5.1 through 7.5.4 inheriting from CustomSapphire. To explore lists of available style templates, PROC TEMPLATE can be used—see Chapter Note 5 in Section 7.8 for a sample and further information.

7.5.2 Computing New Variables in PROC REPORT

While the DATA step is available to compute new variables from existing data, PROC REPORT also allows for new values to be computed with COMPUTE blocks. Program 7.5.6 modifies and extends Program 6.6.9 to include a column that gives the ratio of the mean to median electricity costs in each row.

Program 7.5.6: Computing a New Column in PROC REPORT

```
proc report data= work.ipums2005cost;
  where state in ('North Carolina','South Carolina');
  column state mortgageStatus electric=num electric=mid electric=avg
         ratio❶ electric=std;
  define state / group 'State';
  define mortgageStatus / group 'Mortgage Status' format=$Mort_Status.;
  define electric / '';
  define num / n 'Number of Observations' format=comma8.;
  define mid / median 'Median Electricity Cost' format=dollar10.;
  define avg / mean 'Mean Electricity Cost' format=dollar10.;
  define ratio / computed 'Mean to Median Ratio' format=percent9.2; ❶
  define std / std 'Standard Deviation' format=dollar10.;
  break after state / summarize;
  rbreak after / summarize;
```

```
compute ratio; ❷
  ratio=avg/mid; ❸
endcomp; ❹
run;
```

❶ A new report item, Ratio, is added to the COLUMN statement to create a basis for defining the column where the ratios are to reside. The choice of name for this column must be a legal SAS name and must not already be in use as a data set variable name, alias, or any other column name. The reference must be placed in the COLUMN statement after any columns it references to complete its computation. The DEFINE statement for the ratio column assigns its usage as COMPUTED, along with making typical assignments of a label and a format.

❷ The COMPUTE statement starts a compute block and requires a report item or a location (optionally with a target). Here the report item of Ratio is designated to be computed.

❸ If a COMPUTE block references a report item, at least one statement assigning that report item a value must be present in the COMPUTE block. The single statement inside this compute block is an assignment statement setting a value for ratio using an expression involving the aliases for the mean and median statistic columns. COMPUTE blocks allow for use of many DATA step programming elements; however, not all are supported.

❹ The ENDCOMP statement is required to close any COMPUTE block.

Output 7.5.6: Computing a New Column in PROC REPORT

State	Mortgage Status	Number of Observations	Median Electricity Cost	Mean Electricity Cost	Mean to Median Ratio	Standard Deviation
North Carolina	N/A	8,548	$1,080	$1,327	122.85%	$834
	No	9,382	$1,200	$1,416	117.99%	$803
	Yes	17,303	$1,440	$1,667	115.78%	$892
North Carolina		35,233	$1,320	$1,518	114.98%	$868
South Carolina	N/A	3,807	$1,200	$1,426	118.81%	$862
	No	5,074	$1,320	$1,543	116.92%	$880
	Yes	8,283	$1,560	$1,766	113.18%	$898
South Carolina		17,164	$1,440	$1,624	112.81%	$896
		52,397	$1,320	$1,553	117.63%	$879

While Program 7.5.6 uses aliases to produce multiple statistics on the same variable, and then uses those aliases in a computation, previous examples have shown that aliases are not required to produce these statistics; nesting is also available. It is also possible to produce the same statistic on different variables using nesting (as in Program 6.6.6); however, attempting to reference a statistic keyword is ambiguous in such a case. Program 7.5.7 demonstrates how to reference summary statistics in a PROC REPORT computation when aliases are not in use.

Program 7.5.7: References to Summary Values Not Defined Via an Alias

```
proc report data= work.ipums2005cost;
  where state in ('North Carolina','South Carolina');
  column state mortgageStatus electric,(n median mean std) ratio❶;
  define state / group 'State';
  define mortgageStatus / group 'Mortgage Status' format=$Mort_Status.;
  define electric / '';
  define n / 'Number of Observations' format=comma8.;
  define median / 'Median Electricity Cost' format=dollar10.;
  define mean / 'Mean Electricity Cost' format=dollar10.;
  define std / 'Standard Deviation' format=dollar10.;
  define ratio / computed 'Mean to Median Ratio' format=percent9.2;
```

```
   break after state / summarize;
   rbreak after / summarize;

   compute ratio;
     ratio=electric.mean/electric.median; ❷
   endcomp;
run;
```

❶   Columns used to compute Ratio are created in the nesting. Since the Ratio computation uses values generated in these columns, it must appear after the nested set.

❷   The proper reference to the summary value is of the form *variable.stat-keyword*. This ensures references to the same statistic on different variables, or different variables with the same statistic, are unique.

Output 7.5.7: References to Summary Values Not Defined Via an Alias

State	Mortgage Status	Number of Observations	Median Electricity Cost	Mean Electricity Cost	Standard Deviation	Mean to Median Ratio
North Carolina	N/A	8,548	$1,080	$1,327	$834	122.85%
	No	9,382	$1,200	$1,416	$803	117.99%
	Yes	17,303	$1,440	$1,667	$892	115.78%
North Carolina		35,233	$1,320	$1,518	$868	114.98%
South Carolina	N/A	3,807	$1,200	$1,426	$862	118.81%
	No	5,074	$1,320	$1,543	$880	116.92%
	Yes	8,283	$1,560	$1,766	$898	113.18%
South Carolina		17,164	$1,440	$1,624	$896	112.81%
		52,397	$1,320	$1,553	$879	117.63%

In Program 7.5.7, the columns used in the computation were generated by a combination of an analysis variable and a nested statistic keyword, with those two elements used to build references to their values in the form *variable.stat-keyword*. When using nesting with an across variable, column definitions depend on levels of the across variable and are therefore not explicitly defined by elements provided in the COLUMN statement. Program 7.5.8 is a variation on Program 6.6.12, computing the difference between average electricity costs for North Carolina and South Carolina, separately for metropolitan and non-metropolitan areas. The columns required to complete these computations are all means for Electricity and are distinguished by different combinations of the across variables State and Metro. Those values are data-dependent and use a different method of referencing the columns.

Program 7.5.8: Referencing Columns Implicitly Defined

```
proc format;
  value MetroStatus
    0 = "Unknown"
    1 = "Non-Metro"
    2-4 = "Metro"
  ;
run;

proc report data= work.ipums2005cost;
  where state in ('North Carolina','South Carolina') and metro ge 1;
  column mortgageStatus electric,state,metro NonMetroDiff MetroDiff❶;
  define state / across 'Mean Electricity Costs';
  define metro / across '' format=MetroStatus. order=internal;
  define mortgageStatus / group 'Mortgage Status' format=$Mort_Status.;
  define electric / mean '' format=dollar10.;
  define NonMetroDiff / computed 'Non-Metro Diff (SC-NC)' format=dollar10.2; ❷
  define MetroDiff / computed 'Metro Diff (SC-NC)' format=dollar10.2; ❷
```

```
   compute NonMetroDiff;
     NonMetroDiff=_c4_-_c2_; ❸
   endcomp;
   compute MetroDiff;
     MetroDiff=_c5_-_c3_; ❸
   endcomp;
run;
```

❶ NonMetroDiff and MetroDiff name the two new columns to be computed. As they are computed from columns generated by the nesting of Electricity, State, and Metro, these references must appear after that nested set in the column statement.

❷ The DEFINE statements for each of these two new columns set the usage and format for the variable, and provide a detailed column heading for each.

❸ A column can be referenced by its position in the form: _c#_, where # is the column number. Here, MortgageStatus is the first column, while the next four columns are generated by the combinations of values of State and Metro. Columns 2 and 3 correspond to North Carolina, while columns 4 and 5 are South Carolina, as State is arranged in alphabetical order. Non-Metro values are in columns 2 and 4, while Metro values are in 3 and 5. These are not alphabetical because of the ORDER=INTERNAL option in the DEFINE statement for Metro. The default ordering option of FORMATTED would alter these positions. To use these references effectively, the nesting order of the variables, the set of unique values for each, and the ordering of those values must be known.

Output 7.5.8: Referencing Columns Implicitly Defined

| | Mean Electricity Costs | | | | | |
| | North Carolina | | South Carolina | | | |
Mortgage Status	Non-Metro	Metro	Non-Metro	Metro	Non-Metro Diff (SC-NC)	Metro Diff (SC-NC)
N/A	$1,427	$1,275	$1,529	$1,398	$101.95	$122.61
No	$1,440	$1,389	$1,633	$1,506	$192.98	$117.02
Yes	$1,765	$1,617	$1,898	$1,713	$133.09	$96.51

While the headers for the computed columns in Output 7.5.8 provide useful detail, they do not fit well with the overall structure of the header space. To fit these into the hierarchy, spanning headers can be used in the COLUMN statement, as shown in Program 7.5.9.

Program 7.5.9: Inserting Spanning Headers in the COLUMN Statement

```
proc format;
  value MetroStatus
    0 = "Unknown"
    1 = "Non-Metro"
    2-4 = "Metro"
  ;
run;

proc report data= work.ipums2005cost;
  where state in ('North Carolina','South Carolina') and metro ge 1;
  column mortgageStatus electric,state,metro
         ❶('Diff. (SC-NC)' ❷('' (NonMetroDiff MetroDiff)❸));
  define state / across 'Mean Electricity Costs';
  define metro / across '' format=MetroStatus. order=internal;
  define mortgageStatus / group 'Mortgage Status' format=$Mort_Status.;
  define electric / mean '' format=dollar10.;
  define NonMetroDiff / computed 'Non-Metro' format=dollar10.2;
  define MetroDiff / computed 'Metro' format=dollar10.2;
```

```
compute NonMetroDiff;
  NonMetroDiff=_c4_-_c2_;
endcomp;
compute MetroDiff;
  MetroDiff=_c5_-_c3_;
endcomp;
run;
```

❶ The opening parenthesis defines the start of a spanning header. The general specification is: (*'header1'* *'header2'* ... *'headerN'* report-items), with each header printed on a separate line, spanning the columns generated by the report items given. The report items take their place in the column specification in the same location whether the spanning header is present or not.

❷ A second spanning header is nested under the first header and given an empty value for the literal to match the null value given to the header for Metro in its DEFINE statement. When only a single header spanning NonMetroDiff and MetroDiff is given, it is positioned immediately above them, appearing in the row designated to label the last variable in the nesting hierarchy, Metro. So, this entire header row is re-established and empty cells spanning columns 2 and 3 and columns 4 and 5 appear.

❸ Parentheses enclosing the report items are not necessary but can aid in readability for complex header structures like this one.

Output 7.5.9: Inserting Spanning Headers in the COLUMN Statement

Mortgage Status	Mean Electricity Costs					
	North Carolina		South Carolina		Diff. (SC-NC)	
	Non-Metro	Metro	Non-Metro	Metro	Non-Metro	Metro
N/A	$1,427	$1,275	$1,529	$1,398	$101.95	$122.61
No	$1,440	$1,389	$1,633	$1,506	$192.98	$117.02
Yes	$1,765	$1,617	$1,898	$1,713	$133.09	$96.51

Computations in the REPORT procedure can take advantage of many statements and functions available in the DATA step. While FIRST. and LAST. variables are not available, specific targets for compute blocks are provided to emulate calculation of group level summaries in PROC REPORT. Program 7.5.10 extends and modifies Program 6.6.9, adding another state and computing a percentage of observations at each record. The percentage is constructed so that in each group, the value is the percentage of the group total, whereas on the BREAK lines for each group it is the percentage of the grand total. No percentage is given in the whole-report summary record generated by the RBREAK statement.

Program 7.5.10: Conditional Logic and Various Targets in Compute Blocks

```
proc format;
  value misspct
    .=' '
    other=[percent8.1]
  ; ❶
run;

proc report data= work.ipums2005cost out= work.check❷
            style(column)=[backgroundcolor=grayF3];
  where state in ('North Carolina','South Carolina','Virginia');
  column state mortgageStatus electric=num pct electric=mid
         electric=avg electric=std; ❸
  define state / group 'State';
  define mortgageStatus / group 'Mortgage Status' format=$Mort_Status.;
  define electric / '';
  define num / n 'Number of Observations' format=comma8.;
  define pct / computed '% of Obs.' format=misspct.;
  define mid / median 'Median Electricity Cost' format=dollar10.;
  define avg / mean 'Mean Electricity Cost' format=dollar10.;
  define std / std 'Standard Deviation' format=dollar10.;
```

```
break after state / summarize suppress style=[backgroundcolor=grayD3];
rbreak after / summarize style=[backgroundcolor=grayB9];

compute before;
  tot=num;
endcomp; ❹
compute before state;
  grptot=num;
endcomp; ❺
compute pct;
  if lowcase(_break_) eq 'state' then pct=num/tot;
    else if _break_ eq '' then pct=num/grptot;
      else pct=.; ❻
endcomp;
run;
```

❶ As the percentage is suppressed on the whole-report summary line, this format is constructed to modify the dot that appears for a missing numeric value. To set formatted values using an existing format, the format is given in square brackets in the format label position.

❷ Recall the OUT= option produces a data set corresponding to the columns and records in the report, as seen in Program 6.6.14. That table, not shown here, is available for inspection once PROC REPORT executes successfully and can be helpful in diagnosing problems with computations but may not include all of the desired information.

❸ The new percentage column is inserted after the frequency column—in order to position it here, the column statement uses aliases for the Electric variable rather than nesting of statistic keywords.

❹ To have the correct divisor for the percentages on the break lines, the total frequency must be known. While an RBREAK statement with a target of BEFORE can be used to generate this number in the table, that is neither desired nor necessary. A compute block with a target of BEFORE forces PROC REPORT to produce the same summary values at the start of the report without displaying them. A variable (Tot) is then assigned to that value by referencing the alias for the frequency statistic—the naming of this variable must follow the previously stated rules for naming aliases and/or new columns. In the OUT= data set, Tot is not present as it is not a report item; however, the summary line generated by COMPUTE BEFORE is the first record.

❺ It is also possible to use a target with a variable in a compute block, which allows for the same summaries to be produced as those available with a BREAK statement. The BREAK statements in the code use AFTER, which is too late during processing to get the correct value, so here the total frequency within a group is made available when entering a group with COMPUTE BEFORE and assigned to a variable (GrpTot). Again, this variable is not present in the OUT= data set but the summary lines are.

❻ The final computation of the percentage uses the correct divisor, or sets to missing, by conditioning on the _BREAK_ variable. Note the use of the LOWCASE function to make the comparison robust. The OUT= data set can always be checked for the correct casing of the variable name, but forcing a specific casing using a function is a defensive programming technique.

Output 7.5.10: Conditional Logic and Various Targets in Compute Blocks

State	Mortgage Status	Number of Observations	% of Obs.	Median Electricity Cost	Mean Electricity Cost	Standard Deviation
North Carolina	N/A	8,548	24.3%	$1,080	$1,327	$834
	No	9,382	26.6%	$1,200	$1,416	$803
	Yes	17,303	49.1%	$1,440	$1,667	$892
		35,233	43.4%	$1,320	$1,518	$868
South Carolina	N/A	3,807	22.2%	$1,200	$1,426	$862
	No	5,074	29.6%	$1,320	$1,543	$880
	Yes	8,283	48.3%	$1,560	$1,766	$898

State	Mortgage Status	Number of Observations	% of Obs.	Median Electricity Cost	Mean Electricity Cost	Standard Deviation
		17,164	21.1%	$1,440	$1,624	$896
Virginia	N/A	6,510	22.6%	$960	$1,197	$792
	No	6,525	22.6%	$1,080	$1,318	$822
	Yes	15,816	54.8%	$1,440	$1,635	$912
		28,851	35.5%	$1,200	$1,464	$887
		81,248		$1,320	$1,521	$883

The OUT= data set is a useful check for how a report is being generated; however, it is not a perfect match for what happens during PROC REPORT execution; for example, the variables Tot and GrpTot generated in Program 7.5.10 are not recorded there. Other mismatches between the data generated by OUT= and PROC REPORT execution can cause confusion, as the next example illustrates. Suppose electricity costs were projected to increase from 2005 to 2006 by 3% in North Carolina and 2% in South Carolina. Program 7.5.11 appears to be a reasonable approach to adding columns for the projected mean and median.

Program 7.5.11: Erroneous Conditioning on the Values of a Group Variable

```
proc report data= work.ipums2005cost out= work.check
            style(column)=[backgroundcolor=grayF3];
  where state in ('North Carolina','South Carolina');
  column state mortgageStatus electric=num electric=mid electric=avg projMed projAvg;
  define state / group 'State';
  define mortgageStatus / group 'Mortgage Status' format=$Mort_Status.;
  define electric / '';
  define num / n 'Number of Observations' format=comma8.;
  define mid / median 'Median Electricity Cost' format=dollar10.;
  define avg / mean 'Mean Electricity Cost' format=dollar10.;
  define projMed / computed 'Projected Mean' format=dollar12.2;
  define projAvg / computed 'Projected Mean' format=dollar12.2;
  break after state / summarize suppress style=[backgroundcolor=grayD3];
  rbreak after / summarize style=[backgroundcolor=grayB9];

  compute projAvg;
    if substr(state,1,1) eq 'N' then do; ❶
      projAvg=1.03*avg;
      projMed=1.03*mid;
    end;
    else do; ❷
      projAvg=1.02*avg;
      projMed=1.02*mid;
    end;
  endcomp;
run;
```

❶ Conditioning on the first letter of the state name is sufficient in this case, so the SUBSTR function is used. (See Program 4.5.6.) In this DO block, a value of 103% of the current value is projected for the mean and median.

❷ Other observations are assigned a 2% increase, as done in this DO block.

Output 7.5.11: Erroneous Conditioning on the Values of a Group Variable

State	Mortgage Status	Number of Observations	Median Electricity Cost	Mean Electricity Cost	Projected Median	Projected Mean
North Carolina	N/A	8,548	$1,080	$1,327	$1,112.40	$1,366.64
	No	9,382	$1,200	$1,416	$1,224.00	$1,444.23
	Yes	17,303	$1,440	$1,667	$1,468.80	$1,700.60
		35,233	$1,320	$1,518	$1,346.40	$1,548.09
South Carolina	N/A	3,807	$1,200	$1,426	$1,224.00	$1,454.26
	No	5,074	$1,320	$1,543	$1,346.40	$1,574.21
	Yes	8,283	$1,560	$1,766	$1,591.20	$1,800.86
		17,164	$1,440	$1,624	$1,468.80	$1,656.98
		52,397	$1,320	$1,553	$1,346.40	$1,583.76

Checking Output 7.5.11, the first record correctly projects an increase of 3% (3% of $1080 is $32.40 for a new total of $1,112.40). In the second record, the median electricity cost is $1,200, and is projected to be $1,224— an increase of $24 or 2%. Continuing similar checks, it is apparent that every record except the first projects a 2% increase. Even though the data set from the OUT= option fully populates the state column, the blank cells in the output table are what is actually processed during the execution of PROC REPORT. One way to show that this is what occurs is replacing the `else do;` at ❷ with `else if substr(state,1,1) eq 'S' then do;`, which now has the projections as missing in all but two rows. In addition to trying this (or instead of), remove the SUPPRESS option from the BREAK statement and see how the output changes to the correct values on the break lines.

This problem can be alleviated by using the compute block to target the value of interest at the start of each group and carry it forward through subsequent records, as demonstrated in Program 7.5.12.

Program 7.5.12: Setting an Intermediate Value to Condition on the Values of a Group Variable

```
proc report data= work.ipums2005cost out= work.check
            style(column)=[backgroundcolor=grayF3];
  where state in ('North Carolina','South Carolina');
  column state mortgageStatus electric=num electric=mid electric=avg projMed projAvg;
  define state / group 'State';
  define mortgageStatus / group 'Mortgage Status' format=$Mort_Status.;
  define electric / '';
  define num / n 'Number of Observations' format=comma8.;
  define mid / median 'Median Electricity Cost' format=dollar10.;
  define avg / mean 'Mean Electricity Cost' format=dollar10.;
  define projMed / computed 'Projected Median' format=dollar12.2;
  define projAvg / computed 'Projected Mean' format=dollar12.2;
  break after state / summarize suppress style=[backgroundcolor=grayD3];
  rbreak after / summarize style=[backgroundcolor=grayB9];

  compute before state;
    st=substr(state,1,1);
  endcomp; ❶

  compute projAvg;
    if _break_ ne '_RBREAK_' then do; ❷
      if st eq 'N' then do; ❸
        projAvg=1.03*avg;
        projMed=1.03*mid;
      end;
```

```
        else if st eq 'S' then do;
           projAvg=1.02*avg;
           projMed=1.02*mid;
        end;
        else do; ❹
           projAvg=-1;
           projMed=-1;
        end;
   end;
     else do; ❺
        projAvg=.;
        projMed=.;
     end;
  endcomp;
run;
```

❶ At the top of each State group, the full value of the State variable is present, and the needed portion is extracted and stored as a new variable.

❷ To correctly calculate the value for the projected mean column in the whole-report break line, it must be defined as a weighted mixture of values across both states. That computation is not done here (it is left as an exercise at the end of this chapter), so this condition limits the computation to the other lines of the report.

❸ The conditioning is now based on the value extracted at the top of each group, which makes it correct for all rows in each group and also for the break lines summarizing each group (with or without the SUPPRESS option).

❹ Rather than using a condition for one state with a catch-all ELSE, each state is assigned a specific condition to check, and the final ELSE clause is used to set a value that would be unexpected and/or easily noticeable in the output. Programming IF-THEN-ELSE chains in this fashion is often a useful diagnostic tool, including helping to reveal the issues encountered in Program 7.5.11.

❺ Though this DO block is not necessary to make the final projections missing on the RBREAK line—if unassigned they are missing as they are initialized to missing—it is considered a good programming practice to explicitly set all values rather than rely on default values being correct.

Output 7.5.12: Setting an Intermediate Value to Condition on the Values of a Group Variable

State	Mortgage Status	Number of Observations	Median Electricity Cost	Mean Electricity Cost	Projected Median	Projected Mean
North Carolina	N/A	8,548	$1,080	$1,327	$1,112.40	$1,366.64
	No	9,382	$1,200	$1,416	$1,236.00	$1,458.39
	Yes	17,303	$1,440	$1,667	$1,483.20	$1,717.27
		35,233	$1,320	$1,518	$1,359.60	$1,563.26
South Carolina	N/A	3,807	$1,200	$1,426	$1,224.00	$1,454.26
	No	5,074	$1,320	$1,543	$1,346.40	$1,574.21
	Yes	8,283	$1,560	$1,766	$1,591.20	$1,800.86
		17,164	$1,440	$1,624	$1,468.80	$1,656.98
		52,397	$1,320	$1,553	.	.

7.5.3 Using Compute Blocks to Insert Customized Text and Set Styles

Compute blocks have utility beyond computing values, they can also be used to insert text lines, replace existing values, and set styles. The LINE statement can be used to introduce text rows into the report at various locations. Program 7.5.13 uses this concept to place a custom header into a report.

Program 7.5.13: Inserting and Styling a Text Line with a Compute Block

```
proc report data= work.ipums2005cost
            style(header)=[fontfamily='Arial Black'
                            backgroundcolor=gray55
                            color=white]
            style(column)=[fontfamily='Georgia' backgroundcolor=grayDD fontsize=10pt]
            style(summary)=[backgroundcolor=grayAA fontweight=bold fontstyle=italic];
   where state in ('North Carolina','South Carolina');
   column state mortgageStatus electric,(n median mean std);
   define state / group 'State';
   define mortgageStatus / group 'Mortgage Status' format=$Mort_Status.;
   define electric / '';
   define n / 'N' format=comma8.;
   define median / 'Median' format=dollar10.;
   define mean / 'Mean' format=dollar10.;
   define std / 'Std. Dev.' format=dollar10.;
   break after state / summarize;
   rbreak after / summarize;

   compute before _page_ ❶ / style=[background=gray55 color=white fontsize=14pt
                                    fontfamily='Arial Black' textalign=left] ❷;
      line 'Electricity Costs'; ❸
   endcomp;
run;
```

❶ The _PAGE_ location directs PROC REPORT to the top or bottom of each page, above the header on each page using the BEFORE location, following the last row on each page with AFTER. Given that the result is a single-page report, simply using COMPUTE BEFORE produces the same table.

❷ If a row of text is inserted, it can be styled using a STYLE= option set. Note the use of a slash to separate the style options from the COMPUTE request in this statement. This style can also be set using the LINES location for a STYLE= option in the PROC REPORT statement.

❸ The LINE statement provides a line of text, which can include literals, variables, and expressions. Any direct use of a variable (not inside a function or expression) must include an appropriate format immediately following it. Here the literal text is inserted above the header in the style designated—see Output 7.5.13.

Output 7.5.13: Inserting and Styling a Text Line with a Compute Block

Electricity Costs					
State	**Mortgage Status**	**N**	**Median**	**Mean**	**Std. Dev.**
North Carolina	N/A	8,548	$1,080	$1,327	$834
	No	9,382	$1,200	$1,416	$803
	Yes	17,303	$1,440	$1,667	$892
North Carolina		*35,233*	*$1,320*	*$1,518*	*$868*
South Carolina	N/A	3,807	$1,200	$1,426	$862
	No	5,074	$1,320	$1,543	$880
	Yes	8,283	$1,560	$1,766	$898
South Carolina		*17,164*	*$1,440*	*$1,624*	*$896*
		52,397	*$1,320*	*$1,553*	*$879*

Computations can also be used to insert new values in place of values that already exist in the table. For example, Program 7.5.14 uses two compute blocks to insert new, more descriptive text on all the break lines.

Program 7.5.14: Inserting Replacement Values

```
proc report data= work.ipums2005cost
             style(header)=[fontfamily='Arial Black'
                            backgroundcolor=gray55
                            color=white]
             style(column)=[fontfamily='Georgia' backgroundcolor=grayDD fontsize=10pt]
             style(summary)=[backgroundcolor=grayAA color=black fontweight=bold
                             fontstyle=italic];
  where state in ('North Carolina','South Carolina');
  column state mortgageStatus electric,(n median mean std);
  define state / group 'State' format=$25.;
  define mortgageStatus / group 'Mortgage Status' format=$Mort_Status.;
  define electric / '';
  define n / 'N' format=comma8.;
  define median / 'Median' format=dollar10.;
  define mean / 'Mean' format=dollar10.;
  define std / 'Std. Dev.' format=dollar10.;
  break after state / summarize style=[backgroundcolor=gray77 color=grayEE];
  rbreak after / summarize;

  compute before _page_/ style=[background=gray55 color=white textalign=left
                                fontfamily='Arial Black' fontsize=14pt];
    line 'Electricity Costs';
  endcomp;
  compute after state; ❶
    if substr(state,1,1) eq 'N' then state='Overall: NC';
      else if substr(state,1,1) eq 'S' then state='Overall: SC'; ❷
  endcomp;
  compute after; ❸
    state='All States';
  endcomp;
run;
```

❶ The PROC REPORT includes a BREAK statement targeting State with the AFTER location, and this compute block refers to that same variable and location combination.

❷ The value of State is revised in a manner that is still conditional on the value of State—so it is important in this example that the SUPPRESS option is not in use in the BREAK statement. It is also important to ensure that the replacement text is of the same or lesser length than the original variable.

❸ The RBREAK AFTER and COMPUTE AFTER statements also both refer to the same location. By default, the value of State is null here, but space is still allocated for it, so it can be replaced.

Output 7.5.14: Inserting Replacement Values

Electricity Costs					
State	Mortgage Status	N	Median	Mean	Std. Dev.
North Carolina	N/A	8,548	$1,080	$1,327	$834
	No	9,382	$1,200	$1,416	$803
	Yes	17,303	$1,440	$1,667	$892
Overall: NC		*35,233*	*$1,320*	*$1,518*	*$868*
South Carolina	N/A	3,807	$1,200	$1,426	$862
	No	5,074	$1,320	$1,543	$880
	Yes	8,283	$1,560	$1,766	$898
Overall: SC		*17,164*	*$1,440*	*$1,624*	*$896*
All States		**52,397**	**$1,320**	**$1,553**	**$879**

If both a BREAK and COMPUTE statement target the same variable and location and the compute block includes a LINE statement(s), those lines are placed after the generated break line and before the next group starts.

Compute blocks can be used to set styles using the CALL DEFINE statement and, given the versatility of both the compute block and the CALL DEFINE statement, styles can be changed in a variety of locations conditionally on values in other locations. Program 7.5.15 modifies Program 7.5.6 to show the median value in red text any time the mean exceeds the median by 16% or more.

Program 7.5.15: Styling a Column Conditionally on Values in Another Column

```
proc report data= work.ipums2005cost;
  where state in ('North Carolina','South Carolina');
  column state mortgageStatus electric,(n median mean std) ratio;
  define state / group 'State';
  define mortgageStatus / group 'Mortgage Status' format=$Mort_Status.;
  define electric / '';
  define n / 'Number of Observations' format=comma8.;
  define median / 'Median Electricity Cost' format=dollar10.;
  define mean / 'Mean Electricity Cost' format=dollar10.;
  define std / 'Standard Deviation' format=dollar10.;
  define ratio / computed 'Mean to Median Ratio' format=percent9.2;
  break after state / summarize;
  rbreak after / summarize;

  compute ratio;
    ratio=electric.mean/electric.median;
    if ratio ge 1.16❶
            then call define❷('_c4_'❸, 'style'❹, 'style=[color=cxFF3333]'❺);
  endcomp;
run;
```

❶ The desired style change is conditional on the value of Ratio, and thus must be set after the value of Ratio is computed. So the CALL DEFINE can be established in the same compute block for Ratio, after its assignment statement, or in any compute block executed subsequent to the one for Ratio.

❷ CALL DEFINE requires three arguments: column identifier or _ROW_, attribute name, and attribute value. It may also be useful to think of the three arguments as corresponding to: where to make a change, what to change, and how to change it. For more information about column identifiers in CALL DEFINE, see Chapter Note 6 in Section 7.8 at the end of this chapter.

❸ Here, a column position reference is used to point to the median column—'Electric.Median' is also a legal reference to this same column.

❹ This CALL DEFINE designates the style attribute for modification in the designated column. All attribute names are given as literals.

❺ The specific style alteration to be made is to change the text color to red, and this is written in the same basic syntax as attribute setting in STYLE= options in PROC REPORT. Style attribute values sets are also given as literals; however, this is not universal for all attribute values.

Output 7.5.15: Styling a Column Conditionally on Values in Another Column

State	Mortgage Status	Number of Observations	Median Electricity Cost	Mean Electricity Cost	Standard Deviation	Mean to Median Ratio
North Carolina	N/A	8,548	$1,080	$1,327	$834	122.85%
	No	9,382	$1,200	$1,416	$803	117.99%
	Yes	17,303	$1,440	$1,667	$892	115.78%
North Carolina		35,233	$1,320	$1,518	$868	114.98%
South Carolina	N/A	3,807	$1,200	$1,426	$862	118.81%
	No	5,074	$1,320	$1,543	$880	116.92%

State	Mortgage Status	Number of Observations	Median Electricity Cost	Mean Electricity Cost	Standard Deviation	Mean to Median Ratio
	Yes	8,283	$1,560	$1,766	$898	113.18%
South Carolina		17,164	$1,440	$1,624	$896	112.81%
		52,397	$1,320	$1,553	$879	117.63%

Program 7.5.16 extends the previous example by making multiple style changes based on the mean to median ratio without actually showing Ratio in Output 7.5.16.

Program 7.5.16: Styling Columns and Rows Conditionally

```
proc report data= work.ipums2005cost;
  where state in ('North Carolina','South Carolina');
  column state mortgageStatus electric,(n median mean std) ratio;
  define state / group 'State';
  define mortgageStatus / group 'Mortgage Status' format=$Mort_Status.;
  define electric / '';
  define n / 'Number of Observations' format=comma8.;
  define median / 'Median Electricity Cost' format=dollar10.;
  define mean / 'Mean Electricity Cost' format=dollar10.;
  define std / 'Standard Deviation' format=dollar10.;
  define ratio / computed noprint ❶;
  break after state / summarize;
  rbreak after / summarize;

  compute ratio;
    ratio=electric.mean/electric.median;
    if ratio ge 1.16 and _break_ eq '' then do; ❷
      call define('_c4_','style','style=[color=cxFF3333]');
      call define('_c5_','style','style=[color=cxFF3333]');
      call define(_row_❸,'style','style=[backgroundcolor=grayEE]');
    end;
  endcomp;
  compute after / style=[color=cxFF3333 backgroundcolor=grayEE just=right];
    line 'Mean Exceeds Median by More than 16%';
  endcomp; ❹
run;
```

❶ Ratio is still computed and is available for use as in previous examples, NOPRINT suppresses it in the displayed table only—do not confuse the NOPRINT option and the DISPLAY usage. When using positional references for columns, columns for which the NOPRINT option is active *are included* in the column numbering.

❷ Using a DO block and several CALL DEFINE statements, multiple styles can be set based on this single condition. The first two CALL DEFINE statements set the text color in the median and mean columns.

❸ The third CALL DEFINE uses the _ROW_ location, which applies to the full row for any given observation. This location reference is not (and cannot be) given as a literal.

❹ This COMPUTE block adds a line of text to the end of the report to give some indication as to the meaning of the highlighting—note how it uses styling attributes that are the same as those in the highlighted rows.

Output 7.5.16: Styling Columns and Rows Conditionally

State	Mortgage Status	Number of Observations	Median Electricity Cost	Mean Electricity Cost	Standard Deviation
North Carolina	N/A	8,548	$1,080	$1,327	$834
	No	9,382	$1,200	$1,416	$803
	Yes	17,303	$1,440	$1,667	$892
North Carolina		35,233	$1,320	$1,518	$868
South Carolina	N/A	3,807	$1,200	$1,426	$862
	No	5,074	$1,320	$1,543	$880
	Yes	8,283	$1,560	$1,766	$898
South Carolina		17,164	$1,440	$1,624	$896
		52,397	$1,320	$1,553	$879

Mean Exceeds Median by More than 16%

Changing style is a common use of the CALL DEFINE function, but it is useful for changing other attributes as well—see the SAS Documentation and references for more information. Program 7.5.17 modifies program 7.5.10 to increase the precision shown for the percentages on the break lines only.

Program 7.5.17: Changing a Format in Specific Table Cells Using CALL DEFINE

```
proc report data= work.ipums2005cost out= work.check
                style(column)=[backgroundcolor=grayF3];
  where state in ('North Carolina','South Carolina','Virginia');
  column state mortgageStatus electric=num pct electric=mid
         electric=avg electric=std;
  define state / group 'State';
  define mortgageStatus / group 'Mortgage Status' format=$Mort_Status.;
  define electric / '';
  define num / n 'Number of Observations' format=comma8.;
  define pct / computed '% of Obs.' format=misspct.;
  define mid / median 'Median Electricity Cost' format=dollar10.;
  define avg / mean 'Mean Electricity Cost' format=dollar10.;
  define std / std 'Standard Deviation' format=dollar10.;
  break after state / summarize suppress style=[backgroundcolor=grayD3];
  rbreak after / summarize style=[backgroundcolor=grayB9];

  compute before;
    tot=num;
  endcomp;
  compute before state;
    grptot=num;
  endcomp;
  compute pct;
    if lowcase(_break_) eq 'state' then do;❶
      pct=num/tot;
      call define('_c4_','format','percent9.2')❷;
      end;
      else if _break_ eq '' then pct=num/grptot;
        else pct=.;
  endcomp;
run;
```

❶ Given that the change in precision is to occur on the break lines, the CALL DEFINE that resets the format must be included under that condition. So, the first IF condition in the COMPUTE block for Pct from Program 7.5.10 now contains a DO block to include the computation of Pct and the CALL DEFINE to make the attribute change.

❷ The attribute to be changed is the format, again all attribute references are given as literals. The value for the attribute, also given as a literal in this case, is set to percent9.2. The desired location is the Pct column, which is column 4 in Output 7.5.17, but can also be referenced as 'pct'.

Output 7.5.17: Changing a Format in Specific Table Cells Using CALL DEFINE

State	Mortgage Status	Number of Observations	% of Obs.	Median Electricity Cost	Mean Electricity Cost	Standard Deviation
North Carolina	N/A	8,548	24.3%	$1,080	$1,327	$834
	No	9,382	26.6%	$1,200	$1,416	$803
	Yes	17,303	49.1%	$1,440	$1,667	$892
		35,233	43.36%	$1,320	$1,518	$868
South Carolina	N/A	3,807	22.2%	$1,200	$1,426	$862
	No	5,074	29.6%	$1,320	$1,543	$880
	Yes	8,283	48.3%	$1,560	$1,766	$898
		17,164	21.13%	$1,440	$1,624	$896
Virginia	N/A	6,510	22.6%	$960	$1,197	$792
	No	6,525	22.6%	$1,080	$1,318	$822
	Yes	15,816	54.8%	$1,440	$1,635	$912
		28,851	35.51%	$1,200	$1,464	$887
		81,248		$1,320	$1,521	$883

With the ability to create and manipulate variables, use functions and conditional logic, and target various areas of the report for style changes, styling in COMPUTE blocks is a versatile tool. Program 7.5.18 modifies the previous example by constructing a counter variable and using it to create alternating row striping within the groups, while leaving the break lines in their previously defined styles.

Program 7.5.18: Striping Rows Using CALL DEFINE and Modular Arithmetic

```
proc report data= work.ipums2005cost out= work.check
            style(column)=[backgroundcolor=grayF3];
  where state in ('North Carolina','South Carolina','Virginia');
  column state mortgageStatus electric=num pct electric=mid
         electric=avg electric=std;
  define state / group 'State';
  define mortgageStatus / group 'Mortgage Status' format=$Mort_Status.;
  define electric / '';
  define num / n 'Number of Observations' format=comma8.;
  define pct / computed '% of Obs.' format=misspct.;
  define mid / median 'Median Electricity Cost' format=dollar10.;
  define avg / mean 'Mean Electricity Cost' format=dollar10.;
  define std / std 'Standard Deviation' format=dollar10.;
  break after state / summarize suppress style=[backgroundcolor=grayD3];
  rbreak after / summarize style=[backgroundcolor=grayB9];

  compute before;
    tot=num;
  endcomp;
  compute before state;
    grptot=num;
  endcomp;
```

```
compute pct;
  if lowcase(_break_) eq 'state' then do;
    c=0; ❶
    pct=num/tot;
    call define('_c4_','format','percent9.2');
  end;
    else if _break_ eq '' then do;
      pct=num/grptot;
      c+1; ❷
      if mod(c,2)❸ eq 1 then
        call define(_row_, 'style', 'style=[backgroundcolor=grayF7]');
        else call define(_row_, 'style', 'style=[backgroundcolor=grayE7]'); ❹
    end;
      else pct=.;
  endcomp;
run;
```

❶ A counter is initialized to zero any time a break line is encountered for the State variable. This re-initializes the counter for each of the second and third state groups, but not the first.

❷ For any row within a group, the counter is increased by 1 via a sum statement, using the same syntax as a DATA step sum statement. This sum statement also forces the counter to be initialized to 0 when PROC REPORT execution starts, as it does in the DATA step. Since the sum statement for C appears before the IF-THEN that conditions on it, the value of C is 1 or more each time that condition is checked.

❸ The MOD(a,b) function computes a modulo b, or the remainder when a is divided by b. When computing a whole number modulo 2, the only possible outcomes are 0 and 1. Since the counter starts at 1 and increments by 1, the sequence of possible results is: MOD(1,2) = 1, MOD(2,2) = 0, MOD(3,2) = 1, MOD(4,2) = 0, and so forth.

❹ The first row in any group receives a very light gray background color, the second a slightly darker gray, with the third reverting to the lighter gray—see Output 7.5.18. Resetting the counter at the break lines ensures that the first row in each group is the same color. Applying the CALL DEFINE functions only when the _BREAK_ variable is null ensures the styles only apply within groups and not on break lines.

Output 7.5.18: Striping Rows Using CALL DEFINE and Modular Arithmetic

State	Mortgage Status	Number of Observations	% of Obs.	Median Electricity Cost	Mean Electricity Cost	Standard Deviation
North Carolina	N/A	8,548	24.3%	$1,080	$1,327	$834
	No	9,382	26.6%	$1,200	$1,416	$803
	Yes	17,303	49.1%	$1,440	$1,667	$892
		35,233	43.36%	$1,320	$1,518	$868
South Carolina	N/A	3,807	22.2%	$1,200	$1,426	$862
	No	5,074	29.6%	$1,320	$1,543	$880
	Yes	8,283	48.3%	$1,560	$1,766	$898
		17,164	21.13%	$1,440	$1,624	$896
Virginia	N/A	6,510	22.6%	$960	$1,197	$792
	No	6,525	22.6%	$1,080	$1,318	$822
	Yes	15,816	54.8%	$1,440	$1,635	$912
		28,851	35.51%	$1,200	$1,464	$887
		81,248		$1,320	$1,521	$883

7.6 Connecting to Spreadsheets and Relational Databases

A variety of non-SAS data sources are used as examples in this and previous chapters, all of which have been text files. It is also possible to connect SAS to other data structures, such as spreadsheets and relational databases if the proper SAS/ACCESS products are licensed—see Chapter Note 7 in Section 7.8 for more information. The simplest method of making such connections is via the LIBNAME statement, using the appropriate engine for the type of file to which the connection is made. A variety of engines for spreadsheet products are available. Section 7.6.1 covers using an engine for connecting to Microsoft Excel files.

7.6.1 Connecting to and Working with Data in Excel Spreadsheets

A subset of the IPUMS data is stored in the Excel file Rhode Island.xlsx, which includes both the Basic and Utility data for years 2005, 2010, and 2015 for the state of Rhode Island only. Program 7.6.1 connects to the Excel file with the LIBNAME statement using the XLSX engine and uses the CONTENTS procedure to list the sheets within the workbook.

Program 7.6.1: Connecting to an XLSX File and Viewing Metadata with PROC CONTENTS

```
libname RI xlsx❶ '--path to file--❷\Rhode Island.xlsx❸';

proc contents data=RI._all_ nods;
  ods select members;
run; ❹
```

❶ Following the specification of the *libref*, an engine can be given in a LIBNAME statement. If no engine is provided, the default engine depends on the version of SAS in use and the target of the path if it points directly to a file. XLSX supports connections to Microsoft Excel 2007, 2010, and later files.

❷ Fill in the correct path to the Excel file—or omit this portion, leaving the file name and extension, and set the working directory to the correct path.

❸ For the XLSX engine, the path used in the LIBNAME statement must point directly to the Excel file. If this file does not exist, it is created.

❹ In PROC CONTENTS, the _ALL_ keyword in place of a data set name provides file and directory information for the library, a list of library members, and the metadata for each member. The NODS option suppresses the descriptor portion for each data set, and the Members table includes the list of data set names, as shown in Output 7.6.1.

Output 7.6.1: Connecting to an XLSX File and Viewing Metadata with PROC CONTENTS

#	Name	Member Type
1	IPUMS 2005 BASIC	DATA
2	IPUMS 2005 UTILITY	DATA
3	IPUMS 2010 BASIC	DATA
4	IPUMS 2010 UTILITY	DATA
5	IPUMS 2015 BASIC	DATA
6	IPUMS 2015 UTILITY	DATA

While this data is still native to Excel, the XLSX engine allows SAS to work with it as if the sheets are SAS data sets—there is no need to do any special reading or importing. However, it is important to note that column names and sheet names in Excel do not have to follow SAS naming conventions. In fact, Output 7.6.1 shows that the sheet names do not follow SAS naming conventions for this data. Program 7.6.2 revisits a scenario like the one that generates Output 4.2.2 and demonstrates how to work with these names using name literals.

Program 7.6.2: Using Native XLSX Data in SAS

```
data work.RI05_10_15;
   set RI.'Ipums 2005 Basic'n❶(in = in2005) ❷
       RI.'Ipums 2010 Basic'n(in = in2010)
       RI.'Ipums 2015 Basic'n;

   if in2005 then Year = 2005;
     else if in2010 eq 1 then Year = 2010;
       else year = 2015;
run;

proc sgpanel data= work.RI05_10_15;
   panelby Year / rows=1 novarname headerattrs=(size=12pt);
   histogram MortgagePayment / binstart=250 binwidth=500 scale=proportion
           fillattrs=(color=cx99FF99);
   colaxis label='Mortgage Payment' valuesformat=dollar8. values=(0 to 8000 by 2000)
         fitpolicy=stagger;
   rowaxis display=(nolabel) valuesformat=percent7.;
   where MortgagePayment gt 0;
run;

libname RI clear; ❸
```

❶ For names that do not follow SAS naming conventions, name literals can be used. The general form is 'name'n, where the form of *name* follows an extended set of rules. The quotation marks around *name* can be either single or double, provided the closing and opening quotation marks match. As noted in Chapter Note 9 in Section 1.7, the VALIDMEMNAME= and VALIDVARNAME= system options allow for extensions of these rules. For details about extensions, see Chapter Note 8 in Section 7.8 and the SAS Documentation.

❷ While these data set names are nonstandard and use name literals, they otherwise operate as any other reference to a data set does. Thus, any data set options, such as the IN= option, are written in the same form as they are with any other data set.

❸ Connecting to some file types, such as XLSX files, places a hold on the file so that it cannot be opened in another application. The CLEAR option in a LIBNAME statement cancels the corresponding library assignment in the session, disconnecting the SAS session from the Excel file in this case. The RI library is created in Program 7.6.1.

Output 7.6.2: Using Native XLSX Data in SAS

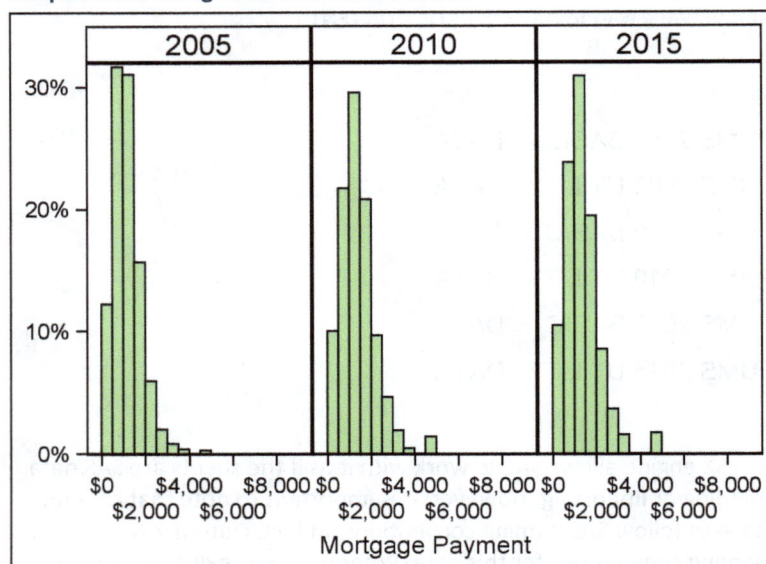

A variety of engines are provided for relational database products as well. Section 7.6.2 reviews using an engine for connecting to Microsoft Access databases.

7.6.2 Connecting to and Working with Data in Access Databases

Similar to the data in Rhode Island.xlsx, a subset of the IPUMS data is stored in the Access database Vermont.accdb, which includes both the Basic and Utility data for years 2005, 2010, and 2015 for the state of Vermont only. Program 7.6.3 connects to the Access database with the LIBNAME statement with the ACCESS engine and uses the CONTENTS procedure to list the tables contained in the database.

Program 7.6.3: Connecting to a Microsoft Access Database and Viewing Metadata

```
libname Vermont access '--path to file--\Vermont.accdb';

proc contents data=Vermont._all_ nods;
  ods select members;
run;
```

Output 7.6.3: Viewing Metadata from a Microsoft Access Database with PROC CONTENTS

#	Name	Member Type	DBMS Member Type
1	ipums2005basic	DATA	TABLE
2	ipums2005utility	DATA	TABLE
3	ipums2010basic	DATA	TABLE
4	ipums2010utility	DATA	TABLE
5	ipums2015basic	DATA	TABLE
6	ipums2015utility	DATA	TABLE

The general form of Program 7.6.3 is the same as Program 7.6.1: a new *libref* is given, an appropriate engine is given, and the path points to the file. For Access databases and the ACCESS engine, if the file does not exist none is created and an error is transmitted to the SAS log.

Program 7.6.4 uses PROC CONTENTS to take a closer look at the Ipums2005Utility table in the Vermont.accdb database, revealing that some of the variable names do not follow standard SAS naming conventions.

Program 7.6.4: Inspecting the Variable Names in an Access Database Table

```
proc contents data=Vermont.ipums2005utility varnum;
  ods select position;
run;
```

Output 7.6.4: Inspecting the Variable Names in an Access Database Table

Variables in Creation Order			
#	Variable	Type	Len	Label
1	SERIAL	Num	8	SERIAL
2	Electric Cost	Num	8	Electric Cost
3	Gas Cost	Num	8	Gas Cost
4	Water Cost	Num	8	Water Cost
5	Fuel Cost	Num	8	Fuel Cost

Name literals are also available for use with variable names. Program 7.6.5 demonstrates how to produce a result similar to Output 6.6.5 from data in the Vermont.accdb file.

Program 7.6.5: Using Access Data Natively in SAS

```
data work.Vermont2005cost;
  merge vermont.ipums2005basic vermont.ipums2005Utility; ❶
  by serial;
run;

proc report data= work.Vermont2005cost;
  column mortgageStatus ('electric cost'n 'gas cost'n) ❷,(mean median);
  define mortgageStatus / group 'Mortgage Status';
  define 'electric cost'n❷ / 'Elec. Cost';
  define 'gas cost'n❷ / 'Gas Cost';
  define mean / 'Mean' format=dollar10.2;
  define median / 'Median' format=dollar10.2;
run;

libname Vermont clear; ❸
```

❶ The Access table names do follow SAS naming conventions, so name literals are not required. However, it is legal to list these names, or any other names, in the literal form even when it is not required.

❷ Electric Cost and Gas Cost are nonstandard variable names, so all references to them must be given as name literals.

❸ Again, the CLEAR option in a LIBNAME statement disconnects the SAS session from the file, Vermont.accdb in this case. The Vermont library is established in Program 7.6.3.

Output 7.6.5: Using Access Data Natively in SAS

	Elec. Cost		Gas Cost	
Mortgage Status	**Mean**	**Median**	**Mean**	**Median**
N/A	$3,137.11	$1,080.00	$6,873.84	$9,993.00
No, owned free and clear	$1,244.62	$960.00	$5,611.14	$9,993.00
Yes, contract to purchase	$1,354.29	$960.00	$6,498.86	$9,993.00
Yes, mortgaged/ deed of trust or similar debt	$1,328.89	$1,200.00	$5,201.76	$2,880.00

Any of the examples or exercises in this book that use the IPUMS Basic and Utility data can be adapted to work with either the Rhode Island.xlsx or Vermont.accdb data subsets directly, no importing or conversions of those files is required.

7.7 Wrap-Up Activity

Use the lessons and examples contained in this and previous chapters to create the results shown in Section 7.2.

Data

This Wrap-Up Activity uses the data from the Wrap-Up Activities for Chapters 5 and 6.

Scenario

Use the skills mastered so far, including those from previous chapters, generate the reports shown in Output 7.2.1 and 7.2.2, which are extensions of Output 6.2.1 and 6.2.2. The tables shown in Output 7.2.3 and 7.2.4 are based on projections used to make the graphs shown in Output 6.2.3, 6.2.4, and 6.2.5. These projections can be done following the DATA step techniques shown in Chapter 6; however, they can also be done inside PROC REPORT itself using COMPUTE blocks demonstrated in this chapter—to get the maximum benefit from this activity, attempt both methods.

7.8 Chapter Notes

1. *Relative Line Pointer Controls.* Recall that an INPUT statement with no line pointer controls or line hold specifiers causes SAS to load the next line of raw text into the input buffer. This action is the basis for the behavior of the INPUT statement demonstrated in Chapters 2 and 3. In fact, the result of the relative line pointer control, /, discussed in Section 7.3.1, is the same as the default action when encountering an INPUT statement. As a result, just as sequential INPUT statements in the same DATA step would read in sequential lines from the raw file, use of the relative line pointer control also results in a sequential reading of lines from the raw file. Neither the default INPUT statement nor the relative line pointer control provides a mechanism for moving to a previous line of raw text—they can only move to the next line.

2. Mixing of Line-Hold Specifiers. The rules for when the INPUT statement releases a line held by a line hold specifier must be applied separately to each INPUT statement in a given DATA step. For example, consider the following DATA step that reads the file shown.

```
0 1 2 3
4 5 6 7
8 9 A B
C D E F
```

```
data LHS;
   infile demo;
   input W $ X $ @@;
   input Y $;
   input Z $ @;
run;
```

In this DATA step, the first INPUT statement reads the string `0 1 2 3` into the input buffer and uses a double trailing @, which instructs SAS to hold the current record in the input buffer until it encounters either an end-of-line marker or an INPUT statement. In this case, the values 0 and 1 are read and the INPUT statement still does not read the end-of-line marker. As such, this line remains in the input buffer.

Next, the DATA step encounters the second INPUT statement. However, the previous line remains, so SAS reads a value of 2 for the variable Y. Now that the DATA step has executed a new INPUT statement, its line hold specifiers (or lack thereof) define how SAS handles the current input buffer. In this case, no line hold specifier is in effect, so the current line is released at the conclusion of this INPUT statement.

When SAS encounters the third INPUT statement, it instructs the DATA step to read a new record —4 5 6 7—into the input buffer causing the value of 3 to remain unread. When SAS reads in a value for Z, it is reading from this new line of data, so Z gets a value of 4. Furthermore, this INPUT statement is now in control of how the DATA step handles the current input buffer. Because it uses a single trailing @, which releases lines at the conclusion of an iteration, this line is reset once SAS encounters the step boundary (a RUN statement in this case).

When multiple INPUT statements appear and use different line hold specifiers—including no line hold specifiers at all—then the release rule is updated after each INPUT statement executes. It is important to understand their actions to simplify the process of reading complex data structures.

3. *GROUPFORMAT.* When using the BY statement in the DATA step, SAS uses the internal, unformatted, value for determining when to change the FIRST.*variable* and LAST.*variable* values for the key variables. If the BY statement includes the GROUPFORMAT option, then SAS uses the formatted values instead of the internal values for determining when to update the values of these variables for each key variable. There is often confusion about the GROUPFORMAT option as several of its aspects are commonly overlooked,: the GROUPFORMAT option is only available in BY statements in a DATA step; the data must still be sorted by the key variables so that the internal values are in the appropriate order; and unlike the DESCENDING option, which is applied to a specific variable, the GROUPFORMAT option applies to all variables in the BY statement.

4. *NOTSORTED.* As with GROUPFORMAT, the NOTSORTED option is commonly misunderstood in that it does not allow a BY statement to include a key variable whose values have not been previously arranged. Instead, it instructs SAS that the values still appear together in a group, but without any expectation that those group values are in ascending or descending order. For example, data indexed only by month would have all the July records together and all the June records together, but the July records would precede the June records alphabetically. The NOTSORTED option would be appropriate here to tell SAS that while the values are not sorted, they are in logical groups. In fact, while the GROUPFORMAT option requires values to be sorted and grouped based on formatted values, the NOTSORTED option does not require sorting but does still require grouping on the internal values. Therefore, a better way to think about the NOTSORTED option is as simply meaning "grouped, but not necessarily sorted."

5. *Explore Available Templates.* To see the current template stores listed in the path, submit the statement:

```
ods path show;
```

The template store(s) named in other ODS PATH statements appear in this list if they were correctly assigned. To see a list of templates available in any template store, use the LIST statement in PROC TEMPLATE with the STORE= option. (Review Chapter Note 8 in Section 1.7.)

```
proc template;
   list /store=BookData.Template;
run;
```

6. *Column Identifier.* In the SAS Documentation, the first argument in CALL DEFINE refers to a column identifier or, in a seeming paradox, the automatic variable _ROW_. The role of the first argument is to give the location for the change called for by the CALL DEFINE statement. Because PROC REPORT builds the report line by line, it is possible to specify any column as the destination; however, it is not possible to point to multiple columns in a single CALL DEFINE statement. Several CALL DEFINE statements can be included in a DO block to create the effect of pointing to multiple columns—but when all columns are desired, _ROW_ is available as a shortcut, pointing to every column in the current row.

7. *Licensed SAS Products.* Submitting the SETINIT procedure with no options sends a list of licensed products and their expiration dates to the log. (Also see Chapter Note 3, Section 2.12.) PROC PRODUCT_STATUS is also available to give more detail on product versions for each.

8. *VALIDVARNAME= and VALIDMEMNAME=.* The default value for the VALIDVARNAME= system option is V7, while UPCASE and ANY are also permitted. The VALIDVARNAME=ANY system option extends the rules for variable names with the following requirements:
    a. The name can begin with or contain any characters, including blanks, national characters, special characters, and multi-byte characters.
    b. The name can be up to 32 bytes in length.
    c. The name cannot contain any null bytes.
    d. Leading blanks are preserved, but trailing blanks are ignored.
    e. The name must contain at least one character. A name with all blanks is not permitted.
    f. The name can contain mixed-case letters. SAS stores and writes the variable name in the same case that is used in the first reference to the variable. However, when SAS processes a variable name, SAS internally converts it to uppercase. Therefore, the same variable names with differing combination of uppercase and lowercase letters **do not** represent different variables.

    If any characters other than the ones that are valid when the VALIDVARNAME system option is set to V7 (letters of the Latin alphabet, numerals, or underscores), then the variable name must be expressed as a name literal and VALIDVARNAME must be set to ANY. If the name includes either the percent sign (%) or the ampersand (&), the name literal must use single quotation marks to avoid interaction with the SAS macro facility.

Setting VALIDMEMNAME to EXTEND creates a similar set of requirements for data sets and data views—see the SAS Documentation for details. Also, when these options are in place, required alterations to variable or data set names follow these rules. For example, when using the ID statement in PROC TRANSPOSE, values of the ID variable are converted to SAS names under the specified value of VALIDVARNAME. Under the V7 rules, the values North Carolina and 2005 are converted to North_Carolina and _2005, respectively, as variable names—if the ANY option is in place no conversion is done.

7.9 Exercises

Concepts: Multiple Choice

1. For a single record, how many times does the following DO loop iterate?

```
do j = 1 to 4;
  j+1;
end;
```

   a. 1
   b. 2
   c. 3
   d. 4

2. Which of the answer choices represents the data set created by the following DATA step?

```
data CondLoop;
  x=5;
  do until(x gt 3);
    x+2;
    output;
  end;
run;
```

   a.

x
7

   b.

x	until
5	7

   c.

x
5

   d. No data set is created since 5 is already larger than 3.

3. Which of the answer choices represents the data set created by the following DATA step?

```
data Reset;
  do j=1 to 4;
    a+j;
    output;
  end;
  a=0;
run;
```

a.

j	a
1	.
2	.
3	.
4	.

b.

j	a
1	1
2	2
3	3
4	4

c.

j	a
1	1
2	3
3	6
4	10

d.

j	a
1	0
2	0
3	0
4	0

4. After submitting the following program, which of the answer choices is a correct representation of the PDV contents when the DATA step concludes?

```
data Loopy;
  do j = 1 to 4;
    x+1;
  end;
run;
```

a.   _N_ = 4, _ERROR_ = 0, J = 4, X = 4
b.   _N_ = 4, _ERROR_ = 0, J = 5, X = 4
c.   _N_ = 1, _ERROR_ = 0, J = 4, X = 4
d.   _N_ = 1, _ERROR_ = 0, J = 5, X = 4

5. Which of the following programs correctly reads in each pair of variables from the raw file shown here?

```
Goat, Billy the Kid, Chicken, Christopher McCluck,
Pig, Petunia, Cow, Cowterbury Tails, Monkey, Inquisitive Igor
```

a. data livestock;
      infile roster dsd;
      input Animal $ Name : $20. @;
    run;

b. data livestock;
      infile roster dsd;
      input Animal $ Name : $20. @@;
    run;

c. data livestock;
      infile roster dsd;
      input Animal $ Name : $20.;
    run;

d. data livestock;
      infile roster dsd;
      input Animal $ Name: @20;
    run;

6. How many raw records are read during each iteration of the following DATA step?

```
data check;
   infile source missover;
   input alpha sampsize @;
   output;
   input power @;
   output;
   input AltMean;
   output;
run;
```

    a. 0

    b. 1

    c. 2

    d. 3

7. The SAS data set Portfolio is represented below. What would be the value of Total for the third observation in the data set Stocks generated by the program below?

Stocks

Stock	Shares	Price
AAA	100	5.00
AAA	100	5.00
BAA	100	2.00
CAA	500	10.00

```
data stocks;
   set portfolio;
   by stock;
   if first.stock then Total = 0;
   total + Shares*Price;
run;
```

    a. . (missing numeric value)

    b. 200

    c. 1000

    d. 1200

8. The SAS data set Portfolio is represented below. What would be the value of Total for the second observation in the data set Stocks generated by the program below?

Stocks

Stock	Shares	Price
AAA	100	5.00
AAA	100	5.00
BAA	100	2.00
CAA	500	10.00

```
data stocks;
  set portfolio;
  by stock;
  if first.stock then Total = 0;
  total = total + Shares*Price;
run;
```

   a. . (missing numeric value)

   b. 500

   c. 1000

   d. 1200

9. If the SAS data set Temperature is properly sorted and appears in a DATA step with the BY statement shown below, which of the following answer choices lists all cases where First.County=1?

```
by State County City;
```

   a. The value of County on the current record in the PDV is different from the previous record and State remains the same.

   b. The value of State on the current record in the PDV is different from the previous record and County remains the same.

   c. The values of both County and State on the current record in the PDV are different from the previous record.

   d. The value of either County or State on the current record in the PDV is different from the previous record.

10. Which of the following programs does not include an error?

   a.

```
proc report data = demo nowd;
  column state perCapita gdp pop;
  define state / group;
  define perCapita / computed format = 4.2;
  compute perCapita;
    perCapita = gdp.sum/pop.sum;
  endcomp;
run;
```

   b.

```
proc report data = demo nowd;
  column state gdp pop perCapita;
  define state / group;
  define perCapita / computed format = 4.2;
  compute perCapita;
    perCapita = gdp.sum/pop.sum;
  endcomp;
run;
```

c.
```
proc report data = demo nowd;
  column state gdp pop;
  define state / group;
  define perCapita / computed format = 4.2;
  compute perCapita;
    perCapita = gdp.sum/pop.sum;
  endcomp;
run;
```

d.
```
proc report data = demo nowd;
  column state gdp pop perCapita;
  define state / group;
  define perCapita / computed format = 4.2;
  compute perCapita;
    perCapita = gdp/pop;
  endcomp;
run;
```

11. Which of the following is not a valid location when modifying a style element with STYLE in PROC REPORT?

   a. Header

   b. Column

   c. Summary

   d. Format

12. Which of the following is legal syntax to associate a *libref* with an Access database?

   a. `libname access mydata 'Utilization.accdb';`

   b. `libname mydata access 'Utilization.accdb';`

   c. `libname 'my data'n access 'Utilization.accdb';`

   d. `libname 'Utilization.accdb';`

13. Which of the following is NOT legal syntax to reference a sheet in an Excel workbook in the Surface library?

   a. `Surface.Model1;`

   b. `Surface.'Model1'n;`

   c. `Surface.'Model 1'n;`

   d. `Surface.'Model 1';`

14. Which of the following is a calculation where a DATA step sum statement is not likely to be of use?

   a. Total of the Expenditure variable across all records in the data set

   b. Sum of the Expenditure and FixedCost variables on each record in the data set

   c. Count the number of records which have a positive value of Expenditure

   d. Count the number of records which have a positive value of Expenditure separately for each department

Concepts: Short-Answer

1. Describe the function of the RETAIN statement and give a scenario in which it is needed. Why is it of no use to put a variable being read in by the DATA step into a RETAIN statement?

2. Suppose BY-group processing in the DATA step is done with the following BY statement.

```
by A B;
```

   a. Is it possible for FIRST.A to be 1 and LAST.A to be 0 on the same record? If not, why not? If so, what condition does this indicate?

   b. Is it possible for FIRST.A to be 0 and LAST.A to be 1 on the same record? If not, why not? If so, what condition does this indicate?

      c.   Is it possible for FIRST.A and LAST.A to both be 0 on the same record? If not, why not? If so, what condition does this indicate?

      d.   Is it possible for FIRST.A and LAST.A to both be 1 on the same record? If not, why not? If so, what condition does this indicate?

3.   Compare and contrast the roles of the double-trailing @ and single-trailing @ in the INPUT statement in the DATA step, and describe the general scenarios expected for the raw records when each is used.

4.   Contrast the use of defining a format and using it in a STYLE= option versus the use of CALL DEFINE statements to set conditional styles. Give scenarios where either method can be made to work and other scenarios where only one of the two methods can be applied.

5.   Assume a REPORT procedure includes a BREAK statement as shown below.

```
break after A / summarize;
```

Further, assume the REPORT procedure also includes the following COMPUTE block.

```
compute after A;
  line 'Text Here';
endcomp;
```

Are either the BREAK line or the text in the LINE statement generated? If so, which ones and where are they placed?

6.   When referencing columns in a compute block, what is an appropriate form of the reference in each of the following cases:

      a.   A column generated via a summary statistic nested within a numeric variable.

      b.   A column generated via a summary statistic applied to the DEFINE statement for an aliased numeric variable.

      c.   A column generated via a summary statistic nested within a numeric variable that is nested within an ACROSS variable.

7.   Is it possible to display a computed value in a column that precedes the display of values it is computed from? If so, how? If not, why not?

Programming Basics

1.   Input Data 7.3.9 showed several records from Utility2005ComplexE.txt before Program 7.3.9 read the file and used line-pointer controls to reshape the file to create the SAS data set UtilityE2005v2. However, in many cases the raw file is not available, and the reshaping must occur based on an already-existing SAS data set.

      a.   The SAS data set Utility2005ComplexE in the BookData folder contains the data as it appears in Utility2005ComplexE.txt. Use a DATA step to re-create the results of Program 7.3.9 by starting with the SAS version of this file.

      b.   Validate the version from Part (a) against the data set resulting from Program 7.3.9.

2.   The BookData library contains two data sets, StudentAnswers and AnswerKey, that contain a set of student responses to 10 quiz questions and the answers to the quiz questions, respectively. Use them to answer the following questions.

      a.   In a single DATA step, use arrays without the _TEMPORARY_ option and read in the answer key once and use arrays to grade the student responses. Include a numeric variable that contains a count of the number of questions the student answered correctly and a character variable that has a value of Pass when students score at least a 7/10 and has a value of Fail otherwise.

      b.   Repeat part (a) using an array with the _TEMPORARY_ keyword.

3.   Update Program 7.5.12 so that it correctly computes the projected mean value on the RBREAK summary line.

4.   Apply the validation process, using PROC COMPARE, to the variable tables from Sashelp.Prdsale and First Sheet from Sales.xlsx. What differences, if any, are present?

5. Read in the data contained in First Sheet in the Sales.xlsx from the RawData directory and validate the results against the Sashelp.Prdsale data set.

6. Apply the validation process, using PROC COMPARE, to the variable tables from Sashelp.Prdsal2 and Sashelp.Prdsal3 against the corresponding variable tables from Sales2.accdb. What differences, if any, are present?

7. Read in the data contained in both tables from Sales2.accdb in the RawData directory and validate the results against Sashelp.Prdsal2 and Sashelp.Prdsal3 for the first and second tables, respectively.

8. Use the Sashelp.Baseball data set to create the report shown (a variation on Programming Basics Exercise 1 from Chapter 6).

League	Division	Team	At Bats	Hits	Home Runs	Salary	$ per HR
American	East	Baltimore	5,077	1,336	164	$6,935,000	$42,286.59
		Boston	4,998	1,378	138	$7,692,500	$55,742.75
		Cleveland	5,478	1,564	153	$5,805,000	$37,941.18
		Detroit	4,786	1,268	183	$5,473,810	$29,911.53
		Milwaukee	4,920	1,285	118	$4,362,500	$36,970.34
		New York	4,802	1,368	183	$8,926,460	$48,778.47
		Toronto	5,155	1,434	167	$6,422,500	$38,458.08
			35,216	**9,633**	**1,106**	**$45,617,770**	**$41,245.72**
	West	California	5,175	1,324	163	$4,864,167	$29,841.52
		Chicago	5,039	1,257	111	$3,895,000	$35,090.09
		Kansas City	5,151	1,315	133	$4,843,000	$36,413.53
		Minneapolis	5,276	1,396	190	$5,272,000	$27,747.37
		Oakland	4,965	1,270	152	$4,315,000	$28,388.16
		Seattle	4,975	1,279	153	$3,107,500	$20,310.46
		Texas	5,087	1,371	177	$3,423,500	$19,341.81
			35,668	**9,212**	**1,079**	**$29,720,167**	**$27,544.18**
National	East	Chicago	4,475	1,188	140	$7,946,665	$56,761.89
		Montreal	4,539	1,216	102	$3,236,500	$31,730.39
		New York	4,901	1,344	139	$7,948,071	$57,180.37
		Philadelphia	4,600	1,217	144	$5,905,833	$41,012.73
		Pittsburgh	4,321	1,132	106	$3,034,500	$28,627.36
		St Louis	4,408	1,089	47	$6,841,667	$145,567.38
			27,244	**7,186**	**678**	**$34,913,236**	**$51,494.45**
	West	Atlanta	4,171	1,055	116	$5,735,000	$49,439.66
		Cincinnati	4,579	1,203	134	$5,338,167	$39,837.07
		Houston	4,386	1,201	114	$5,190,000	$45,526.32
		Los Angeles	4,564	1,192	111	$4,760,000	$42,882.88
		San Diego	4,858	1,313	128	$6,074,167	$47,454.43
		San Francisco	4,918	1,299	109	$3,600,000	$33,027.52
			27,476	**7,263**	**712**	**$30,697,334**	**$43,114.23**

9. Use the Sashelp.Cars data set to create the report shown (a variation on Programming Basics Exercise 2 from Chapter 6).

| | MPG Means | | | | | |
| | Asia | | Europe | | USA | |
Type	City	Highway	City	Highway	City	Highway
Sedan/Wagon	22.8	29.8	19.5	27.0	20.7	28.6
Sports	20.2	26.6	17.7	25.1	16.9	24.2
Truck/SUV	17.5	21.8	14.5	18.7	15.6	20.2
	City Mean Less Than 20, Highway Mean Less Than 25					

Case Studies

For additional practice, multiple case studies are available in addition to the IPUMS CPS case study used in the chapters. See Section 8.7 to apply the skills from this chapter to the Clinical Trials Case Study. For additional case studies, including extensions to the IPUMS CPS case study, see the author pages.

Chapter 8: Clinical Trial Case Study

8.1 Scenario, Learning Objectives, and Introductory Activities

This chapter is a series of activities built around data simulated to create a mock clinical trial, with all data in the Clinical Trial Case Study folder available in the download accompanying the text book (along with some additional information). The trial is a study of safety and efficacy of the test drug in a four-arm crossover design. The test compound is compared to a standard, reference treatment of ibuprofen, both at a 30mg daily dose. Two alternate doses are available: 15mg, twice per day; and 10mg, three times per day (coded as 1, 2, and 3, respectively).

At a recruitment visit to any one of the five sites enrolling patients for the study, demographic information is collected along with informed consent. Those patients continuing on the study are referred for a first clinical visit. The first clinical visit establishes baseline vital signs, lab results, and pain measurements, and provides a final check of screening criteria. Exclusion criteria include: age, under 21 or over 70; pregnant females; high blood pressure at outset or during treatment phase (systolic > 140 or diastolic > 90); unwillingness to sign informed consent; and baseline pain level below threshold (0 on the pain scale).

Patients continuing are randomized to one of four study arms: TRTR, RTRT, RTTR, and TRRT (T-Test, R-Reference). Subsequent visits at 3, 6, 9 and 12 months include vital signs, lab results, and pain measurements. Except for the final visit, each visit also serves to check inclusion criteria are still satisfied and serves as the transition point among treatment elements in the assigned arms (for subjects continuing on the study).

The activities for each section of this chapter are programming tasks that not only work to analyze the clinical trial data, but directly relate to the corresponding book chapter. Therefore, the activities in Section 8.1 are tied to concepts in Chapter 1, activities in Section 8.2 relate to Chapter 2, and so forth. The learning objectives for each section are thus the learning objectives from the corresponding chapter, with the activities given here designed to reinforce those objectives.

Data from different sites comes in different forms, with sites 1, 2, and 3 being used throughout the chapter. Data from sites 4 and 5 is not used until Section 8.7, as site 4 has its data stored in a Microsoft Excel spreadsheet and site 5 has its data stored in a Microsoft Access database—working directly with these files is discussed at the end of Chapter 7.

To start, Program 8.1.1 accesses the metadata for the demographic data for site 1. In order to run this code successfully, the Clinical library must be assigned to the folder where the data for this case study is stored.

Program 8.1.1: Generating Metadata on Demographics Data for Site 1

```
libname Clinical "–place correct path to data folder here–"; ❶

ods trace on; ❷
proc contents data=clinical.demographics_site1;
run;
```

❶ In order for the code to execute, the Clinical library reference must be correctly assigned. Place a correct path reference between the quotations and submit the code. Remember, the path reference can be relative, built from the working directory, or absolute (build from a drive name or letter).

❷ When the code executes successfully, the ODS TRACE ON statement will put the names of all output tables generated into the SAS log. Having this result is essential for successful execution of Program 8.1.2.

Output 8.1.1: Metadata on Demographics Data for Site 1

Data Set Name	CLINICAL.DEMOGRAPHICS_SITE1	Observations	136
Member Type	DATA	Variables	11
Engine	V9	Indexes	0
Created	*Local Information Differs*	Observation Length	112
Last Modified	*Local Information Differs*	Deleted Observations	0
Protection		Compressed	NO
Data Set Type		Sorted	NO
Label			
Data Representation	WINDOWS_64		
Encoding	wlatin1 Western (Windows)		

Engine/Host Dependent Information	
Data Set Page Size	65536
Number of Data Set Pages	1
First Data Page	1
Max Obs per Page	584
Obs in First Data Page	136
Number of Data Set Repairs	0
ExtendObsCounter	YES
Filename	*Local Information Differs*
Release Created	9.0401M4
Host Created	X64_7PRO
Owner Name	*Local Information Differs*
File Size	128KB
File Size (bytes)	131072

Alphabetic List of Variables and Attributes				
#	Variable	Type	Len	Format
6	dob	Num	8	MMDDYY10.
5	dov	Num	8	MMDDYY10.
9	ethnicity	Num	8	

Alphabetic List of Variables and Attributes				
#	Variable	Type	Len	Format
10	ic	Char	1	
4	notif_date	Num	8	MMDDYY10.
8	race	Num	8	
3	screen	Num	8	
7	sex	Char	1	
2	sf_reas	Char	50	
11	site_loc	Char	3	
1	subject	Num	8	

Based on the information in the SAS log generated by ODS TRACE ON, fill in the ODS SELECT statement in Program 8.1.2 to limit the output to the variable information table shown in Output 8.1.2.

Program 8.1.2: Limiting the Metadata to the Variable List

```
proc contents data=clinical.demographics_site1;
  ods select /**--place table name here--**/;
run;
```

Output 8.1.2: Metadata Limited to the Variable List

Alphabetic List of Variables and Attributes				
#	Variable	Type	Len	Format
6	dob	Num	8	MMDDYY10.
5	dov	Num	8	MMDDYY10.
9	ethnicity	Num	8	
10	ic	Char	1	
4	notif_date	Num	8	MMDDYY10.
8	race	Num	8	
3	screen	Num	8	
7	sex	Char	1	
2	sf_reas	Char	50	
11	site_loc	Char	3	
1	subject	Num	8	

Program 8.1.3 includes the VARNUM option in the PROC CONTENTS statement, which alters the variable information table, both in the output generated and its name. Determine the new name of the variable information table and insert it into the ODS SELECT statement in Program 8.1.3.

Program 8.1.3: Limiting the Metadata to a Variable List in Column Order

```
proc contents data=clinical.demographics_site1 varnum;
  ods select /**--place table name here--**/;
run;
```

Output 8.1.3: Metadata Limited to a Variable List in Column Order

#	Variable	Type	Len	Format
	Variables in Creation Order			
1	subject	Num	8	
2	sf_reas	Char	50	
3	screen	Num	8	
4	notif_date	Num	8	MMDDYY10.
5	dov	Num	8	DDMMYY10.
6	dob	Num	8	DDMMYY10.
7	sex	Char	1	
8	race	Num	8	
9	ethnicity	Num	8	
10	ic	Char	1	
11	site_loc	Char	3	

As an exercise, rewrite either Program 8.1.2 or Program 8.1.3 to verify that the set of variables in the demographic data for sites 2 and 3 is the same as the set shown for site 1.

The PRINT procedure is useful for viewing the data records contained in the file; Program 8.1.4 uses it to display the data on a list of variables selected with the VAR statement.

Program 8.1.4: Displaying Demographics Data from Site 1

```
proc print data=clinical.demographics_site1(obs=6) ❶;
  var subject dob dov sex; ❷
run;
```

❶ The OBS= is a data set option that limits the processing to a maximum number of observations or rows, and is commonly used in this book to shorten output tables. (Not all shortened output tables show this option in the code.) Removing the option (including the parentheses) results in the full data set being printed.

❷ The names listed in the VAR statement are a subset of those shown in Output 8.1.2 and Output 8.1.3, and these variables are displayed in the order provided.

Output 8.1.4: Displaying Demographics Data from Site 1 (Partial Listing)

Obs	Subject	Date of Birth	Date of Visit	F: Female, M: Male
1	1	12/02/1954	01/15/2018	M
2	2	01/13/1984	01/16/2018	F
3	3	11/18/1980	01/16/2018	M
4	4	09/09/1959	01/16/2018	M
5	5	01/07/1965	01/16/2018	F
6	6	12/21/1989	01/16/2018	M

Change the set (and/or ordering) of the variables in the VAR statement, see how it affects the output, and generate similar output for each of sites 2 and 3. Also, open a view of each data set and compare the PROC PRINT output to the data in the viewer.

Program 8.1.5 creates a bar chart, which is set up to be delivered to a TIF file in this case; however, it may not be clear where that file will be stored. Before submitting Program 8.1.5, include an appropriate path in the X statement, or remove the X statement and change the working directory to the path where the file is to be placed. (See Program 1.5.1.)

Program 8.1.5: Bar Chart for Sex and Ethnicity Distribution Among Site 1 Enrollees

```
proc format;
  value eth
  1='Non-Hispanic'
  2='Hispanic'
  ;
run;

x 'cd --place path here--'; ❶

ods _all_ close; ❷
ods listing image_dpi=300; ❸
ods graphics/reset width=4in imagename='Sex and Ethnicity Chart-Site 1'
imagefmt=tif; ❹
proc sgplot data=clinical.demographics_site1;
  hbar Sex / group=Ethnicity groupdisplay=cluster;
  xaxis label='Number of Subjects';
  yaxis Display=(nolabel);
  keylegend / title='';
  format Ethnicity Eth.;
run;
```

❶ The X statement accesses the command line for the operating system, and CD is the change directory command. The path placed here becomes the SAS working directory when this command is submitted. This command can also be removed (or commented out) and the working directory can be directly changed in the SAS session.

❷ Graphs from PROC SGPLOT are potentially delivered to a variety of output destinations. ODS _ALL_ CLOSE stops delivery of output to all destinations.

❸ For tables, the ODS LISTING destination is the Output window in the windowing environment. For graphs generated by PROC SGPLOT, the destination is the graphics file itself. IMAGE_DPI is an option that sets the resolution of the image in dots per inch.

❹ The ODS GRAPHICS statement is available to set options for graphs produced, including sizes, names, file types, and various other options.

Output 8.1.5: Bar Chart for Sex and Ethnicity Distribution Among Site 1 Enrollees

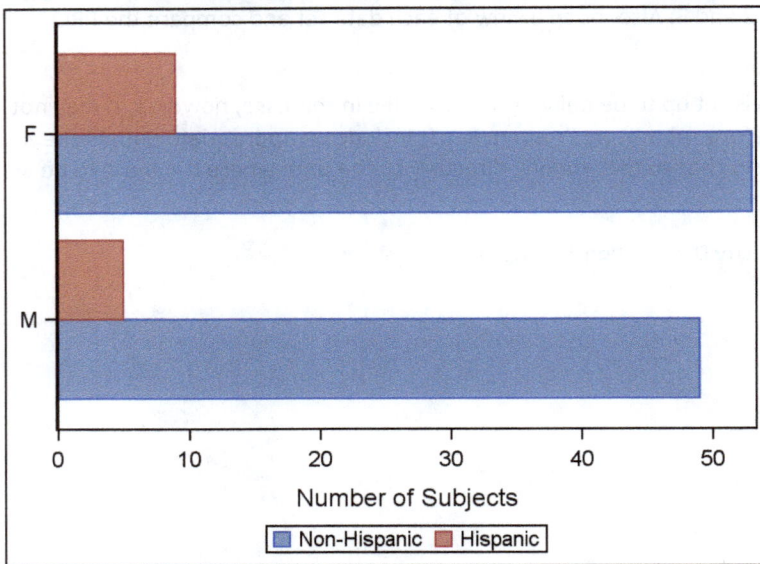

Change the code given in Program 8.1.5 to produce a similar bar chart for sites 2 and 3.

8.2 Reading and Summarizing Visit and Lab Data

Using the techniques covered in Section 2.7, read the Site 1, Baseline Visit.txt and Site 1, Baseline Lab Results.txt files into SAS data sets, and use them to make the tables shown below. Output 8.2.1 is a listing of subjects recruited at site 1 who failed screening at the baseline visit.

Output 8.2.1: Screening Failures at Baseline Visit in Site 1 (Partial Listing)

Subject	Screen Failure Reason	Failure Notification Date
3	LOW BASELINE PAIN	23JAN2018
5	HIGH BLOOD PRESSURE	19JAN2018
7	HIGH BLOOD PRESSURE	25JAN2018
12	LOW BASELINE PAIN	29JAN2018
14	HIGH BLOOD PRESSURE	24JAN2018
15	HIGH BLOOD PRESSURE	25JAN2018

Output 8.2.2 considers subjects who failed screening due to an excessive blood pressure reading at the baseline visit for subjects recruited at site 1, and includes the blood pressure measurements in the listing.

Output 8.2.2: Data for Screen Failures on Blood Pressure at Baseline Visit in Site 1 (Partial Listing)

Subject	Screen Failure Reason	Failure Notification Date	Systolic Blood Pressure	Diastolic Blood Pressure
5	HIGH BLOOD PRESSURE	19JAN2018	144	115
7	HIGH BLOOD PRESSURE	25JAN2018	132	92

Subject	Screen Failure Reason	Failure Notification Date	Systolic Blood Pressure	Diastolic Blood Pressure
14	HIGH BLOOD PRESSURE	24JAN2018	119	98
15	HIGH BLOOD PRESSURE	25JAN2018	124	100
21	HIGH BLOOD PRESSURE	03FEB2018	139	99
22	HIGH BLOOD PRESSURE	04FEB2018	152	120

Output 8.2.3 is an attempt to investigate patterns in pain level across males and females, separated by those who failed initial screening and those who did not.

Output 8.2.3: Sex Versus Pain Level Split on Screen Failure Versus Pass, Site 1 Baseline

Table 1 of Sex by pain						
Controlling for screen=Fail						
Sex	pain					
Frequency Percent Row Pct	0	1	2	3	4	Total
Female	3 21.43 100.00	0 0.00 0.00	0 0.00 0.00	0 0.00 0.00	0 0.00 0.00	3 21.43
Male	11 78.57 100.00	0 0.00 0.00	0 0.00 0.00	0 0.00 0.00	0 0.00 0.00	11 78.57
Total	14 100.00	0 0.00	0 0.00	0 0.00	0 0.00	14 100.00
Frequency Missing = 37						

Table 2 of Sex by pain						
Controlling for screen=Pass						
Sex	pain					
Frequency Percent Row Pct	0	1	2	3	4	Total
Female	0 0.00 0.00	5 6.49 14.29	10 12.99 28.57	10 12.99 28.57	10 12.99 28.57	35 45.45
Male	0 0.00 0.00	7 9.09 16.67	7 9.09 16.67	12 15.58 28.57	16 20.78 38.10	42 54.55
Total	0 0.00	12 15.58	17 22.08	22 28.57	26 33.77	77 100.00

Output 8.2.4 summarizes the distribution of screening failure reasons at the baseline visit across males and females at the baseline visit in site 1.

Output 8.2.4: Sex Versus Screen Failure at Baseline Visit in Site 1

Table of Sex by sf_reason			
Sex	sf_reason(Screen Failure Reason)		
Frequency Row Pct	**HIGH BLOOD PRESSURE**	**LOW BASELINE PAIN**	**Total**
Female	23 88.46	3 11.54	26
Male	14 56.00	11 44.00	25
Total	37	14	51

Considering the results of Output 8.2.3 against the summary in 8.2.4, determine why the first table in Output 8.2.3 has 37 missing values and why the second table has a first column with a zero frequency. Reduce the three tables produced across Output 8.2.3 and 8.2.4 to two tables containing the relevant information.

Output 8.2.5 provides a statistical summary from the baseline visit at site 1 for diastolic blood pressure and pulse rates across combinations of sex and systolic blood pressure (split into two classes).

Output 8.2.5: Diastolic Blood Pressure and Pulse Summary Statistics at Baseline Visit in Site 1

Sex	Systolic Blood Pressure	N Obs	Variable	N	Mean	Std Dev	Minimum	Maximum
Female	Acceptable (120 or below)	37	dbp pulse	37 37	77.4 74.0	8.7 9.6	61.0 57.0	91.0 92.0
	High	24	dbp pulse	24 24	100.6 97.5	9.3 10.2	87.0 79.0	120.0 118.0
Male	Acceptable (120 or below)	55	dbp pulse	55 55	77.1 73.4	10.9 11.8	49.0 42.0	100.0 99.0
	High	12	dbp pulse	12 12	97.6 92.7	10.1 10.8	80.0 73.0	118.0 109.0

Output 8.2.6 shows glucose and hemoglobin statistical summaries from the baseline lab results at site 1.

Output 8.2.6: Glucose and Hemoglobin Summary Statistics from Baseline Lab Results, Site 1

Sex	N Obs	Variable	Minimum	Lower Quartile	Median	Upper Quartile	Maximum
Female	35	c_gluc hemoglob	96.0 11.5	101.0 12.7	105.0 13.6	109.0 14.6	116.0 16.5
Male	42	c_gluc hemoglob	98.0 11.2	101.0 13.0	106.0 13.4	109.0 14.3	118.0 15.8

Produce versions of Output 8.2.1, 8.2.2, 8.2.5, and 8.2.6 for the baseline visits at sites 2 and 3, and also for all other visits at sites 1, 2, and 3.

8.3 Improving Reading of Data; Creating Charts

Using the expanded techniques for reading raw data covered in Chapter 3, re-read the Site 1, Baseline Visit.txt and Site 1, Baseline Lab Results.txt files into SAS data sets. Read all nonstandard values, such as dates, in a manner that maximizes their utility, and use them to create the results shown in this section. Output 8.3.1 is a chart of monthly recruits at site 1.

Output 8.3.1: Recruiting by Month in Site 1

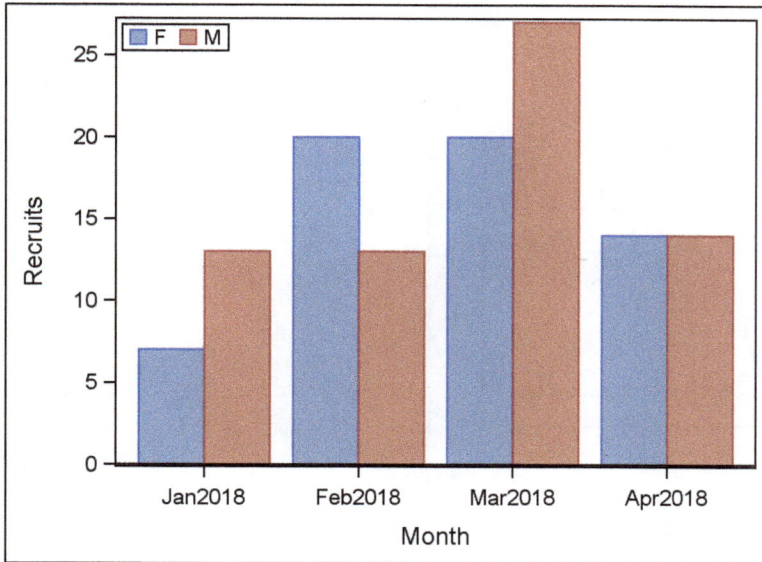

Output 8.3.2 reconstructs Output 8.3.1, removing all recruits who failed initial screening at the baseline visit.

Output 8.3.2: Recruits that Pass Initial Screening, by Month in Site 1

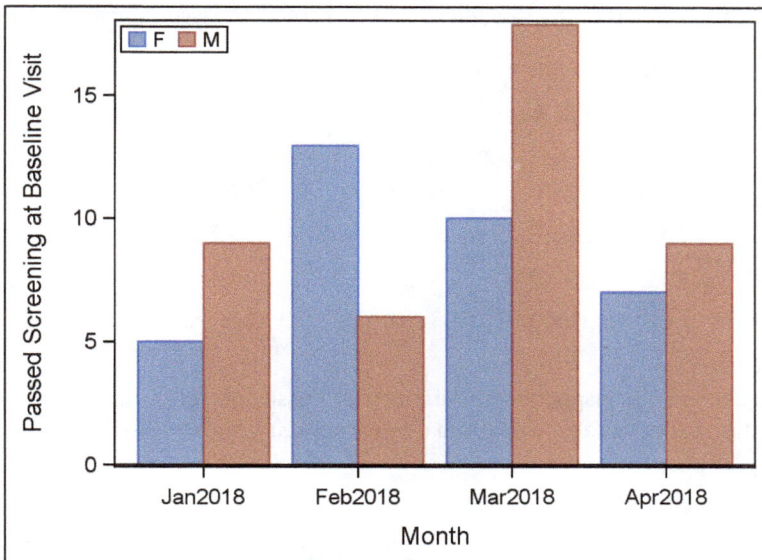

Mean albumin levels are summarized in Output 8.3.3, with error bars representing 95% confidence intervals for the mean.

Output 8.3.3: Average Albumin Results—Baseline Lab, Site 1

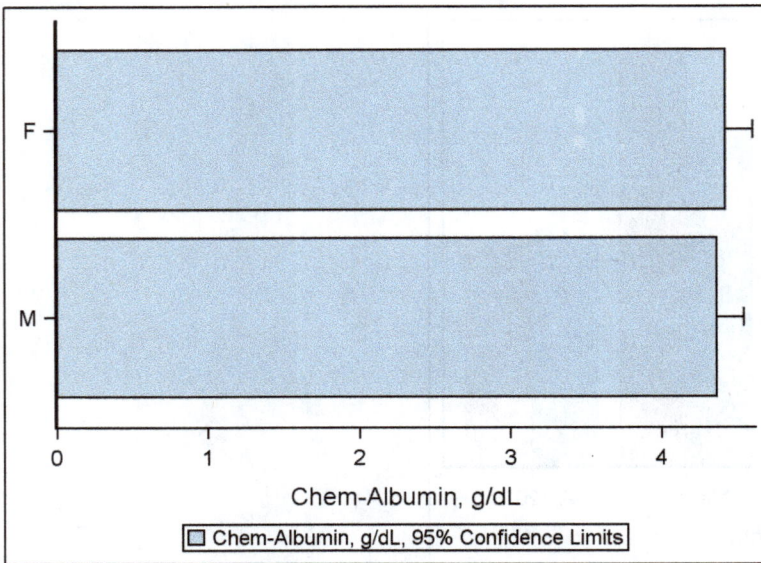

Also construct versions of Output 8.3.1, 8.3.2, and 8.3.3 for each of sites 2 and 3.

Output 8.3.4 represents an investigation of the Pos variable, which records the position of the subject during reading of vital signs. Modify the DATA step that reads in the raw file so that it corrects the issue shown.

Output 8.3.4: Investigation of the Positions Recorded—Baseline Visit, Site 1

Position for VS Reading				
pos	Frequency	Percent	Cumulative Frequency	Cumulative Percent
RECLINED	29	22.66	29	22.66
RECUMBANT	39	30.47	68	53.13
RECUMBENT	60	46.88	128	100.00

Using the techniques like those shown in Section 3.9, inspect other variables for possible errors and correct them, if possible. Complete this investigation for data (labs, vitals, and demographics) from all visits at sites 1, 2, and 3.

8.4 Working with Data Stacked Across Visits (and Sites)

Chapter 4 discusses methods for stacking data sets which, for this case study, allows for lab information across all visits to sites 1, 2, and 3 to be put into a single data set—and the same can be done for the visit data. Output shown in this section relies on assembling these data sets at various levels, with some additional specifications for information to be added for tracking the origin of each record. Specifications for those variables and values are given in Table 8.4.1.

Table 8.4.1: Variables and Values for Visit Tracking

Variable Name	Type (Format or Length)	Value Definition
VisitC	Character, length 20	*Term* Visit *Term* is one of: Baseline, 3 Month, 6 Month, 9 Month, or 12 Month
VisitNum	Numeric, format: best12.	0 for baseline visit, 1 for 3 month visit, 2 for 6 month, and so forth.
VisitMonth	Numeric, format: best12.	0 for baseline visit, 3 for 3 month visit, 6 for 6 month, and so forth.

Output 8.4.1: Subjects Ordered on Visit, Site 1 (Partial Listing)

subject	visitc	visitMonth	visitNum	Date of Visit	Systolic Blood Pressure	Diastolic Blood Pressure	pulse
1	Baseline Visit	0	0	01/22/2018	110	80	75
1	3 Month Visit	3	1	04/24/2018	112	81	76
1	6 Month Visit	6	2	07/25/2018	114	82	77
1	9 Month Visit	9	3	10/25/2018	116	83	78
1	12 Month Visit	12	4	01/25/2019	118	84	79
2	Baseline Visit	0	0	01/23/2018	99	84	81
2	3 Month Visit	3	1	04/26/2018	100	84	80
2	6 Month Visit	6	2	07/28/2018	101	84	79
2	9 Month Visit	9	3	10/29/2018	102	84	78
2	12 Month Visit	12	4	01/30/2019	103	84	77

Reproduce the following graphs, which are built from Lab and Visit information assembled across all visits at site 1.

Output 8.4.2: Albumin Distributions, Post-Baseline Visits, Site 1

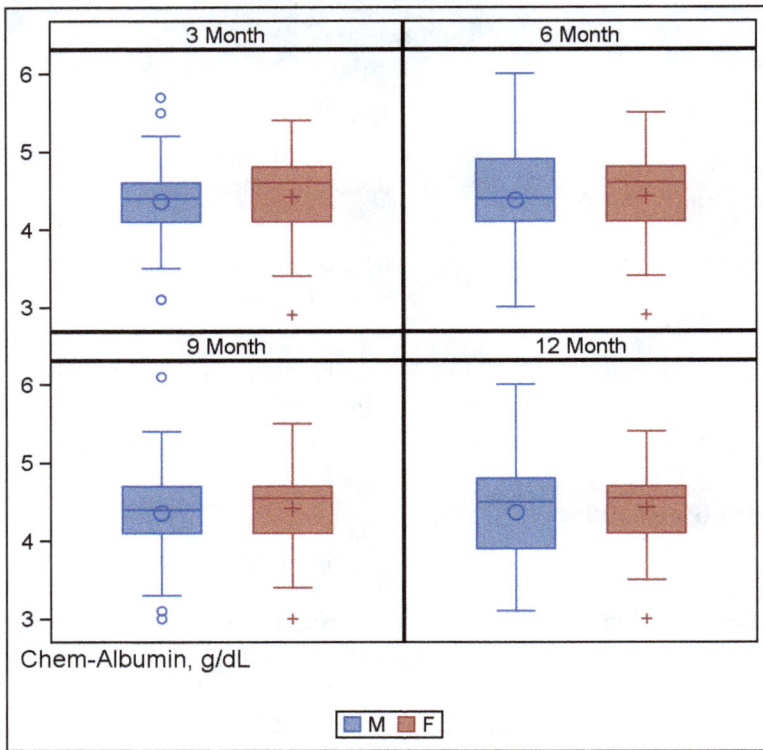

Output 8.4.3: Glucose Distributions, Baseline and 3 Month Visits, Site 1

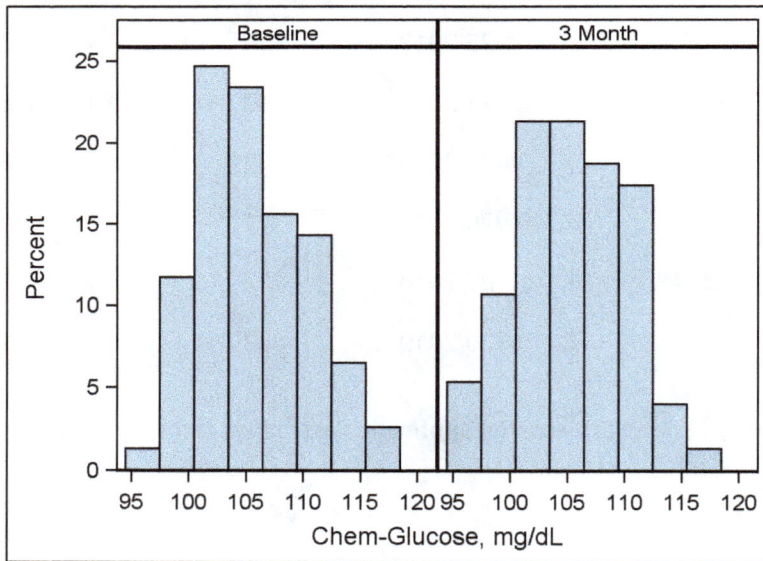

Output 8.4.4: Systolic Blood Pressure Quartiles, Site 1

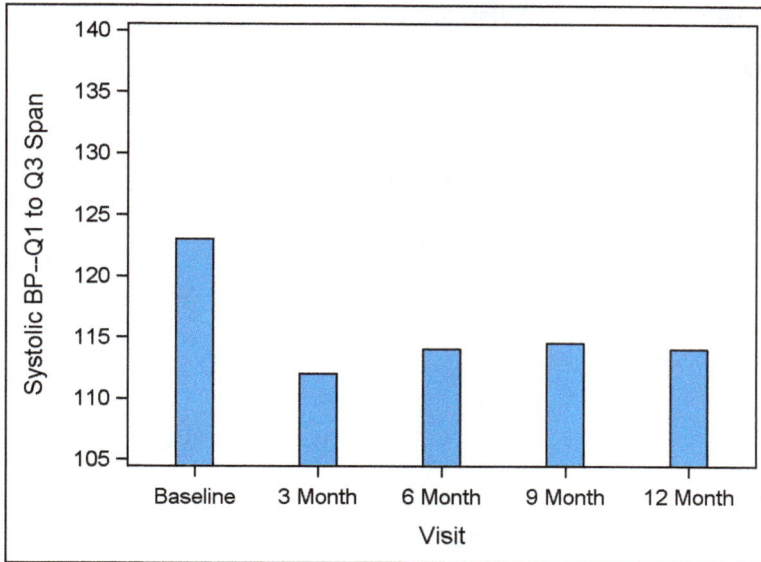

Take the lab data assembled across lab visits for each of site 1, 2, and 3 and put all of those sites together—including a variable that indicates which site a record came from—so all lab records from all visits for sites 1, 2, and 3 are in a single data set with information to identify the site and visit on each record. Do the same for the visit data. Generate variations of Outputs 8.4.2, 8.4.3, and 8.4.4 that summarize similar results across various combinations of sites and visits.

8.5 Assembling and Summarizing Data—Sites 1, 2, and 3

Starting with the data assembled in Section 8.4 for site 1, produce correlations among mean (across visits for each subject) systolic BP, diastolic BP, and pulse. Partial results are given in Output 8.5.1.

Output 8.5.1: Correlations Among BP and Pulse, Site 1, Separated by Sex (Only Females Shown)

sex=F

			Simple Statistics			
Variable	N	Mean	Std Dev	Sum	Minimum	Maximum
sbp_Mean	61	115.73907	17.82507	7060	79.00000	155.00000
dbp_Mean	61	86.90164	14.23967	5301	61.00000	120.00000
pulse_Mean	61	83.77596	14.80678	5110	56.33333	118.00000

Pearson Correlation Coefficients, N = 61
Prob > |r| under H0: Rho=0

	sbp_Mean	dbp_Mean	pulse_Mean
sbp_Mean	1.00000	0.89318 <.0001	0.86459 <.0001
dbp_Mean	0.89318 <.0001	1.00000	0.97953 <.0001
pulse_Mean	0.86459 <.0001	0.97953 <.0001	1.00000

Also create the scatterplot and spline-smoothing plot (shown in Outputs 8.5.2 and 8.5.3, respectively) from the assembled data (the smoothing parameter in the PBSPLINE statement is set to 10,000).

Output 8.5.2: Scatter Plot of BP Values, Site 1, All Visits

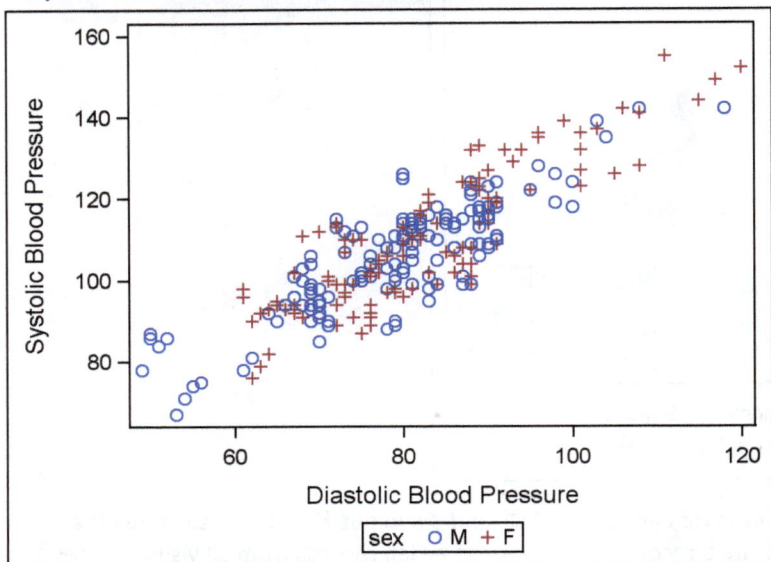

Output 8.5.3: Spline Smoothing on Weight Versus Systolic BP, Site 1, All Visits

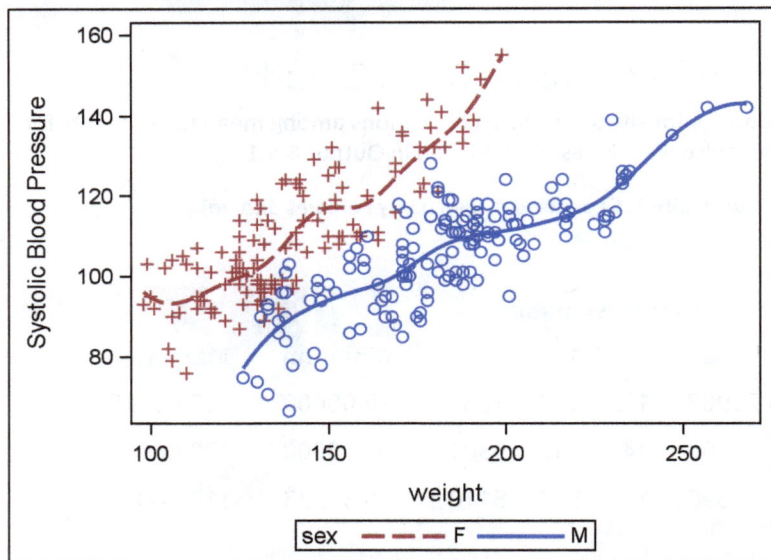

From the visit data sets, create a table of all visit dates for subjects who completed the study (in other words, have a 12 Month visit), computing the number of days from the first visit to the last.

Output 8.5.4: Visits for Subjects Completing the Study for Site 1, with Total Days on Study (Partial Listing)

Obs	subject	dovB	dov3	dov6	dov9	dov12	Days
1	1	01/22/2018	04/24/2018	07/25/2018	10/25/2018	01/25/2019	368
2	2	01/23/2018	04/26/2018	07/28/2018	10/29/2018	01/30/2019	372
3	6	01/24/2018	04/26/2018	07/25/2018	10/22/2018	01/19/2019	360
4	8	01/21/2018	04/25/2018	07/25/2018	10/25/2018	01/25/2019	369

Obs	subject	dovB	dov3	dov6	dov9	dov12	Days
5	9	01/27/2018	04/30/2018	08/02/2018	11/04/2018	02/06/2019	375
6	10	01/24/2018	04/22/2018	07/22/2018	10/23/2018	01/24/2019	365
7	11	01/27/2018	04/27/2018	07/25/2018	10/21/2018	01/17/2019	355
8	13	01/29/2018	04/27/2018	07/25/2018	10/23/2018	01/21/2019	357

Modify Output 8.5.4 to show all subjects with missing visits for non-completing patients, as shown in Output 8.5.5.

Output 8.5.5: Visits for All Subjects, Site 1 (Partial Listing)

subject	vis1	vis2	vis3	vis4	vis5
1	01/22/2018	04/24/2018	07/25/2018	10/25/2018	01/25/2019
2	01/23/2018	04/26/2018	07/28/2018	10/29/2018	01/30/2019
4	01/20/2018	04/19/2018	.	.	.
6	01/24/2018	04/26/2018	07/25/2018	10/22/2018	01/19/2019
8	01/21/2018	04/25/2018	07/25/2018	10/25/2018	01/25/2019
9	01/27/2018	04/30/2018	08/02/2018	11/04/2018	02/06/2019
10	01/24/2018	04/22/2018	07/22/2018	10/23/2018	01/24/2019
11	01/27/2018	04/27/2018	07/25/2018	10/21/2018	01/17/2019

Add a column for days, from first visit to last visit, as shown in Output 8.5.6.

Output 8.5.6: Visits for All Subjects with Days on Study, Site 1 (Partial Listing)

subject	vis1	vis2	vis3	vis4	vis5	Days
1	01/22/2018	04/24/2018	07/25/2018	10/25/2018	01/25/2019	368
2	01/23/2018	04/26/2018	07/28/2018	10/29/2018	01/30/2019	372
4	01/20/2018	04/19/2018	.	.	.	89
6	01/24/2018	04/26/2018	07/25/2018	10/22/2018	01/19/2019	360
8	01/21/2018	04/25/2018	07/25/2018	10/25/2018	01/25/2019	369
9	01/27/2018	04/30/2018	08/02/2018	11/04/2018	02/06/2019	375
10	01/24/2018	04/22/2018	07/22/2018	10/23/2018	01/24/2019	365
11	01/27/2018	04/27/2018	07/25/2018	10/21/2018	01/17/2019	355

From the lab data, create a list of the variable names and labels. Output 8.5.7A shows a possible result; however, these names and labels are chosen during the process of reading the raw data and may differ. Output 8.5.7B shows information from the Lab_Info data set.

Output 8.5.7A: Lab Test Information for Lab Data Sets

Obs	_NAME_	_LABEL_
1	alb	Chem-Albumin, g/dL
2	alk_phos	Chem-Alk. Phos., IU/L
3	alt	Chem-Alt, IU/L
4	ast	Chem-AST, IU/L
5	c_gluc	Chem-Glucose, mg/dL
6	d_bili	Chem-Dir. Bilirubin, mg/dL
7	ggtp	Chem-GGTP, IU/L
8	hematocr	EVF/PCV, %
9	hemoglob	Hemoglobin, g/dL
10	preg	Pregnancy Flag, 1=Pregnant, 0=Not
11	prot	Chem-Tot. Prot., g/dL
12	t_bili	Chem-Tot. Bilirubin, mg/dL
13	u_gluc	Uri.-Glucose, 1=high

Output 8.5.7B: Lab Test Information from Lab_Info Data Set

Obs	lbtestcd	labtest	colunits
1	ALB	ALBUMIN	g/dL
2	ALP	ALK. PHOS.	IU/L
3	ALT	ALT (SGPT)	IU/L
4	AST	AST (SGOT)	IU/L
5	BILDIR	DIRECT BILI	mg/dL
6	GGT	GGTP	IU/L
7	GLUC	GLUCOSE	mg/dL
8	GLUC	GLUCOSE	
9	HCT	HEMATOCRIT	%
10	HGB	HEMOGLOBIN	g/dL
11	BILI	TOTAL BILI	mg/dL
12	PROT	TOTAL PROT	g/dL
13	PREG	PREG	

Determine a mapping/matching from the set of test results given in the lab data sets and the test information in the Lab_Info data set, creating variables in each data set to achieve a match merge. If successful, it is then possible to produce a data set with information like that shown in Output 8.5.8 (for one subject at the baseline visit) and Output 8.5.9 (for one subject and selected labs across multiple visits). Note the RangeFlag variable, which is 1 any time the value is outside the range limits and is 0 otherwise.

Output 8.5.8: Baseline Lab Results for Subject 13, Site 1—Including Flag for Values Outside Normal Range

Subject	Laboratory Test	Collected Units	Value	Normal Range Lower Limit	Normal Range Upper Limit	RangeFlag
13	ALBUMIN	g/dL	3.60	3.4	5.4	0
13	ALK. PHOS.	IU/L	132.00	20.0	140.0	0
13	ALT (SGPT)	IU/L	24.00	5.0	35.0	0
13	AST (SGOT)	IU/L	14.00	10.0	34.0	0
13	DIRECT BILI	mg/dL	0.31	0.0	0.3	1
13	TOTAL BILI	mg/dL	2.19	0.3	1.9	1
13	GLUCOSE	mg/dL	104.00	100.0	110.0	0
13	GGTP	IU/L	39.00	0.0	51.0	0
13	HEMATOCRIT	%	53.00	35.0	49.0	1
13	HEMOGLOBIN	g/dL	14.40	11.7	15.9	0
13	PREG		0.00	.	.	0
13	TOTAL PROT	g/dL	7.20	6.0	8.3	0
13	GLUCOSE		0.00	0.0	0.0	0

Output 8.5.9: Selected Lab Results for Subject 2, Site 1—Including Flag for Values Outside Normal Range

Subject	Date of Visit	visitNum	Laboratory Test	Value	Normal Range Lower Limit	Normal Range Upper Limit	RangeFlag
2	01/23/2018	0	DIRECT BILI	0.10	0.0	0.3	0
2	04/26/2018	1	DIRECT BILI	0.20	0.0	0.3	0
2	07/28/2018	2	DIRECT BILI	0.10	0.0	0.3	0
2	10/29/2018	3	DIRECT BILI	0.20	0.0	0.3	0
2	01/30/2019	4	DIRECT BILI	0.30	0.0	0.3	0
2	01/23/2018	0	TOTAL BILI	0.82	0.3	1.9	0
2	04/26/2018	1	TOTAL BILI	1.49	0.3	1.9	0
2	07/28/2018	2	TOTAL BILI	0.93	0.3	1.9	0
2	10/29/2018	3	TOTAL BILI	1.49	0.3	1.9	0
2	01/30/2019	4	TOTAL BILI	2.06	0.3	1.9	1
2	01/23/2018	0	HEMATOCRIT	48.00	35.0	49.0	0
2	04/26/2018	1	HEMATOCRIT	35.00	35.0	49.0	0

Subject	Date of Visit	visitNum	Laboratory Test	Value	Normal Range Lower Limit	Normal Range Upper Limit	RangeFlag
2	07/28/2018	2	HEMATOCRIT	46.00	35.0	49.0	0
2	10/29/2018	3	HEMATOCRIT	32.00	35.0	49.0	1
2	01/30/2019	4	HEMATOCRIT	28.00	35.0	49.0	1
2	01/23/2018	0	HEMOGLOBIN	13.40	11.7	15.9	0
2	04/26/2018	1	HEMOGLOBIN	12.00	11.7	15.9	0
2	07/28/2018	2	HEMOGLOBIN	12.60	11.7	15.9	0
2	10/29/2018	3	HEMOGLOBIN	11.20	11.7	15.9	1
2	01/30/2019	4	HEMOGLOBIN	9.80	11.7	15.9	1

From the full set of visits at site 1, create a graph to display pain score trends in the various arms, as shown in Output 8.5.10.

Output 8.5.10: Pain Score Trends in Various Arms

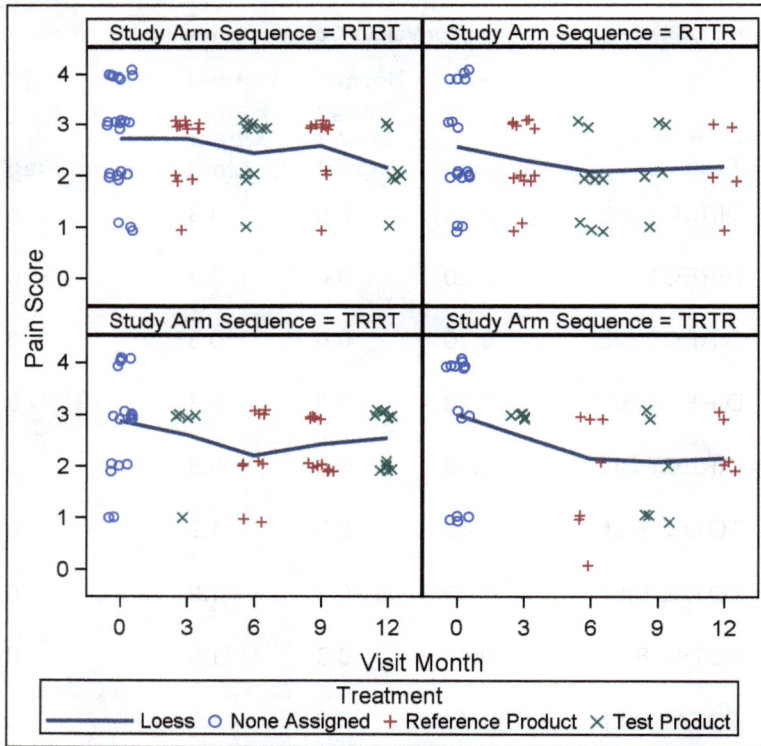

Repeat all of the activities in this section for sites 2 and 3.

8.6 Data Restructuring and Report Writing

Data from the baseline visit for site can be rotated as shown in Output 8.6.1. Once that rotation is achieved successfully, the same code can be extended to rotate the data across all visits.

Output 8.6.1: Rotated Data from Baseline Visit, Site 1 (Partial Listing)

Obs	subject	Date of Visit	name	value	units
1	1	01/22/2018	Systolic BP	110.0	mm/hg
2	1	01/22/2018	Diastolic BP	80.0	mm/hg
3	1	01/22/2018	Pulse	75.0	beats/mi
4	1	01/22/2018	Temperature	98.4	F
5	1	01/22/2018	Weight	192.0	lb
6	2	01/23/2018	Systolic BP	99.0	mm/hg
7	2	01/23/2018	Diastolic BP	84.0	mm/hg
8	2	01/23/2018	Pulse	81.0	beats/mi
9	2	01/23/2018	Temperature	98.5	F
10	2	01/23/2018	Weight	134.0	lb

From the rotated version of the full set of visits at site 1, create the following tables of summary statistics shown in Output 8.6.2 and 8.6.3.

Output 8.6.2: Summary Report on Selected Vital Signs, All Visits, Site 1

Visit	Measurement	Units	Mean	Median	Std. Deviation	Minimum	Maximum
Baseline Visit	Diastolic BP	mm/hg	83.5	82.5	14.09	49.0	120.0
	Pulse	beats/mi	79.9	79.0	14.70	42.0	118.0
	Systolic BP	mm/hg	111.0	110.0	17.03	67.0	155.0
	Temperature	F	98.5	98.5	0.58	97.1	99.8
3 Month Visit	Diastolic BP	mm/hg	77.1	79.0	9.67	49.0	90.0
	Pulse	beats/mi	73.5	75.0	10.46	44.0	89.0
	Systolic BP	mm/hg	103.7	104.0	12.23	70.0	133.0
	Temperature	F	98.4	98.4	0.75	96.9	100.2
6 Month Visit	Diastolic BP	mm/hg	77.7	80.0	9.65	50.0	91.0
	Pulse	beats/mi	74.3	76.0	10.71	43.0	90.0
	Systolic BP	mm/hg	104.9	104.0	12.26	71.0	135.0
	Temperature	F	98.4	98.4	1.04	96.2	100.3
9 Month Visit	Diastolic BP	mm/hg	78.1	80.0	9.53	51.0	91.0
	Pulse	beats/mi	74.8	75.5	10.18	45.0	93.0
	Systolic BP	mm/hg	105.9	105.0	11.99	74.0	138.0

Visit	Measurement	Units	Mean	Median	Std. Deviation	Minimum	Maximum
	Temperature	F	98.5	98.6	1.25	95.7	101.2
12 Month Visit	Diastolic BP	mm/hg	77.6	80.0	9.07	52.0	91.0
	Pulse	beats/mi	74.4	75.0	9.78	45.0	89.0
	Systolic BP	mm/hg	105.4	105.0	11.20	75.0	128.0
	Temperature	F	98.3	98.2	1.43	95.0	102.1

Output 8.6.3: BP Summaries, All Visits, Site 1

	Measurement					
	Diastolic BP			Systolic BP		
Visit	Mean	Median	Std. Dev.	Mean	Median	Std. Dev.
Baseline Visit	83.5	82.5	14.09	111.0	110.0	17.03
3 Month Visit	77.1	79.0	9.67	103.7	104.0	12.23
6 Month Visit	77.7	80.0	9.65	104.9	104.0	12.26
9 Month Visit	78.1	80.0	9.53	105.9	105.0	11.99
12 Month Visit	77.6	80.0	9.07	105.4	105.0	11.20

Repeat each of these for sites 2 and 3, then assemble all data for all sites into a single data set and repeat these activities for that collection of data.

8.7 Advanced Data Reading and Report Writing—Connecting to Spreadsheets and Databases

Read the adverse event data from site 1, create the listing in Output 8.7.1, and repeat the same process for sites 2 and 3.

Output 8.7.1: Adverse Events, Site 1 (Partial Listing)

Obs	subject	aetext	stdt	endt
1	25	TIA	11Jan2018	12Jan2018
2	25	CARPAL TUNNEL IN LEFT HAND	18Jan2018	19Jan2018
3	38	SUICIDAL	24Jan2018	24Jan2018
4	41	MOUTH PAIN	14Jan2018	15Jan2018
5	41	IRRITABLE BOWEL SYNDROME	28Jan2018	29Jan2018
6	41	LOOSE STOOL	30Jan2018	01Jan2018
7	46	RACING HEART BEAT	30Jan2018	01Jan2018
8	46	DELIRIOUS	26Jan2018	27Jan2018
9	50	HEART FLUTTERING	17Jan2018	17Jan2018
10	50	RINGING IN EAR	30Jan2018	01Jan2018

Extend Output 8.6.3 into that shown in Output 8.7.2, repeating this for sites 2 and 3, and for the full data set across all sites.

Output 8.7.2: BP Summaries—Extended, Site 1

| | Measurement | | | | | | | |
| | Diastolic BP | | | Systolic BP | | | Ratios (SBP:DBP) | |
Visit	Mean	Median	Std. Dev.	Mean	Median	Std. Dev.	Mean	Median
Baseline Visit	83.5	82.5	14.09	111.0	110.0	17.03	132.9%	133.3%
3 Month Visit	77.1	79.0	9.67	103.7	104.0	12.23	134.4%	131.6%
6 Month Visit	77.7	80.0	9.65	104.9	104.0	12.26	135.0%	130.0%
9 Month Visit	78.1	80.0	9.53	105.9	105.0	11.99	135.6%	131.3%
12 Month Visit	77.6	80.0	9.07	105.4	105.0	11.20	135.8%	131.3%

Repeat the process leading to Output 8.6.2 to produce the enhanced version shown in Output 8.7.3. Also repeat this for each of sites 2 and 3, and for the full data set across all three sites.

Output 8.7.3: Summary Report on Selected Vital Signs, All Visits, Site 1—Enhanced

Visit	Test	Units	Mean	Median	Std. Dev.	Min.	Max.
Baseline Visit	Diastolic BP	mm/hg	83.5	82.5	14.09	49.0	120.0
	Pulse	beats/mi	79.9	79.0	14.70	42.0	118.0
	Systolic BP	mm/hg	111.0	110.0	17.03	67.0	155.0
	Temperature	F	98.5	98.5	0.58	97.1	99.8
3 Month Visit	Diastolic BP	mm/hg	77.1	79.0	9.67	49.0	90.0
	Pulse	beats/mi	73.5	75.0	10.46	44.0	89.0
	Systolic BP	mm/hg	103.7	104.0	12.23	70.0	133.0
	Temperature	F	98.4	98.4	0.75	96.9	100.2
6 Month Visit	Diastolic BP	mm/hg	77.7	80.0	9.65	50.0	91.0
	Pulse	beats/mi	74.3	76.0	10.71	43.0	90.0
	Systolic BP	mm/hg	104.9	104.0	12.26	71.0	135.0
	Temperature	F	98.4	98.4	1.04	96.2	100.3
9 Month Visit	Diastolic BP	mm/hg	78.1	80.0	9.53	51.0	91.0
	Pulse	beats/mi	74.8	75.5	10.18	45.0	93.0

Visit	Test	Units	Mean	Median	Std. Dev.	Min.	Max.
	Systolic BP	mm/hg	105.9	105.0	11.99	74.0	138.0
	Temperature	F	98.5	98.6	1.25	95.7	101.2
12 Month Visit	Diastolic BP	mm/hg	77.6	80.0	9.07	52.0	91.0
	Pulse	beats/mi	74.4	75.0	9.78	45.0	89.0
	Systolic BP	mm/hg	105.4	105.0	11.20	75.0	128.0
	Temperature	F	98.3	98.2	1.43	95.0	102.1

Use the lab results to produce the table shown in Output 8.7.4 (for potentially any subject at any site, subject 2 at site 1 shown).

Output 8.7.4: Selected Lab Results for Subject 2, Site 1—Enhanced

Subject	Lab Test	Visit	Units	Measured Value	Normal Low	Normal High
2	DIRECT BILI	Baseline Visit	mg/dL	0.1	0	0.3
		3 Month Visit	mg/dL	0.2	0	0.3
		6 Month Visit	mg/dL	0.1	0	0.3
		9 Month Visit	mg/dL	0.2	0	0.3
		12 Month Visit	mg/dL	0.3	0	0.3
	HEMATOCRIT	Baseline Visit	%	48	35	49
		3 Month Visit	%	35	35	49
		6 Month Visit	%	46	35	49
		9 Month Visit	%	32	35	49
		12 Month Visit	%	28	35	49
	HEMOGLOBIN	Baseline Visit	g/dL	13.4	11.7	15.9
		3 Month Visit	g/dL	12	11.7	15.9
		6 Month Visit	g/dL	12.6	11.7	15.9
		9 Month Visit	g/dL	11.2	11.7	15.9
		12 Month Visit	g/dL	9.8	11.7	15.9

Subject	Lab Test	Visit	Units	Measured Value	Normal Low	Normal High
	TOTAL BILI	Baseline Visit	mg/dL	0.82	0.3	1.9
		3 Month Visit	mg/dL	1.49	0.3	1.9
		6 Month Visit	mg/dL	0.93	0.3	1.9
		9 Month Visit	mg/dL	1.49	0.3	1.9
		12 Month Visit	mg/dL	**2.06**	**0.3**	**1.9**

Using the techniques described in Section 7.6, connect to data for each of sites 4 and 5 and repeat each of these results for those sites as well. Combine the site 4 and 5 data with the data from the other three sites and repeat the activities from Sections 8.6 and 8.7 for the full data set.

8.8 Comprehensive Activity

From the data across all sites and all visits, create a table of lab information (LB) and vital signs information (VS) as given in the specifications found in the file SDTM Specs.xls in the Clinical Trial Case Study folder.

Index